Crystallographic Methods and Protocols

Methods in Molecular Biology™

John M. Walker, Series Editor

60. **Protein NMR Protocols,** edited by *David G. Reid, 1996*
59. **Protein Purification Protocols,** edited by *Shawn Doonan, 1996*
58. **Basic DNA and RNA Protocols,** edited by *Adrian J. Harwood, 1996*
57. **In Vitro Mutagenesis Protocols,** edited by *Michael K. Trower, 1996*
56. **Crystallographic Methods and Protocols,** edited by *Christopher Jones, Barbara Mulloy, and Mark Sanderson, 1996*
55. **Plant Cell Electroporation and Electrofusion Protocols,** edited by *Jac A. Nickoloff, 1995*
54. **YAC Protocols,** edited by *David Markie, 1995*
53. **Yeast Protocols:** *Methods in Cell and Molecular Biology,* edited by *Ivor H. Evans, 1996*
52. **Capillary Electrophoresis:** *Principles, Instrumentation, and Applications,* edited by *Kevin D. Altria, 1996*
51. **Antibody Engineering Protocols,** edited by *Sudhir Paul, 1995*
50. **Species Diagnostics Protocols:** *PCR and Other Nucleic Acid Methods,* edited by *Justin P. Clapp, 1996*
49. **Plant Gene Transfer and Expression Protocols,** edited by *Heddwyn Jones, 1995*
48. **Animal Cell Electroporation and Electrofusion Protocols,** edited by *Jac A. Nickoloff, 1995*
47. **Electroporation Protocols for Microorganisms,** edited by *Jac A. Nickoloff, 1995*
46. **Diagnostic Bacteriology Protocols,** edited by *Jenny Howard and David M. Whitcombe, 1995*
45. **Monoclonal Antibody Protocols,** edited by *William C. Davis, 1995*
44. ***Agrobacterium* Protocols,** edited by *Kevan M. A. Gartland and Michael R. Davey, 1995*
43. **In Vitro Toxicity Testing Protocols,** edited by *Sheila O'Hare and Chris K. Atterwill, 1995*
42. **ELISA:** *Theory and Practice,* by *John R. Crowther, 1995*
41. **Signal Transduction Protocols,** edited by *David A. Kendall and Stephen J. Hill, 1995*
40. **Protein Stability and Folding:** *Theory and Practice,* edited by *Bret A. Shirley, 1995*
39. **Baculovirus Expression Protocols,** edited by *Christopher D. Richardson, 1995*
38. **Cryopreservation and Freeze-Drying Protocols,** edited by *John G. Day and Mark R. McLellan, 1995*
37. **In Vitro Transcription and Translation Protocols,** edited by *Martin J. Tymms, 1995*
36. **Peptide Analysis Protocols,** edited by *Ben M. Dunn and Michael W. Pennington, 1994*
35. **Peptide Synthesis Protocols,** edited by *Michael W. Pennington and Ben M. Dunn, 1994*
34. **Immunocytochemical Methods and Protocols,** edited by *Lorette C. Javois, 1994*
33. ***In Situ* Hybridization Protocols,** edited by *K. H. Andy Choo, 1994*
32. **Basic Protein and Peptide Protocols,** edited by *John M. Walker, 1994*
31. **Protocols for Gene Analysis,** edited by *Adrian J. Harwood, 1994*
30. **DNA–Protein Interactions,** edited by *G. Geoff Kneale, 1994*
29. **Chromosome Analysis Protocols,** edited by *John R. Gosden, 1994*
28. **Protocols for Nucleic Acid Analysis by Nonradioactive Probes,** edited by *Peter G. Isaac, 1994*
27. **Biomembrane Protocols:** *II. Architecture and Function,* edited by *John M. Graham and Joan A. Higgins, 1994*
26. **Protocols for Oligonucleotide Conjugates:** *Synthesis and Analytical Techniques,* edited by *Sudhir Agrawal, 1994*
25. **Computer Analysis of Sequence Data:** *Part II,* edited by *Annette M. Griffin and Hugh G. Griffin, 1994*
24. **Computer Analysis of Sequence Data:** *Part I,* edited by *Annette M. Griffin and Hugh G. Griffin, 1994*
23. **DNA Sequencing Protocols,** edited by *Hugh G. Griffin and Annette M. Griffin, 1993*
22. **Microscopy, Optical Spectroscopy, and Macroscopic Techniques,** edited by *Christopher Jones, Barbara Mulloy, and Adrian H. Thomas, 1993*
21. **Protocols in Molecular Parasitology,** edited by *John E. Hyde, 1993*
20. **Protocols for Oligonucleotides and Analogs:** *Synthesis and Properties,* edited by *Sudhir Agrawal, 1993*
19. **Biomembrane Protocols:** *I. Isolation and Analysis,* edited by *John M. Graham and Joan A. Higgins, 1993*
18. **Transgenesis Techniques:** *Principles and Protocols,* edited by *David Murphy and David A. Carter, 1993*
17. **Spectroscopic Methods and Analyses:** *NMR, Mass Spectrometry, and Metalloprotein Techniques,* edited by *Christopher Jones, Barbara Mulloy, and Adrian H. Thomas, 1993*
16. **Enzymes of Molecular Biology,** edited by *Michael M. Burrell, 1993*
15. **PCR Protocols:** *Current Methods and Applications,* edited by *Bruce A. White, 1993*
14. **Glycoprotein Analysis in Biomedicine,** edited by *Elizabeth F. Hounsell, 1993*
13. **Protocols in Molecular Neurobiology,** edited by *Alan Longstaff and Patricia Revest, 1992*
12. **Pulsed-Field Gel Electrophoresis:** *Protocols, Methods, and Theories,* edited by *Margit Burmeister and Levy Ulanovsky, 1992*
11. **Practical Protein Chromatography,** edited by *Andrew Kenney and Susan Fowell, 1992*
10. **Immunochemical Protocols,** edited by *Margaret M. Manson, 1992*
9. **Protocols in Human Molecular Genetics,** edited by *Christopher G. Mathew, 1991*
8. **Practical Molecular Virology:** *Viral Vectors for Gene Expression,* edited by *Mary K. L. Collins, 1991*
7. **Gene Transfer and Expression Protocols,** edited by *Edward J. Murray, 1991*
6. **Plant Cell and Tissue Culture,** edited by *Jeffrey W. Pollard and John M. Walker, 1990*
5. **Animal Cell Culture,** edited by *Jeffrey W. Pollard and John M. Walker, 1990*

Methods in Molecular Biology™ • 56

Crystallographic Methods and Protocols

Edited by

Christopher Jones,
Barbara Mulloy,

National Institute for Biological Standards and Control, Potters Bar, UK

and

Mark R. Sanderson

Department of Biophysics, King's College, London, UK

Humana Press  **Totowa, New Jersey**

999 Riverview Drive, Suite 208
Totowa, New Jersey 07512

For additional copies, pricing for bulk purchases, and/or other information about other Humana titles, contact Humana at the above address or at any of the following numbers: Tel.: 201-256-1699; Fax: 201-256-8341; E-mail: humana@interramp.com

This publication is printed on acid-free paper. ∞
ANSI Z39.48-1984 (American National Standards Institute) Permanence of Paper for Printed Library Materials.

Printed in the United States of America. 10 9 8 7 6 5 4 3 2 1

Library of Congress Cataloging in Publication Data

Main entry under title:
Methods in molecular biology™.

Crystallographic methods and protocols / edited by Christopher Jones,
Barbara Mulloy, and Mark R. Sanderson.
p. cm.
Includes index.
ISBN 0-89603-259-0 (alk. paper)
1. Proteins—Structure. 2. Nucleic acids—Structure. 3. X-ray crystallography—Technique. I. Jones, Christopher, 1954– . II. Mulloy, Barbara. III. Sanderson, Mark R.
QP551.C793 1996
574.19'245—dc20 96-1139
CIP

Preface

The volumes in the series, *Methods in Molecular Biology*, are conceived with the biochemist and molecular biologist in mind. The present book, *Crystallographic Methods and Protocols*, concentrates on the use of X-ray crystallography to solve the detailed three-dimensional structures of proteins, nucleic acids, and their complexes. Such a structure determination is a major undertaking, demands expertise in a range of skills, and requires considerable resources. The biologically trained worker will probably first become involved when identifying an important scientific problem whose solution would benefit from a full structure. The protein or nucleic acid at issue must be sequenced and prepared to high purity in appropriate quantities, probably by either chemical synthesis for nucleic acids or genetic engineering for proteins. *Crystallographic Methods and Protocols* aims to give biologically trained workers an insight into the techniques used to crystallize their proteins, obtain the raw X-ray data, and solve and refine the structure.

The aim of a crystal structure determination is to provide information that will solve biologically relevant problems; that process normally requires a high resolution structure. The preparation of suitable crystals for a high resolution structure remains the major bottleneck in structure determination, and the effective application of appropriate genetic engineering and biochemical techniques in the initial stages pays a handsome dividend later. The production of suitable proteins by recombinant methods and the preparation of crystalline derivatives are covered by Skelly and Madden in Chapter 2. Before beginning the data collection required for a high resolution structure, a preliminary characterization of the crystals is carried out. Abdel-Meguid, Sanderson, and Jeruzalmi explain the procedures involved in Chapter 3.

The past 15 years have seen a revolution in X-ray data collection techniques. This stage in the process, which once required many months of beam time, has become much faster and, with the introduction of area detectors, now needs fewer precious crystals. These instruments, which first became available in a few specialist centers, are now more common, and Garman describes their use in Chapter 4. In Chapter 5, Murthy describes the use of multiple wavelength methods that can greatly reduce the number of heavy atom derivatives to be prepared. Synthesis of these heavy atom derivatives is the second major bottleneck in determining the structure. They are needed to deduce the phase information required in the structure determination, but this information cannot be measured directly. Methods to extract phase information by the use of such heavy atom derivatives are reviewed by Abdel-Meguid in Chapter 6.

X-ray structural studies of macromolecules are carried out on well-characterized samples. The sequence is known, and a wide variety of physical constraints, such as bond lengths, and the structure and chirality of the amino acid or nucleic acid building blocks are available. The continuity of the peptide or nucleic acid chains can also be assumed. In many cases, the crystal structure of a closely related macromolecule is known. The next two chapters in this volume discuss methods to merge this known structural information with the raw X-ray data to improve the initial electron density map. In Chapter 7, Tickle and Driessen introduce molecular replacement methods, and, in Chapter 8, Podjarny, Rees, and Urzhumtev explain the use of density modification methods.

Since X-ray crystallography is a well-defined physical technique, it is possible to reconstruct the expected X-ray diffraction pattern from the electron density map generated by solving the three-dimensional structure. Comparison of this stimulated diffraction pattern with experimental data is a mechanism to estimate the quality of the molecular structure, and to refine atomic coordinates by minimizing the discrepancies. These refinement techniques are described by Westhof and Dumas in Chapter 9, whereas, in Chapter 10, Brünger describes a recently developed approach to crystallographic refinement using stimulated annealing techniques.

The remaining chapters in this volume describe the application of X-ray crystallography to a series of specialist topics that are among the most interesting areas of structural biology. In Chapter 11, Neidle describes the elucidation of the structure of synthetic nucleic acid oligomers, whereas, in Chapter 12, Brown and Freemont extend the discussion to protein–nucleic acid complexes. These are the fundamental regulators of gene expression in living tissue. Recent results on the crystal structures of intact viruses has shown us the beautiful symmetry and organization in these systems, and this topic is covered by Fry, Logan, and Stuart. Finally, Newman outlines the progress made in the X-ray crystallography of membrane-bound proteins, which were long considered one of the most difficult structural problems to tackle.

We would like to emphasize that a large number of the most interesting biological structures that have been reported recently have been solved primarily by single individuals. These people, not trained as crystallographers, have a good understanding of the molecular biological and biochemical techniques demanded, and obtained suitable crystals for X-ray study. From then on, they were guided through the processes of data collection, structure determination, and refinement by experts familiar with these specialist techniques. Good Luck!

We would like to thank all of the authors for the care and detail with which they prepared their manuscripts, and for their forbearance in updating chapters during the gestation of this book.

Christopher Jones
Barbara Mulloy
Mark R. Sanderson

Contents

Contributors

SHERIN S. ABDEL-MEGUID • *Department of Macromolecular Sciences, SmithKline Beecham, King of Prussia, PA*
DAVID G. BROWN • *Randall Institute, King's College, University of London, UK*
AXEL T. BRÜNGER • *Howard Hughes Medical Institute and Department of Molecular Biophysics and Biochemistry, Yale University, New Haven, CT*
HUUB P. C. DRIESSEN • *Department of Crystallography, Birkbeck College, London, UK*
PHILIPPE DUMAS • *Institut de Biologie Moléculaire et Cellulaire, Centre National de la Recherche Scientifique, Strasbourg, France*
PAUL S. FREEMONT • *Protein Structure Laboratory, Imperial Cancer Research Fund, London, UK*
ELIZABETH FRY • *Laboratory of Molecular Biophysics, University of Oxford, UK*
ELSPETH F. GARMAN • *Laboratory of Molecular Biophysics, University of Oxford, UK*
DAVID JERUZALMI • *Department of Molecular Biophysics and Biochemistry, Yale University, New Haven, CT*
DEREK LOGAN • *Laboratory of Molecular Biophysics, University of Oxford, UK*
C. BERNADETTE MADDEN • *CRC Biomolecular Structure Unit, The Institute of Cancer Research, University of London, UK*
H. M. KRISHNA MURTHY • *Fels Institute for Cancer Research and Molecular Biology, Temple University School of Medicine, Philadelphia, PA*
STEPHEN NEIDLE • *CRC Biomolecular Structure Unit, The Institute of Cancer Research, Sutton, UK*

RICHARD NEWMAN • *Imperial Cancer Research Fund, London, UK*
ALBERTO D. PODJARNY • *UPR de Biologie Structurale, Université Louis Pasteur, Illkirch, C.U. de Strasbourg, France*
BERNARD REES • *UPR de Biologie Structurale, Université Louis Pasteur, Illkirch, C.U. de Strasbourg, France*
MARK R. SANDERSON • *Department of Biophysics, King's College, London, UK*
JANE V. SKELLY • *CRC Biomolecular Structure Unit, The Institute of Cancer Research, University of London, UK*
DAVID STUART • *Laboratory of Molecular Biophysics, University of Oxford, UK*
IAN J. TICKLE • *Department of Crystallography, Birkbeck College, London, UK*
ALEXANDRE G. URZHUMTSEV *UPR de Biologie Structurale, Université Louis Pasteur, Illkirch, C.U. de Strasbourg, France*
ERIC WESTHOF • *Institut de Biologie Moléculaire et Cellulaire, Centre National de la Recherche Scientifique, Strasbourg, France*

CHAPTER 1

Introduction

Mark R. Sanderson

This chapter is intended to give an overall view of the process of structure solution with some of the basic theory behind it. It is possible to skip the most mathematical section, at any rate, on a first reading. There is a bibliography at the end of this chapter that should provide further reading matter for readers at every level of crystallographic experience.

1.1. Fundamentals of X-Ray Diffraction

X-rays are a form of electromagnetic radiation, with a shorter wavelength than radio waves or visible light. X-rays are used in crystal studies because their wavelength (1.542×10^{-10} m for copper K α radiation) is comparable to the planar separation of atoms in a crystal lattice, if the Bragg description of diffraction from a crystal is considered. The Ångstrom unit, where 1 Å = 10^{-10} m, is still widely used in diffraction circles. Measurements in these units, rather than their SI equivalents, can be spoken in fewer syllables (e.g., 1.547 Å, compared with 0. 1547 nm).

Safety: It must be stressed that X-ray equipment must under no circumstances be used by an untrained operator. Training in its use must be received from an experienced worker.

1.1.1. X-Ray Generation

X-rays are generated when a beam of electrons at a potential of approx 10,000 eV is accelerated from a small tungsten filament (the cathode) to strike an anode (usually a copper target for macromolecular studies). The deceleration of these electrons, which is known by its German name *bremsstrahlung,* causes electrons to be knocked out of the inner *K* and *M*

From: *Methods in Molecular Biology, Vol. 56: Crystallographic Methods and Protocols*
Edited by: C. Jones, B. Mulloy, and M. Sanderson Humana Press Inc., Totowa, NJ

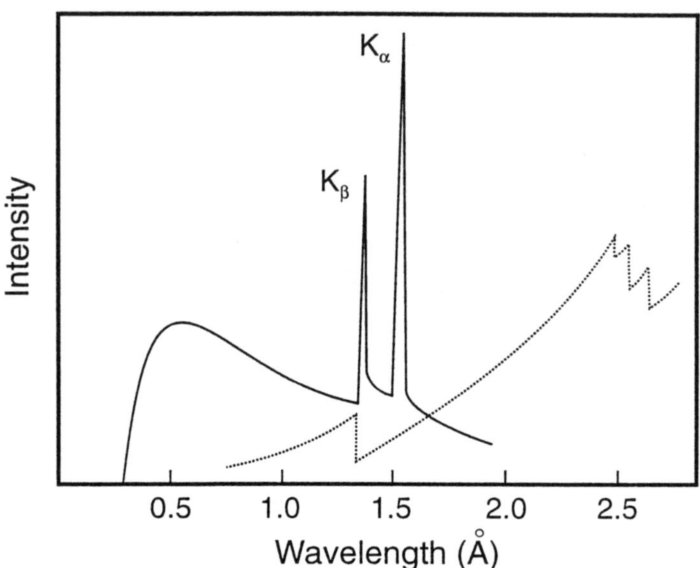

Fig. 1. X-ray spectrum of MoKa, 50 kV, and CuKa, 35 kV. The absorption spectrum of nickel is shown by the dotted line.

atomic shells and dissipates a large amount of heat. When the electrons in higher levels fall back to these inner shells, emission of X-ray radiation occurs. When the transitions are from K to L, then K α1 and K α2 radiations are produced, whereas the transition from M to K leads to K β1 and K β2 radiation. Since the electrons are involved in multiple collisions, these defined lines are superimposed on a background of white radiation. Figure 1 shows a typical X-ray emission spectrum. In macromolecular studies, copper K α radiation is usually used with the K β filtered out either by a graphite monochromator or by nickel filters. Molybdenum radiation of wavelength 0.71 Å is often used in small organic and inorganic molecule diffraction studies, but has also been used for several high-resolution protein data collections.

An alternative source of X-rays is synchrotron radiation, which is generated tangentially to a ring of accelerating electrons. This source of X-radiation is available at various centers throughout the world, such as the Daresbury Laboratories (Warrington, UK), Brookhaven National Laboratories (Long Island, NY), The Photon Factory (Japan), L.U.R.E. (Paris, France), and the E.S.R.F. (Grenoble, France). Synchrotron radiation offers the possibility of tuning the X-ray wavelength to suit the problem being studied, as discussed by Krishna Murthy in Chapter 5, and it has a beam with narrow divergence, resulting in small spot sizes, which is a great advantage when studying viral crystals as discussed by Elizabeth Fry et al. in Chapter 13. The X-ray flux attainable at synchrotron rings is also much higher than that generated in a conventional X-ray laboratory, often allowing higher-resolution data to be collected in a shorter time. Research groups apply for "beam time" at these centers, travel to the synchrotron with their crystals, and collect data during their allocated period. Two types of generators are in general use in X-ray diffraction laboratories, known as sealed-tube generators and rotating anode generators.

1.1.1.1. Sealed-Tube Generators

These X-ray sources consist of a sealed evacuated glass tube containing a filament and a fixed hollow target anode, which is cooled by water. Generators fitted with these tubes produce X-rays of up to 3 kW, corresponding to a current of 50 mA and voltage of 60 kV. Heat generated by the decelerating electrons means that these tubes cannot be operated at very high powers since the anode will melt. The advantage of sealed-tube generators is that they require less maintenance than the rotating anode generators described below, and the sealed tube may easily be replaced at the end of its lifetime. The major disadvantage of these systems is the limit on the operating power of a fixed target source, which results in lower X-ray fluxes compared with those from rotating anode generators.

1.1.1.2. Rotating Anode Generators

Rotating anode generators were developed in order to increase the X-ray flux. The filament is mounted in a focal cup in the electron gun, and the electron beam is directed at a rotating anode (usually copper). The anode is spun so that a cooler region of the copper anode is continually brought into the path of the X-ray beam. This allows higher powers to be used without melting the target. Here again, the rotating copper wheel is water-cooled, often on an internal circuit that is heat-exchanged against an external cooling loop. Figure 2 shows a Rigaku RU-200 X-ray generator, with the rotating anode mounted on top of the stainless-steel column. In this generator, X-rays can exit from two ports (to the left and to the right), sealed by air-tight beryllium windows, which are transparent to X-rays. In the figure, only the right-hand port is in use and has an

Fig. 2. Rigaku RU-200 X-ray generator with a mirror system and an R-AXIS II image plate detector mounted against the right port (courtesy of Dr. Paul Freemont, I.C.R.F.).

X-ray mirror system and image plate detector mounted against it (Rigaku Raxis II, image plate detector; X-ray mirrors developed by Z. Otwinowsky and marketed by Molecular Structure Corporation). The electron gun is evacuated to 10^{-5} Pa by a turbomolecular pump, which is backed on to an oil diffusion pump. These generators typically operate at a power of 5.4 kW when a small filament (300 μm) is used and 12 kW when a broad focus (500 μm filament) is used. Recently, X-ray sources have become available with more compact, high-voltage generators. The older instruments have oil immersed high-voltage tanks, which take up much more floor space, an important consideration when laboratory space is limiting.

1.2. Crystals and Symmetry

A crystal may be thought of as a three-dimensional lattice of molecules. An early study of crystal morphology of quartz in 1669 by a Danish physician, Nicolaus Steno, concluded that the angles between similar crystal faces were the same. At the end of the 18th century, Abbé Hauy and Romé de l'Isle extended these observations to other crystals, and found that the interfacial angles were the same even though the overall morphology of the crystals may be very different. Bravais showed that symmetry criteria limited the number of lattices to the 14 lattices shown in Fig. 2 of Chapter 3. It was known even before the discovery of X-rays, through the mathematical studies of Federov in Russia, Schoenflies in Germany, and Barlow in Britain at the turn of the century, that there is only a finite number of ways of arranging objects symmetrically within a crystal lattice. This gives rise to the 230 possible space groups, which are listed in *International Tables for Crystallography*, published by Rediel NE/Kluwer Academic Publishers, Norwell, MA. A copy of these tables should be available to anyone wishing to work in crystallography.

For biological studies, we need only consider 65 out of the 230 possible space groups, because macromolecules are chiral and therefore only those space groups lacking a center of symmetry need be considered. The subject of crystal symmetry is discussed more fully in Chapter 3.

1.2.1. Miller Indices

The crystal may be thought of as sectioned into planes as shown below (Fig. 3). Miller indices are the three intercepts that a plane makes with the cell axes, in units of the cell edge. For example, if the plane intersects the axes of a cell with lengths a, b, and c at coordinates a', b', and c', then the Miller indices are given by $h = a/a'$, $k = b/b'$ and $l = c/c'$.

1.2.2. Diffraction from Lattices

The crystal may be viewed, by analogy with the diffraction of visible light, as a three-dimensional grating, with the diffracted rays interfering in phase and out of phase to produce a diffraction pattern. The spacing of the resulting pattern is inversely proportional to the lattice spacing as given by Bragg's law:

$$n\lambda = 2d\sin\theta \quad (1)$$

where λ = wavelength, θ = diffraction angle, d = lattice spacing, and n = diffraction order. Figure 4 shows the derivation of Bragg's law. Two incident rays are shown with a path difference given by $\Delta(path) = PQ + QR = n\lambda$.

1.2.3. Resolution

Having crystals that diffract X-rays to large values of θ is vital to being able to solve a structure so that biological detail may be extracted. When a crystallographer is found talking about a new crystal form diffracting to the edge of the film (on a precession camera with a crystal-to-film

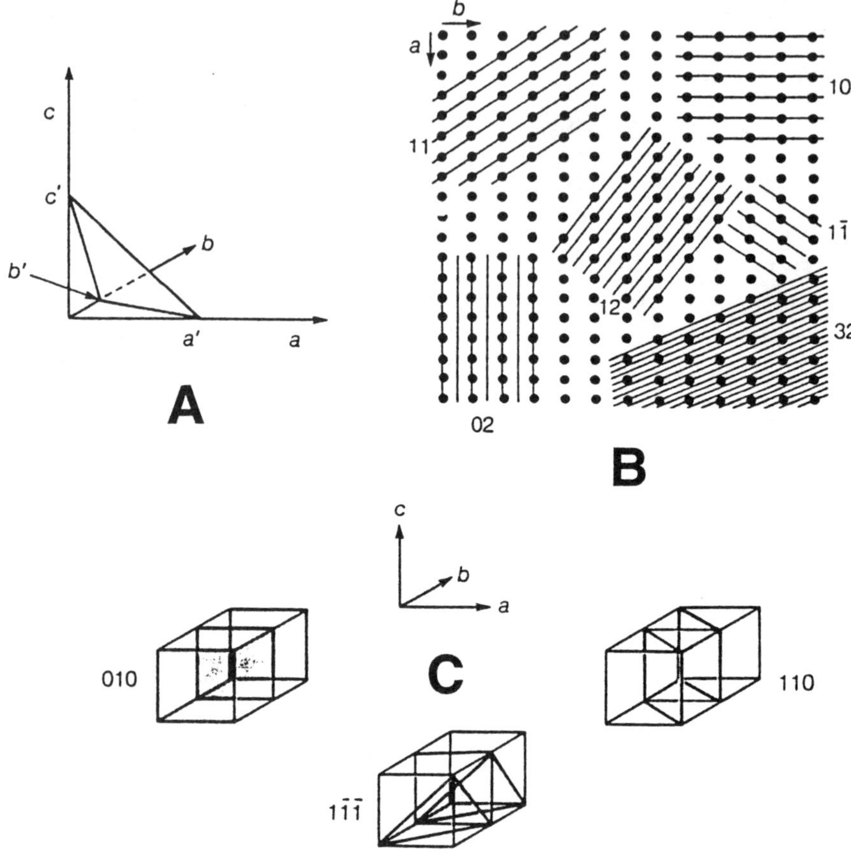

Fig. 3. Miller indices of lattice planes within a crystal. **(A)** A Lattice plane with intercepts *a*', *b*', and *c*' along the *a, b,* and *c* axes. **(B)** Lattice planes in a two-dimensional lattice. **(C)** Lattice planes in a three-dimensional lattice. (Reproduced with permission from ref. *1*.)

distance of 10 cm), this is often a cause for celebration, since the data once collected and processed from this crystal form will allow the polypeptide backbone to be traced (for a protein) or unambiguous positioning of the backbone and bases (for a nucleic acid). Equation 1 may be rearranged as $d = \lambda/2 \sin \theta$, since we are considering first-order diffraction with $n = 1$. Substituting for the diffracting angle θ gives the useful form of the equation $d = \lambda/2 \sin [(1/2)\tan^{-1} (r/F)]$ where r is the distance of a diffraction intensity from the center of the film and F is the crystal-

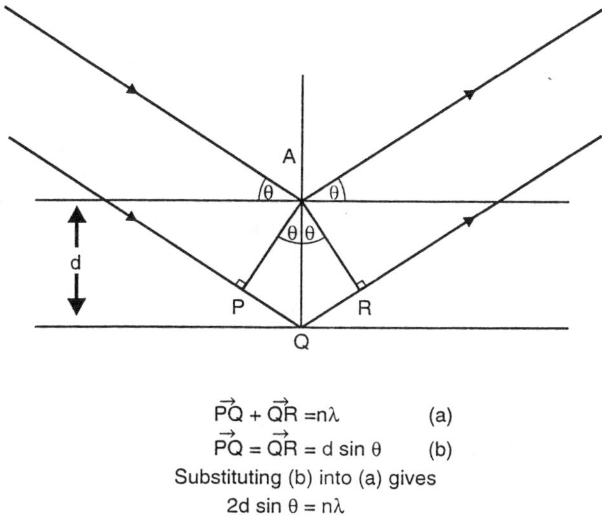

Fig. 4. The derivation of Bragg's law.

to-film distance (10 cm for many precession cameras). Further details of preliminary crystal characterization are discussed in Chapter 3 by Sherin Abdel-Meguid et al. Figure 5 shows the diffraction pattern from a crystal of the thymidine kinase from herpes simplex virus type 1, which has been mounted together with a small amount of buffer in an X-ray capillary tube (Fig. 6) and irradiated with X-rays. Since water is an integral part of the crystal lattice, crystals must be mounted and kept hydrated, a very important observation first made by Hodgkin and Bernal *(2)*. Flash-freezing crystals to liquid nitrogen temperatures may also be used to maintain the lattice hydration as described in Chapter 3. The reflections recorded in this 2° oscillation photograph may be assigned indices h, k, and l and their intensities $I(hkl)$ measured by using integration software. The photograph shows a distorted picture of the reciprocal lattice. In the past, precession X-ray cameras were used to give an undistorted view of the reciprocal lattice, which facilitated space group assignment, and indexing of the reflections, when this was done by hand.

1.3. An Overview of Macromolecular Crystal Structure Solution

This section shall give a brief nonmathematical overview of macromolecular structure solution, leaving a more detailed treatment for later in the chapter (Section 1.4.).

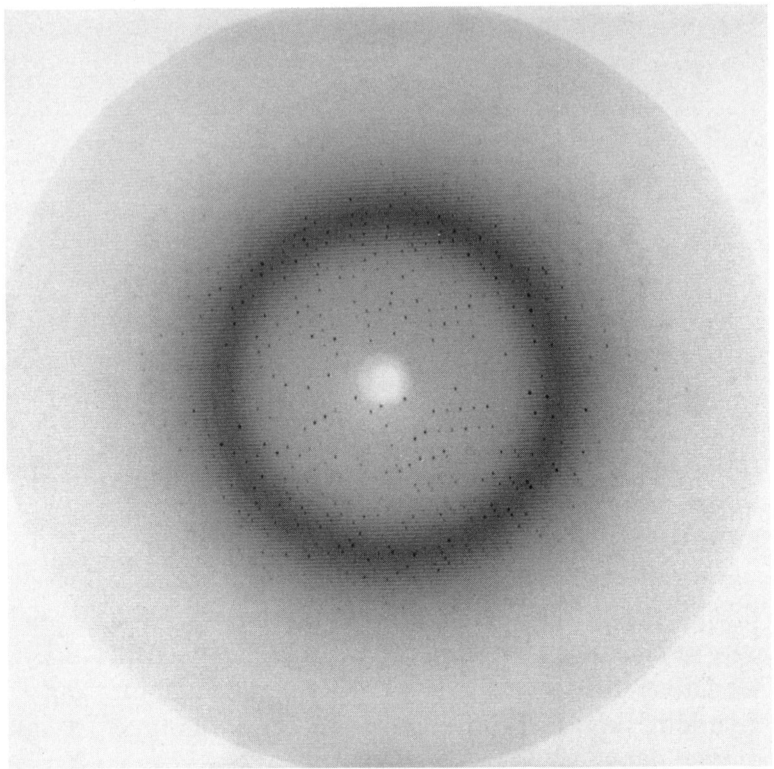

Fig. 5. Diffraction of thymidine kinase from herpes simplex virus type 1 recorded on an MAR image plate detector. (M. R. Sanderson and W. C. Summers, unpublished results.)

1.3.1. Stage 1: Protein Preparation and Crystal Growing

1. The first stage in a crystallographic study is to obtain tens of milligrams of the macromolecule (or macromolecules when the structure of a complex is being undertaken) in a very pure form, either from:
   a. A natural source rich in the protein;
   b. The use of cloning techniques to engineer a vector that will overexpress the desired macromolecule in large amounts; or
   c. Chemical methods, as in the case of DNA synthesis for DNA crystallization.

   Chapter 2 covers aspects of genetic engineering. Biochemical techniques are used to purify the macromolecule; this can usually be achieved in fewer steps with cloned material. An affinity "tag" is often attached in order to aid purification, although cleaving the tag away from the molecule

Fig. 6. Crystal of thymidine kinase mounted in a glass capillary tube and attached to a goniometer head using plasticine. The arcs and sledges on the goniometer head allow the crystal to be centered in the X-ray beam.

of interest may introduce heterogeneity, which hampers crystallization. The knowledge of solubility in different buffer solutions at different salt concentrations gained by biochemical manipulation of the protein can often be very useful when crystallizations are set up.

2. Crystallization of proteins is discussed in Chapter 2, of DNA and protein–DNA complexes in Chapter 12, and for membrane proteins in Chapter 14.

1.3.2. Stage 2: Symmetry Determination

The symmetry of the macromolecular crystals is determined as discussed in Chapter 2. If the crystals are found to be sensitive to radiation damage in initial experiments, then cooling techniques, also discussed in Chapter 2, may be used to extend the crystal lifetime. Macromolecular crystals are formed of molecules that are chiral, so only the 65 space groups that lack a center of symmetry need be considered.

1.3.3. Stage 3: The Strategy for Structure Solution

1. The strategy for structure solution will depend on whether or not a similar macromolecule, or fragment of it, has been solved before, and the coordinates are available.
2. If coordinates are obtainable, then the structure may be solved by molecular replacement using the phase information from the previously solved structure, and only a native X-ray diffraction data set needs to be collected. "Native data" are crystallographic jargon describing data collected from crystals in their native state, unmodified by, for example, heavy-atom derivatization.
3. If a structurally related macromolecule has not been solved, then the phase information has to be obtained *"de novo"* from either several heavy atom derivatives with the technique of multiple isomorphous replacement (MIR, described in Chapter 6), or by using a single heavy-atom derivative and the multiple wavelength methods covered in Chapter 5.
4. Once native X-ray diffraction data and phase information are available, then the electron density map is calculated and the chemical structure of the macromolecule fitted into the electron density map using a computer graphics system, and refinement may begin. In refinement, the best fit between the X-ray diffraction data and the fitted model is achieved computationally, either using the more traditional technique of conjugate gradient energy minimization discussed in Chapter 9 by Eric Westhof and Phillippe Dumas, or by using the recent technique of molecular dynamics discussed in Chapter 10 by Axel Brunger.

1.4. Diffraction Theory

This section shall discuss diffraction theory. The reader may wish to skip this section on a first reading.

Most crystallographic computer programs use as input the structure factor amplitudes F_{hkl}. These structure factor amplitudes are proportional to the square roots of the intensities *(I)*, $| F_{hkl} | = \sqrt{(kI_{hkl}/Lp)}$ where L is the Lorentz correction, which is dependent on the geometry of the camera used to collect the data and arises because the different reflections spend varying times in a reflecting position dependent on their location within reciprocal space and their angle of approach to the reflecting position. The constant p is a correction for the polarization that X-rays experience on reflecting from a crystal. The component of the electric vector parallel to the crystal plane will only be affected by the electron density parallel with the plane, whereas the electric vector perpendicular to the plane will be dependent on electron density in the vertical plane and on the incident angle. The remaining constant k is dependent on beam intensity, crystal size, and other fundamental constants. Its use is avoided by using a relative F, with $| F_{rel} | = c | F_{obs} | = \sqrt{(I_{hkl} / Lp)}$ where F_{obs} is the observed F and is scaled relative to F_{rel} at the refinement stage, once the structure is determined, and a calculated value of F has been derived by back-transformation of the structure model.

The final objective during crystal structure solution is the calculation of an electron density map so that the atomic model may be fitted into it. In order to calculate the electron density, both the amplitudes and phases of the reflections must be known. Since only the square of the amplitude of the waves is recorded, the phase information is lost. The regaining of phase information (known in crystallography as the Phase Problem) is therefore central to structure solution and is discussed below.

The total scattering by the crystal is given by the ratio of the sum of the atomic scattering amplitudes of the atoms in the lattice to the scattering by a point electron at the origin. For N atoms, the structure factor is defined by Eq. (2).

$$F_{hkl} = \sum_{j=1}^{N} f_j \exp (\mathrm{r}_j \cdot s) \tag{2}$$

$$\mathrm{y} = a \sin (\omega t \pm kx) \tag{3}$$

where s is the scattering vector and r_j is the position vector of the jth atom and is given by Eq. (4) in terms of fractional atomic coordinates.

$$r_j = x_j a + y_j b = z_j c \tag{4}$$

The structure factor equation is similar to the wave equation (Eq. 3) encountered in physics in having an amplitude term a and a phase term $(\omega t \pm kx)$. For an explanation of dot (scalar) $r_j \cdot s$ and cross (vector) products, and the vector notation and Fourier transforms given below, the reader is referred to one of the number of excellent mathematical texts listed at the end of this chapter.

The reciprocal lattice vector G_{hkl} is defined by Eq. (5) in terms of the lattice planes of the real lattice; *hkl* are the reflection indices:

$$G_{hkl} = ha^* + kb^* + lc^* \tag{5}$$

where a^*, b^*, and c^* are base vectors in reciprocal space related to the real space vectors a, b, and c for a right handed system:

$$a^* = \frac{b \times c}{a.b \times c},\ b^* = \frac{c \times a}{a.b \times c},\ c^* = \frac{a \times b}{a.b \times c} \tag{6}$$

Equation 7 defines the scattering vector with respect to the reciprocal lattice vector:

$$s = 2\pi G_{hkl} \tag{7}$$

An expression for the scalar or dot product $r_i \cdot s$ may be derived by substituting Eqs. (4) and (5) into Eq. (7):

$$r_j \cdot s = 2\pi(x_j a + y_j b + z_j c) \cdot (ha^* + kb^* + lc^*) = 2\pi(hx_j + ky_j + lz_j) \tag{8}$$

Substituting this expression into Eq. (3) gives an alternative expression for Eq. (3) in terms of fractional atomic coordinates and Miller indices:

$$F_{hkl} = \sum_{j=1}^{N} f_j \exp 2\pi i(hx_j + ky_j + lz_j) \tag{9}$$

The form of the scattering vector is complex, and hence, may be resolved into real and imaginary components:

$$F_{hkl} = A_{hkl} + iB_{hkl} \tag{10}$$

$$A_{hkl} = \sum_{j=1}^{N} f_j \cos 2\pi r_j \cdot s \tag{11}$$

$$B_{hkl} = \sum_{j=1}^{N} f_j \sin 2\pi r_j \cdot s \tag{12}$$

The phase angle may be given as

$$\phi_{hkl} = \tan^{-1}(B_{hkl}/A_{hkl}) \tag{13}$$

1.4.1. Electron Density

If one imagines the crystal divided up into small volumes *dv* with point charges where ρ is the electron density distribution, then an expression for the total scattering amplitude is:

$$F(s) = \int_V \rho(r) \exp(2\pi i s \cdot r) dv \tag{14}$$

The interesting expression for X-ray crystallography is Eq. (15), the inverse transform of Eq. (14), since we are interested in solving the structure by calculating its electron density.

$$\rho(r) = \int_{v^*} F(s)\exp(-2\pi i s \cdot r)dv^* \tag{15}$$

Equation 16 is used for computing the electron density:

$$\rho(r) = (1/v) \sum_{h=-\infty}^{\infty} \sum_{k=-\infty}^{\infty} \sum_{l=-\infty}^{\infty} F_{hkl}\exp - 2\pi(hx_j + ky_j + lz_j) \tag{16}$$

If Friedel's law [$I(h,k,l) = I(-h, -k, -l)$] holds, then Eq. (16) above simplifies to:

$$\rho(xyz) = (1/v) \sum_{h=-\infty}^{\infty} \sum_{k=-\infty}^{\infty} \sum_{l=-\infty}^{\infty} |F_{hkl}|\cos[2\pi(hx_j + ky_j + lz_j) - \phi_{hkl}] \tag{17}$$

The theory of diffraction is covered in greater depth in a number of excellent texts, some of which are listed in the bibliography.

1.4.2. Phasing the Macromolecular Structure

1.4.2.1. The Phase Problem

As discussed above, in order to compute an electron density map, the phase information must be recovered. A solution to this problem for macromolecular crystallography was achieved by Max Perutz and coworkers *(3)*, who showed that if heavy atoms (such as mercury in a compound, which may bind to a cysteine group in a protein) were soaked into the crystal lattice, and they bound to the protein without disturbing the crystal cell dimensions, then the positions of these heavy atoms may be used to regain phase information. Figure 7 shows the perturbation of amplitude and phase induced in a hypothetical triangular molecule on binding a heavy-atom compound. Data from such crystals are called "heavy-atom derivative data" or simply "derivative data" by crystallographers. The technique for structure solution using heavy atom derivatives is known as multiple isomorphous replacement (MIR). In order to overcome the phase ambiguity, the heavy-atom positions for two or more heavy-atom derivatives are used to determine the phase. In practice, the more derivatives that can be used, the better, since the overall phase may be calculated with greater certainty. Chapter 6 provides a full discussion of phasing using heavy-atom derivatives. Chapter 5 discusses the use of the anomalous contribution with the isomorphous contribution to calculate

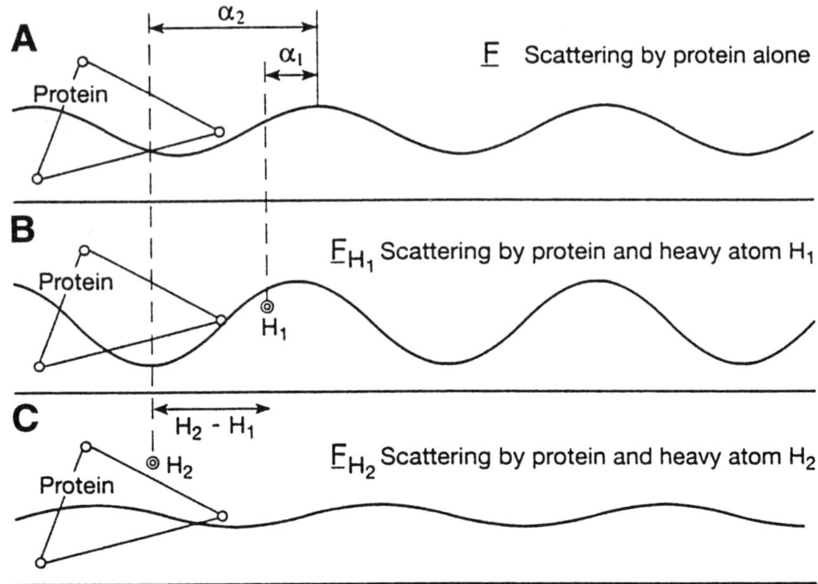

Fig. 7. Wave diffracted by triangle of atoms representing a protein. **(A)** Scattering by the protein alone. **(B)** and **(C)** Changes in amplitude and phase of diffracted wave caused by the heavy atoms H1 and H2. (Reproduced with permission from Protein Structure by Max Perutz.)

phases using only one derivative. The first stage in this phasing process is being able to locate the heavy-atom positions, which may be achieved by either calculating a Patterson map based on the difference between the derivative and native data or by using Direct methods on this difference data.

A very active area of research is the development of techniques to phase structures directly from the intensity data using probabilistic methods. This area has been pioneered primarily by Bricogne *(4–6)*. Gilmore and Bricogne have now written a program based on these methods called MICE *(7)*. Structure solution for small organic and inorganic molecules directly from intensity data is now routine. The problem for macromolecules is much more difficult, since crystals of macromolecules do not diffract to atomic resolution. Direct methods techniques cannot therefore be used to phase macromolecular data at present, though in the future this may become possible using probabilistic methods.

1.4.2.2. The Patterson Method

Patterson developed this method in 1934 initially to locate heavy-atom positions in small organic and inorganic molecules, so that their positions may be used in phasing these structures. He derived an equation *(8)*, now named the Patterson function, using as coefficients the phaseless square of the structure factor amplitudes:

$$P(r) = 1/v \sum_{hkl} | F_{hkl} |^2 \exp(-2\pi ih.r) \quad (18)$$

where $h = G_{hkl}$ is the reciprocal lattice vector. Since $| F_{hkl} |^2 = | F_{\overline{hkl}} |^2$, the Patterson is a real expression and may be expressed as:

$$P(r) = 1/v \sum_{hkl} | F_{hkl} |^2 \cos(-2\pi ih.r) \quad (19)$$

Using convolution theory and Fourier transformation, the expression in terms of electron density is:

$$P(r) = v \int_{V} \rho(u)\rho(u + r) \quad (20)$$

The Patterson function has the following important features:

1. There are $N^2 - N$ nonorigin peaks in a calculated map, so peak overlap makes the Patterson map hard to interpret.
2. The distance of the peaks from the origin is the interatomic vectors between the *i*th and *j*th atoms $(r_i - r_j)$.
3. The heights of the peaks in the Patterson are proportional to the products of the atomic weights of the *i*th and *j*th atoms Z_iZ_j.
4. Space group symmetry introduces simplification into Patterson interpretation. For example, in space group $P2_1$, the vectors between general equivalent positions x, y, z and $\overline{x}, 0.5 + y, \overline{z}$ produces the Harker section $(2x, 0.5, 2z)$ with all the vectors of this type in the plane $y = 0.5$.

The Patterson vectors *u, v, w* of the peaks high in the peak height listing are analyzed for correspondence with the Harker vectors derived from the crystal space group. This latter analysis is often called "hand solution."

The Patterson technique has now been widely applied to protein difference data in order to locate the heavy-atom positions within a macromolecular crystal. The differences $(F_{deriv} - F_{native})$ are calculated for derivative and for the native protein, once they have been scaled together. If one takes the case of a protein soaked in a mercury compound, one may think of the difference data as containing only the contribution from the mercury atom, since the contribution from the protein has been removed by taking the difference. In order for this to be true, the derivative data must be very well scaled against the native; time spent making sure this is true often pays dividends. An example of the difference Patterson map for an osmium derivative of the porcine growth hormone is given in Chapter 6.

In addition to solving the difference Patterson by hand as described above, a range of software is now available, such as HASSP *(9)* written by Terwilleger et al. and RSPS by Knight, which is integrated into the CCP4 package. These programs take as input $(F_{deriv} - F_{native})$ and will solve the Patterson map automatically. It is advisable to check the automatic solution against the hand solution, and to compare these results with those determined from Direct methods. Finally, crossdifference Fourier maps calculated using phases determined from one heavy-atom derivative should solve the positions of other heavy-atom derivatives. Finally, the positions determined by Patterson and Direct methods and crossdifference Fourier maps should be self-consistent.

1.4.3. Direct Methods

Direct methods are used routinely for solving small organic/inorganic structures and are based on the inequality and probability relationships between structure factors that arise from the impossibility of negative electron density.

Because of the fact that the structure factor F_{hkl} is dependent on $\sin\theta/\lambda$ and space group symmetry, normalized structure factors E_h with these contributions removed are used in inequality and probability relationships.

$$|E_h| = \frac{|F_h|}{\varepsilon^{1/2}\left(\sum_{i=1}^{N} f_i^2\right)^{1/2}} \tag{21}$$

where $h = hkl$, and ε is a term that varies for certain groups of reflections in given space groups (these conditions are tabulated in International tables). For the case of solving heavy-atom positions within a protein, it is the scaled structure factor differences $(F_{deriv} - F_{native})$ that are normalized; the Wilson plot from the normalization routine should be linear. If it is not, this may be because of poor scaling and may result in failure to obtain a Direct methods solution. In order to solve the heavy-atom positions in derivatives of macromolecules, one needs only to consider the noncentrosymmetric space groups, discussion will be limited to these. For reflections in noncentrosymmetric space groups, a phase angle can take any value from 0 to 2π. Once origin and enantiomorph reflections have been defined, it is then possible to build up a "tree" of phased reflections from this starting set by using the expression in an equation known as the $\Sigma 2$ relationship:

$$\phi_h = (\phi_k + \phi_{h-k}) \tag{22}$$

where the parentheses represent summation over reciprocal space. SHELXS *(10)* and MULTAN 80 *(11)* are the Direct methods programs generally used, and these will automatically select starting reflections and use the $\Sigma 2$ relationship with each reflection being given a phase value of ($\pi/4$, $3\pi/4$, $5\pi/4$, $7\pi/4$) in turn. The phases are then refined by the weighted tangent formula:

$$\langle \tan_h \rangle = \frac{\sum_k w_k w_{h-k} | E_k E_{h-k} | \sin (\phi_k + \phi_{h-k})}{\sum_k w_k w_{h-k} | E_k E_{h-k} | \cos (\phi_k + \phi_{h-k})} \tag{23}$$

where $w_h = \tanh [(\alpha_h)/2]$ and $\alpha_h = N^{-1/2} | E_h E_k E_{h-k} |$. α_h is a test for the validity of a phase and N is the number of atoms in a unit cell.

The programs analyze the probability that a given starting phase set gives rise to a correct solution on the basis of several figures of merit criteria, and then calculate and peak-pick Fourier maps for the highest ranked solutions. The solution to the heavy-atom positions often corresponds to the highest peaks within this map *(12)*.

1.4.4. Multiple Isomorphous Replacement

Once the positions of the heavy atoms have been located, the phases α_{best} determined from them may then be used to calculate the electron density map using Eq. (24), where m is the figure of merit, as discussed fully in Chapter 6:

$$\rho(r) = \Sigma m F_p \exp | i\alpha_{best} | \exp (-2\pi i s \cdot r) \tag{24}$$

In the case of a protein, the amino acid sequence is then fitted into the electron density map using a graphics program, such as O *(13,14)*, and the structure refined as discussed in Chapters 9 and 10.

Acknowledgments

I thank all the authors for their contributions and all the subsequent revisions. I thank Drs. Max Perutz and Don Crothers for permission to reproduce figures from their books, and Kate Kerwin and Mark Simon for photographic and graphical work.

References

1. Eisenberg, D. and Crothers, D. (1979) *Physical Chemistry with Applications to the Life Sciences,* Benjamin-Cummings, Redwood City, CA.
2. Bernal, J. D. and Crowfoot, D. (1934) Use of the centrifuge in determining the density of small crystals. *Nature* **134,** 809,810.
3. Green, D. W. Ingram, V. M., and Perutz, M. F. (1954) The structure of haemoglobin IV. Sign determination by the isomorphous replacement method. *Proc. Roy. Soc.* **A225,** 287–307.
4. Bricogne, G. (1984) Maximum entropy and the foundations of Direct methods. *Acta Cryst.* **A40,** 410–445.
5. Bricogne, G. (1988) A Bayesian statistical theory of the phase problem. I. A multichannel maximum-entropy formalism for constructing generalized joint probability distributions of structure factors. *Acta Cryst.* **A44,** 517–545.
6. Bricogne, G. (1991) Maximum entropy as a common statistical basis for all phase determination methods, in *Crystallographic Computing 5* (Moras, D., Podjarny, A. D., and Thierry, J. C., eds.), Oxford University Press, Oxford, UK.
7. Gilmore, C. J. and Bricogne, G. (1991) Maximun entropy, likelihood, and the phase problem in single crystal and powder diffraction, in *Crystallographic Computing 5* (Moras, D., Podjarny, A. D., and Thierry, J. C., eds.), Oxford University Press, Oxford, UK.
8. Patterson, A. L. (1935) A direct method for the determination of the components of interatomic distances in crystals. *Z. Krist.* **90,** 517–542.
9. Terwilliger, T. C., Kim, S.-H., and Eisenberg, D. (1987) Generalized method of determining heavy-atom positions using the difference Patterson function. *Acta Cryst.* **A43,** 1–5.
10. Sheldrick, G. M. (1990) Phase annealing in SHELX-90: Direct methods for larger structures. *Acta Cryst.* **A46,** 467–473.
11. Germain, G., Main, P., and Woolfson, M. M. (1971) The application of phase relationships to complex structures III. The optimum use of phase relationships. *Acta Cryst.* **A21,** 410–445.
12. Sheldrick, G. M. (1991) Chapter 13, in *Crystallographic Computing 5* (Moras, D., Podjarny, A. D., and Thierry, J. C., eds.), Oxford University Press, Oxford, UK.
13. Jones, T. A., Zou, J.-Y., Cowan, S. W., and Kjeldegaard, M. (1991) Improved methods for building protein models in electron density maps and the location of errors in these models. *Acta Cryst.* **A47,** 110–119.
14. Jones, T. A. and Kjeldegaard, M. (1994) Chapter 1, in *From First Map to Final Model* (Bailey, S., Hubbard, R., and Waller, D., eds.), CCP4 Workgroup.

Bibliography

Mathematical Texts

Arfken, G. (1970) *Mathematical Methods for Physicists.* Academic, New York.

Bamberg, P. and Sternberg, S. (1991) *A Course in Mathematics for Students of Physics*, vols. 1 and 2. Cambridge University Press, New York.

Boas, M. L. (1983) *Mathematical Methods for the Physical Sciences.* Wiley, New York.

DuChateau, P. C. (1992) *Advanced Math for Physicists and Engineers.* Harper Collins outline series, Harper Collins, New York.

Fitts, D. D. (1974) *Vector Analysis in Chemistry.* McGraw Hill, New York.
Hirst, D. M. (1994) *Mathematics for Chemists.* Macmillan, New York. (This covers vector manipulation and Fourier transformation and is particularly recommended as an introduction.)
Janin, J. (1985) Chapter 5, in *Methodes Biophysiques pour l'etude des Macromolecules,* Hermann, Paris.
Margeneau, H. and Murphy, G. M. (1961) *The Mathematics of Physics and Chemistry.* van Nostrand, Princeton, NJ.
Prince, E. (1994) *Mathematical Techniques in Crystallography and Material Science,* 2nd ed., Springer Verlag, New York.
Stephenson, G. (1979) *Mathematical Methods for Science Students*, 2nd ed., Longman, London.
Stephenson, G. (1985) *Worked Examples in Mathematics for Scientists and Engineers.* Longman, London.

Books on Crystallography

Some of the older books give a very clear introduction to the subject. Unfortunately many of these are out of print and may only be obtainable from libraries.

Basic Introductions

Dressier, D. and Potter, H. (1991) *Discovering Enzymes.* W. H. Freeman, New York.
Matthew, C. K. and van Holde, K. E. (1990) *Biochemistry*, Benjamin-Cummings, Redwood City, CA.
Stryer, L. (1995) *Biochemistry*, 4th ed., W. H. Freeman, New York.

Short Introductions to X-Ray Structure Determination

Branden, C.-I. and Tooze, J. (1991) Chapter 17, in *Introduction to Protein Structure*, Garland, New York.
Cantor, C. R. and Schimmel, P. R. (1980) Part II of *Biophysical Chemistry*, W. H. Freeman, New York.
Eisenberg, D. and Crothers, D. (1979) Chapters 16 and 17, in *Physical Chemistry with Applications to the Life Sciences.* Benjamin-Cummings Publishing Company, Redwood City, CA. (Contains clear description of X-ray diffraction, which includes interesting short biographies of J. D. Bernal and J.-B. J. Fourier.)
Holmes, K. C. and Blow, D. M. (1965) *Methods of Biochemical Analysis*, vol. 13, Wiley, New York, 113–239.
Janin, J. (1985) Chapters 1–4, in *Methodes Biophysiques pour l'etude des Macromolecules.* Hermann, Paris. (This is a very good introduction for the French reader.)
Perutz, M. (1992) Chapter 1 and Appendix 1, in *Protein Structure, New Approaches to Disease and Therapy.* W. H. Freeman, New York.
Sawyer, L. and Turner, M. A. (1992) Chapter 12, in *Crystallization of Nucleic Acids and Proteins* (Ducruix, A. and Geige, R., eds.), IRL, New York.
Stuart, D. and Jones, Y. (1993) Chapter 9, in *Protein Engineering* (Sternberg, M., ed.), IRL, New York.
van Holde, E. (1985) *Physical Biochemistry*, 2nd ed., Prentice-Hall, Englewood Cliffs, NJ.

Texts on X-Ray Structure Determination

Blundell, T. B. and Johnson, L. H. (1976) *Protein Crystallography.* Academic, New York. (This is an excellent, indispensible guide to the subject, although the data collection sections are now dated.)

Buerger, M. J. (1959) *Vector Space.* Wiley, New York.

Buerger, M. J. (1976) *Contemporary Crystallography.* McGraw Hill, New York.

Bunn, C. W. (1961) *Chemical Crystallography*, 2nd ed., Oxford University Press, New York.

Drenth, J. (1994) *The Principles of Protein X-ray Crystallography.* Springer-Verlag, New York.

Dunitz, J. D. (1979) *X-ray Analysis and the Structure of Organic Molecules.* Cornell University Press, Ithaca, N. Y. (This is a thorough treatment of small molecule crystallography.)

Giaccovazzo, C., Monaco, H. L., and Viterbo, B. (1992) *Fundamentals of Crystallography.* Oxford University Press, New York.

Glazer, A. M. (1987) *The Structure of Crystals,* Adam Hilger, Bristol, UK.

Glusker, J. P. and Trueblood, K. N. (1985) *Crystal Structure Analysis, A Primer,* 2nd ed., Oxford University Press.

Ladd and Palmer, R. (1989) *X-ray structure determination: A Practical Guide*, 2nd ed., Wiley, New York. (This is a very good on symmetry and space group derivations, covers both small-molecule and macromolecular crystallography.)

Lifson, H. and Taylor, C. A. (1958) *Fourier Transforms and X-ray Diffraction.* G. Bell, London.

Lipson, H. S. (1970) *Crystals and X-rays.* Wykeham Publications Ltd., London (This is a clear elementary introduction.)

Lipson, H. and Cochran, W. (1957) *The Determination of Crystal Structures.* G. Bell, London.

McRee, D. E. (1993) *Practical Protein Crystallography.* Academic, New York.

Rhodes, C. (1993) *Crystallography Made Crystal Clear: A Guide for Users of Macromolecular Models.* Academic, New York.

Sherwood, D. (1976) *Crystals, X-rays and Proteins.* Wiley, New York. (This is a very understandable treatment, which derives all the mathematical aspects of the subject.)

Stout, G. H. and Jensen, L. H. (1989) *X-ray Crystal Structure Determination*, 2nd ed. Wiley, New York. (This provides good coverage of the basics of X-ray crystallography, primarily from a small-molecule perspective.)

Wilson, H. R. (1966) *Diffraction of X-rays by Proteins, Nucleic Acids, and Viruses.* Edward Arnold, London. (This gives a description of diffraction and especially of helical diffraction by a member of the King's College DNA group.)

Woolfson, M. M. (1978) *An Introduction to X-ray Crystallography.* Cambridge University Press, Cambridge, UK. (This develops diffraction theory from a scattering theory perspective, rather than starting with the Bragg equation. It is an excellent treatment of Direct methods by a leader in the field and recommended for readers stronger in physics.)

Woolfson, M. M. (1961) Direct Methods in Crystallography, Oxford University Press, New York.

Advanced Texts

Dodson, G., Glusker, J. P., and Sayre, D. (eds.) (1981) *Structural Studies on Molecules of Biological Interest.* Oxford University Press, New York.

Moras, D., Podjarny, A. D., and Thierry, J. C. (eds.) (1991) *Crystallographic Computing 5.* Oxford University Press, New York. (This is an extensive series of articles on all aspects of macromolecular structure solution.)

Rollett, J. S. (1965) *Computing Methods in Crystallography.* Pergamon Press, Oxford.

Rossman, M. G. (ed.) (1972) *The Molecular Replacement Method.* Gordon and Breach, New York.

Wyckoff, H. W., Hirs, C. H. W., and Timasheff, S. N. (eds.) (1985) *Diffraction Methods for Biological Macromolecules, Methods in Enzymology*, vols. 114 and 115. Academic, New York.

CCP4 Weekend Workshops

The contributions to these workshops (which are organized by the CCP4 workgroup) are written up and circulated to the participants. They provide an invaluable source of up-to-date methods and applications. Below are listed the titles since 1987.

Helliwell, J. R., Machin, P. A., and Papiz, M. Z. (1987) Computational Aspects of Protein Crystal Data Analysis.

Bailey, S., Dodson, E., and Phillips, S. (1988) Improving Protein Phases.

Goodfellow, J., Hendrick, K., and Hubbard, R. (1989) Molecular Simulation and Protein Crystallography.

Hendrick, K., Moss, D. S., and Tickle, I. J. (1990) Accuracy and Reliability of Macromolecular Crystal Structures.

Wolf, W., Evans, P. R., and Leslie, A. G. W. (1991) Isomorphous Replacement and Anomalous Scattering.

Dodson, E. J., Gover, S., and Wolf, W. (1992) Molecular Replacement.

Sawyer, L., Isaacs, N., and Bailey, S. (1993) Data Collection and Processing.

Bailey, S., Hubbard, R., and Walter, D. (1994) From First Map to Final Model.

Computer Packages for Macromolecular Structure Solution

Information on crystallographic software is obtainable on the World Wide Web address http://www.unige.ch/crystal/w3vlc/crystal.index.html

Software	Source
CCP4	SERC Daresbury Laboratories, Warrington, Cheshire WD4 4AD, UK
Phases	Bill Furey, VA Medical Centre, Pittsburgh, PA
Protein	Wolfgang Steigemann, Max Planck Institut fur Biochimie, Martinsreid, Germany
X-plor	Axel Brunger, Department of Molelcular Biophysics and Biochemistry, Yale University, CT 06511
Xtal	S. Hall, Crystallographic Centre, University of Western Australia, Nedlands 6009, Australia

CHAPTER 2

Overexpression, Isolation, and Crystallization of Proteins

Jane V. Skelly and C. Bernadette Madden

1. Introduction

Rapid developments in recombinant technology have made it possible to overproduce selected proteins of specific interest to the levels required for structural analysis by X-ray crystallography. High-level gene expression has facilitated the purification of many proteins that are normally only expressed at low concentrations, as well as those that have proven difficult to purify to homogeneity from natural sources. Furthermore, advances in oligonucleotide site-directed mutagenesis have enabled proteins to be engineered so as to possess certain features that may confer stability or assist in their isolation. There are several examples of proteins that, despite rigorous purification from their natural source, have defied crystallization attempts, e.g., human growth hormone, but have been successfully crystallized from recombinant sources *(1)*. The lack of posttranslational processing in bacterial expressed proteins can often be an advantage to the crystallographer where microheterogeneity presents a problem. Indeed, certain features or residues of a protein that are believed to impede crystal formation by preventing a close-packing arrangement may be successfully deleted by genetic manipulation without destroying its essential functionality *(2)*.

2. Overexpression

Many factors influence the selection of an appropriate expression system for providing a protein suitable for structural studies. Probably the

From: *Methods in Molecular Biology, Vol. 56: Crystallographic Methods and Protocols*
Edited by: C. Jones, B. Mulloy, and M. Sanderson Humana Press Inc., Totowa, NJ

simplest and least expensive method for production is in bacteria, usually *Escherichia coli*, but if the protein requires further processing for its stability and activity, then it may be necessary to select a eukaryotic based system. These include yeast, fungi, insect, and mammalian cells. Other factors to be considered include protein size, the presence of disulfide bonds, and whether the foreign gene product is likely to be toxic to the host cell.

The methodology for the overexpression of recombinant genes is ever-expanding. It is possible here merely to provide a limited overview of some of the expression systems at our disposal together with a brief rationale as to their selection. It is assumed throughout this discussion that the gene coding for the protein to be overproduced has already been cloned. For detailed laboratory protocols, *see* Sambrook et al. *(3)*.

2.1. Overexpression in E. coli

An understanding of the genetics of *E. coli* has enabled the design and construction of expression vectors and selection of host strains to achieve the maximum possible expression of virtually any cloned gene. The *E. coli* promoter sequence that provides the signal for transcription, i.e., recognition by the σ factor of RNA polymerase, consists of two consensus sequences situated some –10 and –35 bases upstream from the initiation codon. Expression vectors based on *E. coli* are designed to contain a promoter region supplied by the upstream region of an appropriate *E. coli* gene. This is sited before a unique restriction site into which the gene to be expressed may be inserted. The new gene is then placed under the control of the *E. coli* promoter. Minor differences between the consensus promoter sequences are effective in determining the level of transcription of the gene, i.e., the frequency with which RNA polymerase initiates transcription. The most effective way to maximize transcription is to locate the gene downstream from a strong regulatable promoter. A number of plasmid vectors containing such strong promoters have been designed for use with suitable host strains *(4–6)*.

Levels of expressed gene product are normally measured as a percentage of the total soluble cell protein. This can vary widely from <1% to >50% depending on several factors, including:

1. The vector-host system used;
2. The stability of the mRNA;
3. The stability of the expressed gene product;
4. The possible adverse effects of the accumulated product on the host; and
5. The conditions of fermentation and induction, as detailed for each vector.

Some examples of the more frequently used *E. coli* promoters are given below.

2.1.1. *The* lac *Operon*

The *lac* operon is probably the best example of regulatory gene expression in bacteria *(7)* and has therefore been extensively used in the construction of expression vectors. It has the disadvantage of requiring a chemical inducer, which can be prohibitively expensive if used for large-scale fermentation. The *lac* promoter contains the sequence that controls transcription of the lacZ gene coding for β-galactosidase, one of the enzymes that converts lactose to glucose and galactose. It also controls transcription of lacZ', which codes for an α-peptide fragment of the same enzyme. Certain strains of *E. coli* that lack this fragment are only able to synthesize a functional β-galactosidase enzyme when harboring vectors carrying the lacZ' sequence, e.g., pUC and M13. This can be used advantageously as a means of selecting for recombinants. The *lac* promoter is induced by either allolactose, a naturally occurring isomeric form of lactose, or isopropyl β-D-thiogalactoside (IPTG), a nondegradable substrate, at a concentration of <1 m*M* in the growth medium.

2.1.2. *The* trp *Promoter*

The *trp* promoter is located upstream of several genes coding for enzymes responsible for the biosynthesis of tryptophan. The *trp* promoter is repressed in the presence of tryptophan, but induced by either 3-indolylacetic acid or the absence of tryptophan in the growth medium (in a defined minimal medium, such as M9CA). A series of plasmids containing the *trp* promoter have been described *(8,9)*.

2.1.3. *The* tac *Promoter*

The *tac* promoter, a synthetic hybrid containing the –35 sequence derived from the *trp* promoter and –10 from *lac*, is regulated by the lac repressor and is therefore induced in the presence of IPTG. The *tac* promoter is several times stronger than either *lac* or *trp*, and has been found to be extremely successful for high-level expression. A series of plasmid vectors containing the *tac* promoter together with the appropriate restriction sites for cloning have been constructed by Amann et al. *(10)*.

2.1.4. *Bacteriophage λpL*

Bacteriophage λpL is an extremely powerful promoter responsible for the transcription of λ DNA. The product of the λcI gene, i.e., λ repressor, represses the λ promoter at an adjacent operator site in the plasmid. Selected *E. coli* host strains synthesize a temperature-sensitive defective form of the cI protein. At temperatures <30°C, the mutant cI is able to repress the λpL promoter, but, at a higher temperature, this protein becomes inactive, thus allowing the cloned gene to be transcribed. The host cells are usually grown at 28–30°C to midlog phase when the temperature is rapidly adjusted to 42°C. The temperature shift can also induce heat-shock genes, some of which encode proteases. However, this problem can be avoided by using a λcI^+ lysogen and inducing with nalidixic acid, which results in inactivation of the repressor at low temperatures *(11)*.

2.1.5. Bacteriophage T7 Promoter

E. coli host cells are lysogenized with bacteriophage carrying the gene for T7 RNA polymerase, which specifically recognizes the bacteriophage T7 gene 10 promoter carried on the plasmid vector upstream of the gene to be expressed. The bacteriophage RNA T7 polymerase is usually more efficient than the *E. coli* RNA polymerase and consequently often enables very high levels of expression, which cannot be attained in other systems *(12)*. The target protein can represent up to 50% of the total cell protein in only a few hours after induction. The target genes, in the first instance, are cloned using hosts that do not contain the T7 RNA polymerase gene. The plasmids, once established, are then transferred into the expression host harboring the T7 RNA polymerase gene under the control of a suitable inducible promoter. The specificity of the T7 polymerase is important. The host-cell RNA polymerase does not recognize the T7 promoter, and so provides a better control of expression, which is important for plasmid stability. This tightly controlled system prevents basal or leaky expression of the foreign protein, which may prove toxic to the cell, and thus helps to reduce plasmid instability. A series of derivatives of the original pET vectors constructed by Studier and Moffatt *(12)* have been developed by Novagen, Inc. (Madison, WI) and are now available in kit form for convenient cloning and expression. Expression is induced by addition of IPTG to the medium.

2.2. Expression of Eukaryotic Genes in E. coli

Fundamental differences in the mechanism and control of gene expression between eukaryotes and prokaryotes present certain difficulties when using prokaryotic expression systems for expressing eukaryotic cDNA *(13)*. Nevertheless, there are numerous examples of biologically active eukaryotic proteins overexpressed in *E. coli.* The problems that need to be addressed are the following.

2.2.1. The Absence of a Suitable Strong Ribosome Binding Site in Eukaryotes

In *E. coli*, the ribosome binding site consists of the initiation codon AUG and the Shine-Dalgano sequence located several bases upstream. It is possible to provide a synthetic oligonucleotide fragment containing a suitable binding site or, alternatively, to clone into a specific vector that possesses a translation initiation region. This entails insertion of the cloned gene so that the second codon is adjacent to the *E. coli* initiation codon. More sophisticated vectors have been developed that provide all the necessary signals for gene expression, i.e., strong regulatable promoter, ribosome binding site, and termination sequence derived from a cloned *E. coli* gene from which the reading frame has been removed. These, commonly known as translation vectors, are also referred to as cassette vectors.

2.2.2. The Presence of Introns in Eukaryotic DNA

cDNA can be generated by use of an mRNA template reverse transcriptase, a viral enzyme that produces a strand of DNA. DNA polymerase allows synthesis of the second strand, and S1 nuclease cuts any loops. Alternatively, the entire gene can be synthesized. In the case of large genes, fragments can be synthesized and ligated.

2.2.3. Termination Sequences

The foreign DNA may contain sequences that are recognized as termination sequences in *E. coli*, resulting in premature termination of transcription. Differences in codon preference between eukaryotes and *E. coli* may affect translation, resulting in low levels of expression and/or premature termination.

2.2.4. Induced Cell Fragility

The expression of the foreign gene at high levels may also be harmful to the host cell, often resulting in cell fragility, which therefore places the recombinant cell at a disadvantage with respect to the nonrecombinant cell.

2.2.5. Posttranslation Modification

Many eukaryotic proteins become fully functional and active only after specific posttranslational modifications have been carried out. Glycosylation, phosphorylation, and specific proteolytic cleavage will not be executed by the bacterial cells. Furthermore, many eukaryotic proteins expressed in bacterial systems are incorrectly folded with inappropriately formed disulfide bonds.

2.2.6. Plasmid Instability

During rapid cell growth, recombinant plasmids may be lost or the copy number may be reduced, allowing the nonrecombinant cells to take over. Plasmid instability can arise when the product of the cloned gene is toxic to the host. This frequently occurs when the cells have reached their saturation level, but is less likely to occur if antibiotic selection is maintained in culture. It should be noted that ampicillin is inactivated by β-lactamases secreted by *E. coli* into the medium. The use of carbenicillin instead of ampicillin can help to avoid this, or alternatively, use another antibiotic for selection.

2.2.7. Inclusion Bodies

Recombinant proteins expressed at high levels in *E. coli* often form insoluble, inactive aggregates known as inclusion bodies, which can be separated out by centrifugation. The formation of inclusions can be advantageous for stabilization and may be used as a primary separation step, provided the protein can be solubilized and renatured into its active form *(14)*. This usually involves treatments with detergents and/or various denaturants, such as guanidinium chloride and urea, followed by renaturation in a suitable buffer. Although commonly used for extracting small proteins and polypeptide hormones, such as insulin, many larger proteins do not successfully refold to their active native forms, and therefore, inclusions should be avoided. The formation of insoluble inclusion bodies may be circumvented by engineering the secretion of the product into the periplasm or medium (*see* Section 2.4.).

2.3. Eukaryotic Expression Systems

The difficulties associated with expression of eukaryotic genetic material in *E. coli* have led to the adaptation of higher organisms as the hosts for overexpression. Microbial eukaryotes, such as yeast and filamentous fungi, process their gene products in a way that is more closely related to higher organisms. Brewer's yeast *(Saccharomyces cerevisiae)* is particularly favored for large-scale production of eukaryotic gene products, since it is nonpathogenic and its fermentation characteristics are well documented. A number of expression vectors based on yeast promoters and translation initiation signals have been designed *(15)*. Recently, an expression system has been developed based on *Pichia pastoris,* a yeast that is able to metabolize methanol as a sole carbon source. Expression is regulated by the AOX1 promoter, which controls the expression of alcohol oxidase, the enzyme involved in the first stage of methanol metabolism *(16)*. *Pichia pastoris* has been used for the expression of a range of heterologous proteins, which can either be secreted (if provided with an appropriate leader sequence) or expressed intracellularly. This novel yeast expression system is available in kit form (Invitrogen BV, San Diego, CA).

Whereas yeast export protein to the cell vacuole, the filamentous fungi *Aspergillus nidulans* and *Aspergillus niger* secrete their gene products into the growth medium *(17)*. Secretion of the protein is sometimes preferred, since it facilitates the recovery and isolation of the product. High yields of active soluble protein may be obtained by secretion, which intracellularly would otherwise form insoluble inclusion bodies. DNA can be inserted in the fungal genome at a high copy number without giving rise to problems of instability.

Recently, much attention has been given to the exploitation of cultured insect cells for the expression of foreign genes using recombinant baculovirus vectors *(18–20)*. Baculoviruses are a group of viruses that specifically infect insect cells. Included in the baculovirus genome is a gene that encodes polyhedrin, the protein that acts as a protection shield for the viral particles outside the host. It is possible to replace the polyhedrin gene with foreign DNA and maintain the level of expression. In this way, high-level expression of cytoplasmic, secretory, or cell-surface proteins can be achieved. Also, the recombinant gene products will be processed with respect to glycosylation, phosphorylation, and so forth. Complete baculovirus expression systems that provide all the materials required to generate recombinant virus stocks are now available commercially (Invitrogen BV).

2.3.1. Overexpression in Higher Eukaryotic Cells

For very large molecules that present problems with accurate folding and disulfide bridge formation, expression in a mammalian host-vector system may be the only feasible way of obtaining correctly folded active material. For example, Factor VIII, 180 kDa with 17 disulfides, has been successfully expressed at low levels in hamster cells. Mammalian expression vectors are usually hybrids containing elements derived from prokaryotic plasmids that facilitate construction and amplification, and controlling sequences from eukaryotes, i.e., promoters, transcription enhancers and polyadenylation signals required for the expression of foreign DNA. Some of the most efficient regulatable promoters used for expression in mammalian cells include (1) the mouse metallothionein promoter, which is switched on by the addition of Zn^{2+} to the culture medium *(21)*, and (2) the human hsp 70 heat-shock promoter *(22)*.

Recombinant animal viruses containing the gene of interest can be used to infect a variety of host cells, e.g., Simian virus (SV40), adenovirus, and mouse retroviruses. Epstein-Barr virus-based vectors produce high levels of recombinant proteins in a wide range of mammalian cells. The use of vaccinia virus (pox virus) for this purpose has received much attention *(23)*. Recombinant vaccinia can accommodate foreign DNA to a high degree, thereby enabling multiple genes to be expressed simultaneously at a high level. However, since the virus replicates in the cytoplasm, intron splicing, which takes place in the nucleus, will not occur. Therefore, only cDNA may be cloned in vaccinia.

It should be emphasized that, for some proteins, obtaining high levels of expression is not straightforward. Considerable time, effort, and cost can be devoted to the construction of a suitable expression system and subsequently optimizing the yields. As stated above, large-scale fermentation involving high cell densities may simply result in a loss of the expression vector through selection, and/or the product may prove toxic to the host when expressed in large amounts. Errors in translation frequently occur when overexpressing foreign proteins in *E. coli.* Growth media, antibiotics, and chemical inducers, such as IPTG, can also be costly. For the uninitiated, therefore, it is advisable to seek the advice of an experienced molecular biologist or fermentation technologist before embarking on a strategy for overproduction.

2.4. Engineering Proteins for Purification

Fusion proteins may be specifically engineered for overexpression *(24)* and to facilitate ease of purification and detection *(25)*. An N-terminal signal sequence may be used to direct the recombinant protein product into the periplasm. The signal peptide sequence derived from a secreted bacterial protein may be appended to the 5' end of the coding sequence to enable the protein to be recognized by the host as a secretory product *(26,27)*. In this way, the initiating fMet, which is often retained in *E. coli*, but not in eukaryotes, can be removed, if desired, by engineering an appropriate signal sequence. The fMet will then be cleaved off during translocation across the bacterial membrane. Expression vectors that incorporate the *ompT* and *pelB* leaders upstream of the 5' cloning sites are commercially available. Through the direction of an appropriate leader sequence, antibody fragments can be secreted into the periplasm and through the outer membrane of *E. coli* into the culture medium, where they may be reconstituted *(28)*. Fusion at the N-terminus with a coding sequence derived from an *E. coli* gene, as supplied by a cassette vector, may confer stability on the protein, thus providing some resistance to host-cell proteases. Proteins expressed as fusions may be purified directly from crude cell lysates by affinity chromatography, and fusions with staphylococcal protein A can be purified to near-homogeneity by chromatography on an IgG affinity column *(29)*. Immunological detection is often complicated by immunoaffinity procedures, a problem that can be avoided by fusion of the protein to the C-terminus of *Schistosoma japonicum* glutathione S-transferase (GST), which is subsequently purified on immobilized glutathione *(30)*. A series of vectors is available (Pharmacia-Biotec, Sweden) that already provides the GST under the control of the *tac* promoter (pGEX) together with a specific site for proteolytic cleavage. Another commercial system is based on the *mal*E gene, which encodes maltose binding protein (MPB) under the control of the *tac* promoter. The cloned gene is inserted downstream of the *mal*E, and the expressed fusion protein can be purified in one step on amylose resin *(31)*.

A characteristic physical property of the protein, such as its pI, may also be altered by designing a fusion protein to assist its separation by ion-exchange or isoelectric focusing *(25,32)*. Fusion proteins may also be designed for raising antibodies to aid detection. If an antibody has been previously raised against the fusion sequence, the product may fold to provide an epitope for antibody recognition. One example is the use of the tripeptide Glu-Glu-Phe motif for the immunoaffinity purification of HIV enzymes. This motif is recognized by the YL1/2 monoclonal antibody to α-tubulin, which is available commercially *(33)*. Immunoblotting can then be used for the detection of the fusion product *(34)*. Another example features the HSV·Tag antibody. This is a mouse monoclonal

IgG having high specificity for an 11-amino acid peptide derived from herpes simplex glycoprotein D. This peptide is encoded by the HSV·Tag sequence in pET vectors (Novagen, Inc.) enabling detection of the target protein by immunoblotting. Alternatively, a specific assay for the fusion protein may exist. For example, GST fusions produced in *E. coli* may be assayed using the GST substrate 1-chloro-2,4-dinitrobenzene *(35)*. Subsequent posttranslational removal of the fusion sequence can be provided for by engineering the incorporation of sites for specific peptidase cleavage. These come already supplied in many of the commercial vectors. One major disadvantage of fusions is the relatively high cost of the proteases required. Thrombin, factor Xa, and enterokinase are commonly used. It is also worth noting that although specific, some proteases, particularly thrombin, can behave promiscuously, cleaving elsewhere in the protein. It is important, therefore, to optimize the digestion conditions before embarking on a large-scale preparation. Another disadvantage of fusions is the fact that, in order to provide the site for proteolytic cleavage, an alteration to the terminal sequence of the fused protein is often required. This may alter some physical characteristic of the protein that may otherwise affect its ability to crystallize. Nonetheless, there are several examples where the incorporation of a polyhistidine tag to either the N or C terminus of the target protein has proven positively beneficial in this respect. The sequence of consecutive histidine residues facilitates purification by immobilization on a metal chelation resin *(36)*. If desired, the polyhistidine tag may then be removed proteolytically, but good-quality crystals have been obtained with the tag remaining in place *(37)*.

3. Extraction and Isolation

3.1. Cell Disruption

A variety of chemical and mechanical methods can be used to rupture the cell wall; for reviews, *see* refs. *38* and *39*. The choice of method depends not only on the rigidity and architecture of the host cell wall, but also the physical and chemical properties of the protein to be extracted together with the method and scale of extraction. For bacterial cells, enzymic digestion with hen egg white lysozyme, which specifically catalyzes the hydrolysis of 1,4 glycosidic bonds in the peptideoglycan cell wall of gram-positive bacteria, is a selective and gentle method that minimizes denaturation of the product. For gram-negative bacteria, e.g., *E. coli*, metal chelators, such as EDTA, are additionally required to remove the cations that maintain the integrity of the outer lipopolysaccharides *(40)*. Chemical disruption methods mainly involve the use of a combination of anions, nonionic detergents, and chaotropic agents. The choice of detergent is critical in order to avoid irreversible denaturation of the protein and possible interference in the subsequent purification.

Methods that involve physical disruption include sonication and high-speed homogenization. These can be used for both microbial and higher eukaryotic cells. Sonication disrupts the cells by a combination of cavitation and shearing. Because sonication generates a considerable amount of heat and it is often difficult to judge the point at which the majority of cells are disrupted, all operations should be carried out in an ice bath. Bead milling, a technique by which small glass beads are agitated at high speeds in cell suspension, is especially suitable for cells that are difficult to break, e.g., yeast. The Bead-Beater cell disrupter (Biospec Products, Bartsville, OK) provides a totally sealed, air-free system that eliminates frothing and aerosol generation.

Whereas chemical methods may create the problem of contamination, the chief disadvantage of physical and mechanical methods is the generation of heat and the hazardous production of aerosols. This is especially true of sonication, which is suitable only for resuspension volumes of <100 mL, restricting its application to small-scale preparations. To avoid proteolytic degradation, it is advisable to include protease inhibitors in the resuspension buffer. The serine protease inhibitors, aprotinin or 1 m*M* phenylmethanesulfonyl fluoride (PMSF), are commonly used and may be adequate. Some proteins that appear to be particularly susceptible to proteolysis may require a "cocktail" of inhibitors to cope with each type of protease *(41)*. For bacterially expressed proteins where proteolytic degradation presents a serious problem, the use of protease-deficient host strains, e.g., *lon⁻* *(42)*, should be used. Proteolysis can be minimized by carrying out all procedures at low temperatures.

3.1.1. Removal of Cell Debris and Nucleic Acids

Much of the cell debris generated after disruption can be removed by centrifugation at 10,000*g* for 30 min, although some fine particulate matter may still remain in suspension. The nucleic acid component, a major contaminant that increases the viscosity of the cell extract, can be removed by precipitation with a positively charged polymer, such as polyethyleneimine (typically 0.5–1% of a 10% solution) or by application to a cellulose-based adsorption media (Cell Debris Remover) supplied by Whatman, Ltd. UK. It is important to note that some loss of protein may occur by coprecipitation. This is especially true of some DNA binding proteins and may be overcome by prior 1:1 dilution of the crude extract with an appropriate buffer.

3.2. Primary Isolation

Before embarking on the isolation of a novel protein, a specific assay for its activity or a detection method, such as immunoblotting, is an essential prerequisite.

The first stage in the recovery of a soluble protein is the preparation of a concentrated extract. If the physicochemical properties of the protein are known, then ion exchange on a rigid matrix, such as Sephacel (Pharmacia Biotech), can be used both for concentration and as preliminary clean-up step. To reduce the volume of extract, ultrafiltration by either pressure or vacuum can be used, for which membranes with a broad range of mol wt "cut offs" are available (Amicon Corp., Danvers, MA). Alternatively, selective precipitation with salts, organic solvents, or long-chain polymers, e.g., polyethylene glycol (PEG), may be employed as an initial fractionation step. Conditions of high protein concentration, low temperatures, and pH close to neutrality are generally used in salt fractionation. Ammonium sulfate is the most commonly used salt, chiefly because of its high solubility and stabilizing properties. A nomogram and various charts have been devised *(43)* for determining the amount of salt to be added to the protein fraction to attain the required saturation. A number of organic precipitants, including ethanol and acetone, have proven successful in the fractionation of many extracellular proteins, e.g., plasma proteins, histones, and polypeptide hormones, but are not to be recommended for unstable proteins. Polyethylene glycol, on the other hand, shows little tendency to cause denaturation and can be easily removed either by ion-exchange or size-exclusion chromatography, obviating the need for extensive dialysis as in the case of ammonium sulfate *(44)*.

3.3. Chromatographic Methods

The principal chromatographic methods and the properties they exploit are given in Table 1. For protein purification, a vast range of chromatographic media is now available. Functional properties have been combined with matrices of different strengths and/or porosities to optimize

Table 1
Chromatographic Methods

Chromatographic method	Physical/chemical property	Resolution
Ion exchange	Charge	High
Chromatofocusing	Charge/pI	High
Size exclusion (gel filtration)	Size/shape	Low
Hydrophobic interaction	Surface hydrophobicity	Medium
Reversed-phase chromatography	Polarity	High
Affinity chromatography	Biological affinity	High
Adsorption chromatography	Dipole interactions Hydrogen bonding van der Waals forces	High

flowrates and selectivity. Detailed descriptions and methodologies for these media can be found in a number of dedicated texts *(45,46)*. The selection of media and the number and combination of chromatographic steps are determined by the physical, chemical, and functional properties of the protein to be recovered. If the pI is unknown, the amino acid sequence may provide an indication of the acid, basic, or hydrophobic nature of the protein, which may assist the design of a purification strategy.

Ideally, buffer exchange, dialysis to reduce ionic strength, and ultrafiltration should be avoided where possible between chromatographic stages. The sequence of separations can usually be manipulated to avoid such processes. For example, size exclusion, which desalts and dilutes the sample, would normally follow a concentration step, such as ion exchange. On the other hand, ion exchange would not be a suitable step to follow ammonium sulfate fractionation because of the extensive subsequent dialysis necessary to reduce the ionic strength and enable the protein to bind to the column. Instead, hydrophobic interaction chromatography (HIC) could be used, thereby binding proteins preferentially at high ionic strengths. The sample can be eluted selectively with a descending salt gradient. Because of its relatively low capacity, size-exclusion or gel-filtration chromatography is more appropriately used in the final "polishing" stages of purification. It is often employed as a convenient alternative to dialysis for desalting and buffer exchange, for which dedicated prepacked columns are commercially available, such as

Table 2
Affinity Chromatography Applications

Ligands	Target molecules	Examples
Lectins	α-D-glucopyranosyl	Adenosine deaminase
Concanavalin A	α-D-mannopyranosyl	γ-Glutamyl transferase
Heparin	Nucleic acid binding proteins	DNA helicases Restriction endonucleases DNA/RNA polymerase
	Growth factors	Endothelial cell growth factor
Calmodulin	Calmodulin-dependent enzymes	Adenylate cyclase ATPase
ssDNA	DNA-dependent enzymes	DNA-polymerases alkyl transferase
Triazine dyes:	NAD^+, $NADP^+$	Dehydrogenases
Cibacron blue	dependent enzymes	Quinone reductases
5'-AMP	ATP-dependent enzymes	cAMP-dependent protein kinase
Protein A	Immunoglobulins	Fc region of IgGs
MAB	Antigens	Protein A fusions
Glutathione	Glutathione S-transferases	Glutathione S-transferase fusions
Amylose	Maltose binding protein	MBP fusions

PD-10 (Pharmacia-Biotech). To reduce the loss of protein and optimize yields, purity should be achieved by the least number of chromatographic steps. To a large extent, this has been made possible by the development of affinity techniques *(47)*.

Affinity chromatography provides high specificity by exploiting the biological properties of the molecule, which is reversibly adsorbed to an immobilized ligand coupled to an insoluble matrix. Some applications of affinity chromatography are given in Table 2. For immunoaffinity chromatography, antibodies raised against the protein are coupled to the activated adsorbant, e.g., cyanogen bromide-activated Sepharose (Pharmacia-Biotech). Immunoadsorbants offer high binding capacities and extremely high specificity, but the method used for desorption or elution is critical. Sometimes such harsh conditions are required that the protein becomes irreversibly denatured. However, with the development of dedicated media for purifying monoclonal antibodies (MAb), e.g., Protein A (Pharmacia-Biotech), immunoaffinity is becoming an increasingly popular and powerful separation method. Of the adsorption methods, hydrated calcium phosphate or hydroxyapatite chromatography deserves a mention here. The mode of binding to this gel is not readily understood, but adsorption is believed to occur by dipole–dipole interactions between Ca^{2+} and PO_4^{3-} groups on the adsorbant and the protein. Because of its unique binding mechanism, hydroxyapatite can sometimes provide selectivity not always achievable by alternative methods. In the past, lengthy preparation times and low flowrates have restricted the use of hydroxyapatite, but more recently, gel beads coated with calcium phosphate have become available (HA-Ultragel IBF, Columbia, MD), which provide much greater mechanical stability.

For membrane proteins or protein aggregates, where detergents and high-salt buffers are used, metal chelate chromatography (MCAC) has proven effective *(48)*. This technique is a refinement of ligand chromatography, where the protein interacts with the matrix via a complex-forming metal ion. Histidine imidazoles, thiol groups, and possibly tryptophan residues supplied by the protein are involved in metal ion complex formation. The column is precharged with the chosen metal-ion, e.g., Cu^{2+} or Zn^{2+}, and equilibrated at neutral pH to achieve maximum adsorption of the sample. The protein is then eluted by lowering the pH to <6; strongly bound proteins can be eluted with chelating agents, such as EDTA.

Ion-exchange and affinity media can now be obtained in convenient prepacked cartridges of 1–5 mL (Econo-Pac, from Bio-Rad Laboratories Ltd., Richmond CA; HiTrap, from Pharmacia-Biotech). These are ideal for small-scale preparations and method development.

The speed and resolution of protein separations have been assisted enormously by the developments in low-pressure high-performance liquid chromatography (HPLC) and fast-protein liquid chromatography (FPLC), pioneered by Pharmacia-LKB *(49)*. A range of prepacked columns based on rigid mono disperse beads 10 ± 0.2 µm (Mono beads) has been developed for use at high flowrates relative to conventional "soft gel" media. Accordingly, dedicated biocompatible HPLC systems have been designed by manufacturers for protein purification applications by replacing the conventional stainless-steel components with inert fluoroplastics and titanium (Waters 650 Advanced Protein Purification System, Millipore Corp., Miford, MA). Halide-containing eluting solvents can then be used confidently without risk of corrosion and metal-ion contamination. Also, developments in the applications of electrophoretic techniques have led to the design of systems that can provide a one-step purification based on either pI or charge, on a semipreparative scale, e.g., The Rotophor Isoelectric Focusing Cell and Model 491 Prep Cell Continuous Elution Electrophoresis, Bio-Rad Laboratories Ltd. The Isoprime Multi-Chambered Electrofocusing Unit, recently introduced by Hoefer Scientific Instruments (San Francisco, CA) is an integrated system that uses immobilized pH gradients to purify micrograms to hundreds of milligrams of protein in a single run *(50)*.

Advances in membrane-based chromatography have produced a new concept of membrane convective liquid chromatography (MCLC). Using membrane cartridges (MemSep MCLC Cartridges, Millipore Corp.), mass transport of the molecules to be separated is achieved through convection rather than diffusion, which permits higher flowrates without compromising resolution. Many of the problems inherent in conventional column chromatography, such as bed collapse and the entrapment of air bubbles, are eliminated with membrane-based chromatography. The MemSep units are also compatible with FPLC and HPLC systems.

3.4. Criteria of Purity

One-dimensional SDS-PAGE stained with Coomassie blue provides a reasonable indication of purity in the first instance *(51)*. Minor impurities (<5%) can be detected by ultrasensitive stains, e.g., silver and colloidal gold *(52,53)*. Protein contaminants that differ by >1% in molecular weight can be detected. For protein contaminants that differ in charge and for the detection of enzymes with several isomeric forms, analytical isoelectric focusing and/or two-dimensional PAGE is recommended. Also for proteins that have an optical chromophore, e.g., cytochromes, a ratio of the absorption peak in the visible region to the absorption at 280 nm can be used as a criterion of purity. Microheterogeneity is frequently encountered as an impediment to crystallization and/or can adversely affect crystal quality. It may be caused by incomplete posttranslational processing, chemical modification, or partial denaturation during purification. More sophisticated analytical methods for detecting the presence of microheterogeneity now exist and should be used wherever possible to monitor preparations. This is particularly important where batch-to-batch variability occurs. The molecular size of proteins may now be checked by electrospray mass spectrometry (ESI-MS) to within the mass of a single amino acid residue. This sensitive technique permits the measurement of relative mass in the range of 5–40 kDa with a precision of better than 0.01%, and may reveal the presence of proteolytically degraded and/or incorrectly processed proteins that differ in mass *(54,55)*. N-terminal sequence determination combined with mass determination can be used to provide an absolute confirmation of the sequence. Matrix-assisted laser desorption mass spectrometry (MALD-MS) is a less sensitive technique for mass analysis, but unlike ESI-MS, it can be used in the presence of buffers and detergents *(56,57)*. Capillary electrophoresis (CE) has now become a powerful analytical tool for the detection of chemical modifications in proteins *(58,59)*. The technique provides rapid, high-efficiency, and reproducible separations of peptides and proteins under both native and denaturing conditions. CE is capable of separating and quantitating a single-charge difference. Incorrectly bonded disulfides, which do not differ in size or charge, can usually be detected by high-resolution peptide mapping *(60)*. Recent advances in instrumentation technology have now improved the availability of all these techniques, and it is likely that they become increasingly used for the monitoring of protein preparations on a routine basis.

4. Crystallization

4.1. Introduction

Owing to the present inadequate level of understanding of the complex physical processes involved, procedures for growing crystals of macromolecules are generally empirical. Crystallization is usually achieved by varying those physical parameters affecting solubility of the macromolecule in order to attain a state of supersaturation. This will usually involve a systematic search to determine those conditions that indicate crystallinity and then "fine-tuning" them for subsequent experiments. Nucleation sites for crystal growth are formed by chance collisions of molecules in solution forming molecular aggregates, and the probability that these molecules will occur simultaneously in a small volume will be greater in a saturated solution. A supersaturated solution of protein can be achieved by varying those factors that affect its solubility, namely pH, ionic strength, and temperature, as well as the presence of precipitating agents, such as organic solvents, which alter the overall dielectric constant. These parameters, together with the initial concentration of the protein, may be varied independently in a controlled fashion so as to bring the system slowly to a state of minimum solubility and thus allow the molecules to order themselves into a crystalline lattice. If approached too rapidly, the molecules may aggregrate to form an amorphous precipitate. The theoretical aspects of protein solubility and crystal formation are discussed more fully in refs. *61–63*.

A degree of familiarity with the protein is usually an important advantage. To limit the number of experiments needed, and thus avoid unnecessary and sometimes irreversible loss of material through denaturation, it is advisable to acquire some knowledge of the behavior of the protein in solution, i.e., stability over a range of pH and ionic strengths, any tendency toward aggregation, and the importance of cofactors, such as FAD, metal-ions, ligands, and so forth. Occasionally, one can predict the chances of a protein crystallizing from its behavior during isolation. For instance, the formation of a crystalline precipitate (biochemist's crystals) may occur on salt fractionation or during dialysis while desalting.

Before embarking on a course of crystallization experiments, it is necessary to have determined the level of purity of the sample. Although a single band on SDS-PAGE stained with Coomassie blue may provide a good indication of purity in the first instance, it is no longer considered adequate for crystallization. It is usually the case that a homogeneous preparation has a greater chance of yielding crystals than material that contains some degree of impurity. However, if a protein shows a strong inclination to crystallize, it may do so while tolerating a high percentage, e.g., <10%, impurity. Moreover, different levels of purity may result in different crystal forms, some of which may be crystallographically more desirable than others. Microheterogeneity may be detected by isoelectric focusing (IEF) and two-dimensional PAGE or other methods (*see* Section 3.4.). Difficulties in reproducing good-quality crystals are frequently experienced, especially from different batches of material. For this reason, it is important not to mix protein from the different batches, and it is always advisable to keep a detailed record of each purification. To minimize denaturation on storage, lyophilization and repeated freezing and thawing should be avoided. It is strongly recommended that only highly purified analytical-grade reagents be used for all experiments and that all buffers should be prepared only with a reliable source of ultrapure water.

Methods for obtaining cocrystals of protein with nucleic acid and crystals of membrane proteins are dealt with specifically in Chapters 12 and 14, respectively, of this volume.

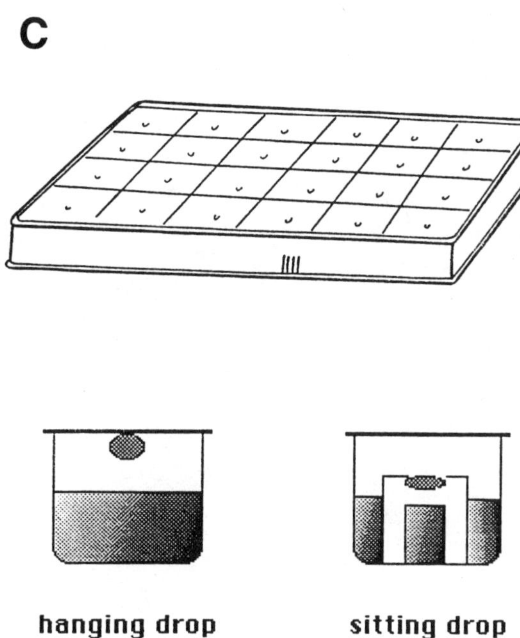

Fig. 1. **(C)** Vapor-diffusion crystallizations using the hanging drop and sitting drop technique. Twenty-four-well tissue-culture plates (ICN-Flow) are ideally suited for initial screening to determine the appropriate starting conditions.

Repetition Engineers UK are available in 10–50-μL volumes. A degree of manual dexterity is required when handling these buttons to avoid trapping air bubbles, which would otherwise interfere with the progress of equilibration *(69)*. Zeppezauer *(70)* introduced the use of capillary tubes (Fig. 1B), which can also be sealed with a plug of acrylamide in place of dialysis tubing. The percentage of crosslinkage of acrylamide can then be used to control the rate of diffusion.

4.2.2. Vapor Diffusion

This has become a popular method for screening a wide range of conditions especially when material is limited. It relies on the principle of controlled equilibration through the vapor phase to bring about a state of supersaturation in the sample. The vapor diffusion method also lends itself to automation *(71)*, with robotic crystallization machines now available (*see* Fig. 2) (Douglas Instruments, UK; ICN Flow Labs, McLean, VA). The most popular variation is the hanging drop technique, which

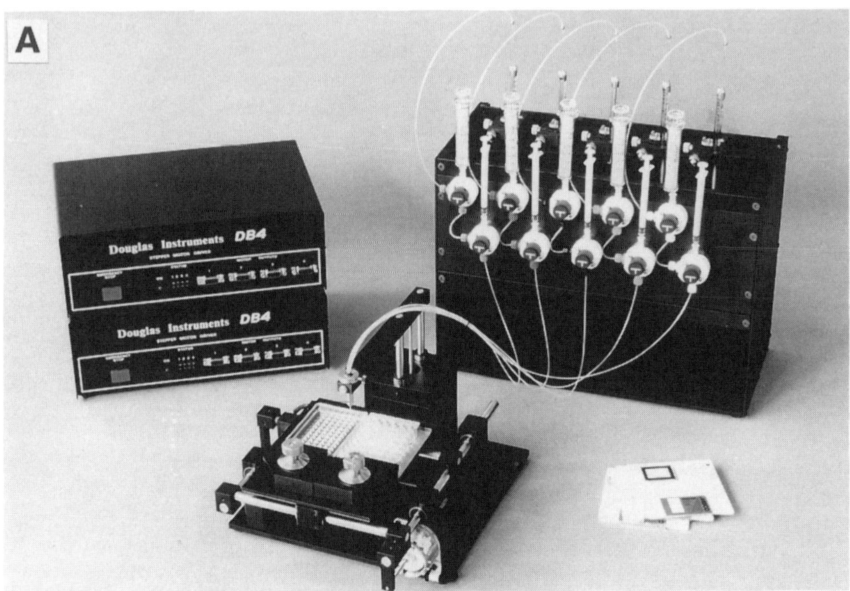

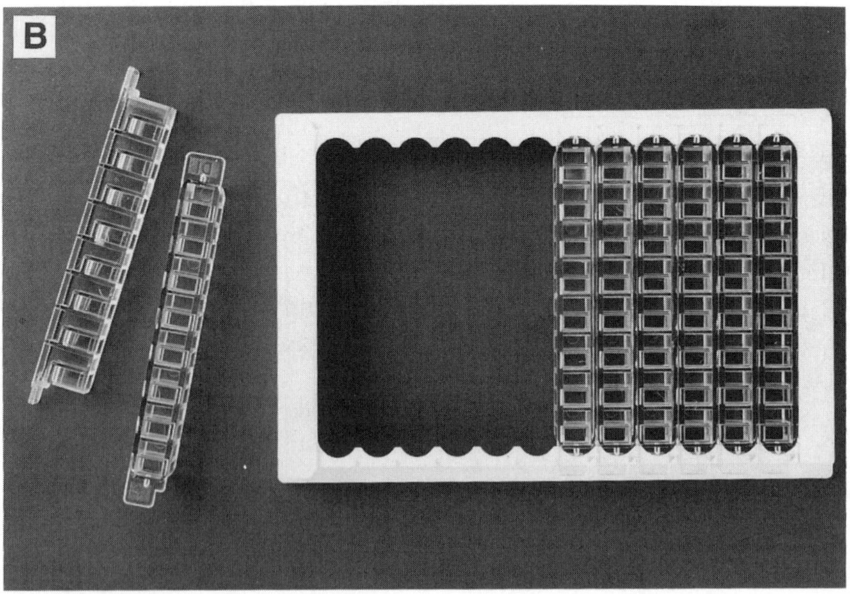

Fig. 2. **(A)** Robotic Crystallization Apparatus DB4 marketed by Douglas Instruments. **(B)** Crystallization plate for setting up sitting drops for use with the robot.

by using microdroplets of 2–20-μL vol permits a large number of conditions to be tested simultaneously. Essentially, a microdroplet of protein solution is deposited on a glass coverslip that has been treated previously with dichlorodimethylsilane to reduce surface tension. The coverslip is then inverted and placed over a sealed reservoir containing a precipitant solution. The droplet contains precipitant initially at a lower concentration than in the reservoir: a range of gradients can be set up. Vapor diffusion is a useful method for determining the solubility of a concentrated protein solution at various pHs by equilibration with a range of volatile buffers. This information may then provide a good starting point for subsequent experiments. Linbro 24-well tissue-culture plates (ICN Flow Laboratories) are generally favored for hanging drop experiments (Fig. 1C). A variety of crystallization plates and growth chambers have been specifically designed which allow for changes of the reservoir conditions during an experiment as well as facilitating so-called sitting and sandwiched drops (American Crystallographic Association; Hampton Research, Riverside, CA; Cryschem, Inc., Riverside, CA).

4.2.3. Free Interface Diffusion

A supersaturated protein solution may be achieved simply by means of free diffusion between the protein solution and a solution of precipitant. The two solutions are layered one on top of the other in a capillary tube. As diffusion proceeds to equilibrium, the interface is exposed to a transient supersaturating concentration of precipitant, which may promote the formation of nuclei. This method has been adapted for microliter volumes by Salemme *(72)*.

4.3. Strategies for Crystallization Screening

The number of variables that influence crystal growth are too great to enable exhaustive screening to be carried out, especially if the amount of protein is limited. To restrict the number of experiments, it may be necessary therefore to make a choice between a "knowledge-based" approach, where only the physical properties of the protein in question are considered, or sparse matrix sampling, where a range of trial conditions is preselected. Jancarik and Kim *(73)* have chosen a set of 50 different conditions for random sampling. These conditions have been largely determined using a data base of successful protein crystallizations. In this way a wide range of pHs, counterions, and precipitating agents can be rapidly screened with minimal material. These reagents are now commercially available as kits (Crystal Screen, Hampton Research).

Alternatively, statistical and systematic approaches to crystallization screening can be applied by designing a set of factorial experiments *(74,75)* intended to identify the effects of a number of different possible "factors." The information obtained from a factorial design can then be analyzed, scored, and used in subsequent experiments to optimize the conditions for crystal growth. Computer programs have been written to assist in the design of factorial experiments, e.g., INFAC *(76)*.

4.3.1. A Strategy for the Crystallization of an Intracellular Enzyme Using Vapor Diffusion

1. Adjust the concentration to 5–15 mg/mL in a relatively low ionic-strength buffer at a pH at which the enzyme is optimally active. It is often important to include cofactors, metal ions, or reducing agents known to be essential for activity at the appropriate concentration in the buffer. The sample can be concentrated using an Amicon microconcentrator with the appropriate mol-wt size exclusion. Ensure that the sample is free from particulate matter by filtration through a 0.22-μ filter or alternatively, by centrifugation in a microcentrifuge (12,000*g* for 5 min).
2. Select suitable buffers in an incremental range of pHs. Ensure that the buffers chosen do not react with essential cofactors or metal ions. Buffers should be filtered prior to use, and antimicrobial agents, such as thiomersal or sodium azide (0.02%), should be added where appropriate.
3. Add precipitating agents, such as polyethylene glycol 4000 or 6000, in a range of concentrations, ideally 5–20%, to each of the reservoirs of a 4 × 6 well tissue-culture plate.
4. Deposit accurate volumes of 2–10 μL on siliconized glass coverslips.
5. Set up a gradient by adding an equal volume of precipitant from each reservoir to the droplet. The coverslip may then be inverted and sealed by means of vacuum grease or mineral oil over the reservoir chamber.
6. Store the plate in a thermostatically controlled incubator and/or cold room, undisturbed, initially for 1–2 d, and then inspect the experiment at periodic intervals. Preferably, the experiments should be set up at more than one temperature. A low-power stereo zoom microscope with polarizer and analyzer attachments is an essential tool for examining crystals.

4.4. Crystal Quality

Although the conditions for crystallization may be established, the crystals may not be entirely suitable for crystallographic data collection. Frequently, they grow twinned, are of poor diffraction quality, or are of inadequate size. Twinning can sometimes be prevented by the presence of an organic solvent, e.g., the addition of 1% of 1,4-dioxan to the crystallization buffer. Large single crystals may be encouraged to grow by simply increasing the scale of the experiment. The relationship between crystal size and the volume of the crystallization sample has now been demonstrated *(77)*. An increase in crystal volume appears to be the result of the increased amount of available protein and the slower equilibration rates associated with the larger sample volumes. Alternatively, the quality and size may be improved by fine-tuning the conditions, so as to approach saturation over a longer period and reducing the number of possible nucleation sites. To minimize the number of nuclei formed, it is essential to ensure that glass and plastic surfaces are clean and free of contaminating dust particles. Compressed air is a useful standby for removing offending particles prior to setting up an experiment.

Nonionic detergents have been used with some success to improve crystal quality and size presumably by reducing hydrophobic interactions between molecules, which may give rise to nonspecific aggregation. Several examples of crystals grown in the presence of detergents, which result in improved resolution of diffraction, are given by McPherson *(78)*. The presence of up to 1% β–octyl glucoside in the equilibration buffer has also been shown to induce different crystal forms. The use of detergents for crystallizing membrane proteins has been extensively researched by Michel et al. *(79)*, who succeeded in crystallizing and elucidating the structure of the photosynthetic reaction center *(80)* (*see also* Chapter 14).

Excessive nucleation may be reduced by growing crystals in hydrogels. Silica and agarose have both proven successful for increasing the size and quality of protein crystals *(81)*. Silica gel, in particular, is compatible with most precipitating agents and is stable over a wide temperature range (4–60°C). The experiments can be set up as sitting drops using the vapor diffusion method or in glass capillaries (liquid–liquid diffusion).

4.4.1. Seeding

Large single crystals may sometimes be grown by "seeding" once the preliminary conditions have been determined. Microcrystals can be introduced into a pre-equilibrated supersaturated protein solution by means of a thin glass fiber or fine hair, although it is not easy to control the number of seeds entering the solution. The result may therefore be a "shower" of microcrystals. It may be possible to overcome this, however, by introducing seeds that have been crushed and serially washed in a buffer solution containing a low concentration of precipitant. Thaller et al. *(82)* reported a method for producing large single crystals of globular proteins by "repeated seeding." A small, carefully washed crystal is used as the primary seed. After growth has ceased, the crystal is removed and transferred to a fresh protein solution of a slightly lower concentration. This is then repeated until the crystal has reached the required dimensions. These two methods are commonly referred to as micro- and macroseeding, respectively, and are described in some detail by Stura and Wilson *(83)*.

Should all these measures continue to fail to produce good-quality crystals, a re-examination of the purification procedure is advisable in the first instance. More fruitful results are quite often obtained from crystallizing protein derived from a closely related species. Where proteins have proven persistently difficult to crystallize, dynamic light scattering may be used to provide some indication of "crystallizability." Quasielastic light scattering (QELS) has been applied to the investigation of nucleation processes in macromolecular systems and shown that aggregation patterns that lead to crystal formation differ from those that result in amorphous precipitate *(84,85)*. Thus, QELS may be used to assess the suitability of conditions for crystallization for a given protein system.

Acknowledgments

We wish to thank our colleagues Steve Hobbs, David Suter, and Sandy Carne for their advice and helpful comments, and we gratefully acknowledge the support of the Cancer Research Campaign.

References

1. Abdel-Meguid, S. S., Smith, W. W., Violand, B. N., and Bentle, L. A. (1986) Crystallisation of methionyl porcine somatotropin, a genetically engineered variant of porcine growth hormone. *J. Mol. Biol.* **192,** 159,160.
2. Jancarik, J., de Vos, A., Kim, S.-H., Miura, K., Ohtsuka, E., Noguchi, S., Nishimura, S. (1988) Crystallisation of human c-H-*ras* oncogene products. *J. Mol. Biol.* **200,** 205–207.
3. Sambrook, J., Fritsch, E. F., and Maniatis, T. (1989) *Molecular Cloning: A Laboratory Manual*, 2nd ed. Cold Spring Harbor Laboratory, Cold Spring Harbor, NY.
4. Das, A. (1990) Overproduction of proteins in *Esherichia coli*: Vectors, hosts, and strategies. *Methods Enzymol.* **182,** 93–112.
5. Reznikoff, W. and Gold, L. (eds.) (1986) *Maximizing Gene Expression.* Butterworth, Boston.
6. Pouwels, P. H., Enger-Valk, B. E., and Brammar, W. J. (1985) *Cloning Vectors.* Elsevier, Amsterdam.

7. Jacob, F. and Monod, J. (1961) Genetic regulatory mechanisms in the synthesis of proteins. *J. Mol. Biol.* **3,** 318–356.
8. Tacon, W. C. A., Cary, N., and Emtage, S. (1980) The construction and characterization of plasmid vectors suitable for the expression of all DNA phases under the control of the *E. coli* tryptophan promoter. *Mol. Gen. Genet.* **177,** 427–438.
9. Tacon, W. C. A., Bonass, W. A., Jenkins, B., and Emtage, J. S. (1983) Plasmid expression vectors containing tryptophan promoter transcriptional regulons lacking the attenuator. *Gene* **23,** 255–265.
10. Amann, E., Brosius, J., and Ptashne, M. (1983) Vectors bearing a hybrid *trp-lac* promoter useful for regulated expression of cloned genes in *Escherichia coli. Gene* **25,** 167–178.
11. Shatzman, A. R. and Rosenberg, M. (1987) Expression, identification and characterisation of recombinant gene products in *E. coli. Methods Enzymol.* **152,** 661–673.
12. Studier, F. W. and Moffatt, B. A. (1986) Use of bacteriophage 7 RNA polymerase to direct selective high level expression of cloned genes. *J. Mol. Biol.* **189,** 113–130.
13. Harris, T. J. R. (1983) Expression of eukaryotic genes in *E. coli*, in *Genetic Engineering 4* (Williamson, R., ed.), Academic, New York, pp. 127–185.
14. Marston, F. A. O. (1986) The purification of eukaryotic polypeptides synthesized in *Escherichia coli. Biochem. J.* **240,** 1–12.
15. Bitter, G. A., Egan, K. M., Koski, R. A., Jones, M. O., Elliott, S. G., and Giffin, J. C. (1987) Expression and secretion vectors for yeast. *Methods Enzymol.* **153,** 516–544.
16. Cregg, J. M., Vedvick, T. S., and Raschke, W. C. (1993) Recent advances in the expression of foreign genes in *Pichia pastoris*. *Bio/Technology* **11,** 905–910.
17. Saunders, G. (1989) Heterologous gene expression in filamentous fungi. *Trends Biotechnol.* **7,** 283–287.
18. Yong Kang, C. (1988) Baculovirus vectors for expression of foreign genes. *Adv. Virus Res.* **35,** 177–192.
19. Fraser, M. J. (1989) Expression of eukaryotic genes in insect cell cultures in vitro. *Cell. Devel. Biol.* **25,** 225–235.
20. Luckow, V. A. and Summers, M. D. (1988) Trends in the development of baculovirus expression vectors. *Biotechnology* **6,** 47–55.
21. Lee, F., Mulligan, R., Berg, P., and Ringold, G. (1982) Glucocorticoids regulate expression of dihydrofolate reductase cDNA in mouse mammary tumour virus chimaeric plasmids. *Nature* **294,** 228–235.
22. Pelham, H. and Benz, M. (1982) A synthetic heat-shock promoter element confers heat-inducibility on the herpes simplex virus thymidine kinase gene. *EMBO J.* **1,** 1473–1477.
23. Piccini, A., Perkus, M. E., and Paoletti, E. (1987) Vaccinia virus as an expression vector. *Methods Enzymol.* **153,** 545–563.
24. Brewer, S. J. and Sassenfeld, H. J. (1989) Engineering proteins for purification, in *Protein Purification Applications: A Practical Approach* (Harris, E. L. V. and Angal, S., eds., IRL, Oxford, UK, pp. 91–111.
25. Uhlen, M. and Moks, T. (1991) Gene fusions for purpose of expression: an introduction. *Methods Enzymol.* **185,** 129–143.

26. Stader, J. A. and Silhavy, T. J. (1990) Engineering *Escherichia coli* to secrete heterologous gene products. *Methods Enzymol.* **185,** 166–187.
27. Nilsson, B., Holmgren, E., Josephson, S., Gatenbeck, S., Philipson, L., and Uhlen, M. (1985) Efficient secretion and purification of human-like growth factor I with a gene fusion vector in staphylocci. *Nucleic Acids Res.* **13,** 1151–1162.
28. Yarranton, G. T. and Mountain, A. (1992) Expression of proteins in prokaryotic systems-principles and case studies, in *Protein Engineering: A Practical Approach* (Rees, A. R., Sternberg, M. J. E., and Wetzel, R., eds.), IRL, Oxford, UK, pp. 303–305.
29. Nilsson, B., Abrahmsen, L., and Uhlen, M. (1985) Immobilisation and purification of enzymes with staphylococcal protein A gene fusion vectors. *EMBO J.* **4,** 1075–1080.
30. Smith, D. B. and Johnson, K. S. (1988) Single step purification of polypeptides expressed in *Escherichia coli* as fusions with glutathione S-transferasc. *Gene* **67,** 31–40.
31. Riggs, P. (1992) Expression and purification of maltose binding protein fusions, in *Current Protocols in Molecular Biology* (Ausebel, F. M., ed.), Wiley Interscience, New York, pp. 16–21 to 16–27.
32. Sassenfeld, H. M. and Brewer, S. J. (1984) A polypeptide fusion designed for the purification of recombinant proteins. *Biotechnology,* **2,** 76–81.
33. Stammers, D. K., Tisdale, M., Court, S., Parmar, V., Bradley, C., and Ross, C. K. (1991) Rapid purification and characterization of HIV-1 reverse transcriptase and RNase H engineered to incorporate a C-terminal tripeptide alpha-tubulin epitope. *FEBS Lett.* **283,** 298–302.
34. Towbin, H., Staehelin, T., and Gordon, J. (1979) Electrophoretic transfer of proteins from polyacrylamide gels to nitrocellulose sheets: procedure and some applications. *Proc. Natl. Acad. Sci. USA* **76,** 4350–4354.
35. Habig, W. H., Pabst, M. J., and Jakoby, W. B. (1974) The first enzymatic step in mercapturic acid formation. *J. Biol. Chem.* **249,** 7130–7139.
36. Smith, M. C., Furman, T. C., Ingolia, T. D., and Pidgeon, C. (1988) Chelating Peptide-immobilized metal ion affinity chromatography. *J. Biol. Chem.* **263,** 7211–7215.
37. Zhang, F., Robbins, D. J., Cobb, M. H., and Goldsmith, E. J. (1993) Crystallisation and preliminary X-ray studies of extracellular signal-regulated kinase-2/MAP kinase with an incorporated His-tag. *J. Mol. Biol.* **233,** 550–552.
38. Naglak, T. J., Hettwer, D. J., and Wang, H. Y. (1989) in S*eparation Processes in Biotechnology.* (Asenjo, J. A., ed.), Marcel Dekker, New York.
39. Scawen, M. D., Atkinson, T., Hammond, P. M., and Sherwood, R. F. (1988) Downstream processing: large scale protein purification, in *Molecular Biology and Biotechnology* (Walker, J. M. and Gingold, E. B., eds.), Royal Society of Chemistry, Cambridge, pp. 295–322.
40. Cull, M. and McHenry, C. S. (1990) Preparation of extracts from prokaryotes. *Methods Enzymol.* **182,** 147–153.
41. North, M. J. (1989) Prevention of unwanted proteolysis, in *Proteolytic Enzymes: A Practical Approach* (Beynon, R. J. and Bond, J. S., eds.), IRL, Oxford, UK, pp. 105–123.

42. Gottesman, S. (1990) Minimising proteolysis in *Escherichia coli*: genetic solutions. *Methods Enzymol.* **185,** 119–129.
43. Dawson, R. M. C., Elliott, D. C., Elliott, W. H., and Jones, K. M. (eds.) (1986) *Data For Biochemical Research*, 3rd ed. Clarendon Press, Oxford.
44. Hao, Y. L., Ingham, K. C., and Wickerhauser, M. (1980) in *Methods of Protein Fractionation* (Curling, J. M., ed.), Academic, New York.
45. Scopes, R. K. (1987) *Protein Purification: Principles and Practice*, 2nd. ed. Springer-Verlag, New York.
46. Harris, E. L. V. and Angal, S. (eds.) (1989) *Protein Purification Methods: A Practical Approach.* IRL, Oxford, UK.
47. Lowe, C. R. (1979) An Introduction To Affinity Chromatography in *Laboratory Techniques in Biochemistry and Molecular Biology*, vol. 7, part II (Work, T. S. and Work, E., eds.), Elsevier/North-Holland, Amsterdam, pp. 267–518.
48. Rijken, D. C. and Collen, D. (1981) Purification and characterisation of the plasminogen activator secreted by human melanoma cells in culture. *J. Biol. Chem.* **256,** 7035–7041.
49. Pharmacia AB. (1985) *FPLC Ion Exchange and Chromatofocusing-Principles and Methods.* Laboratory Separations Division, Pharmacia AB, Uppsala.
50. Righetti, P. G., Wenisch, E., and Faupel, M. (1989) Preparative protein purification in multi-compartment electrolyser with immobiline membranes. *J. Chromatogr.* **475** 293–309.
51. Laemmli, U. (1970) Cleavage of structural proteins during the assembly of the head of bacteriophage T4. *Nature* **227,** 680–685.
52. Merril, C. R. (1990) Gel-staining techniques. *Methods Enzymol.* **182,** 477–488.
53. Bristow, A. F. (1990) Purification of proteins for therapeutic use, in *Protein Purification Applications: A Practical Approach* (Harris, E. L. V. and Angal, S. eds.), IRL, Oxford, UK, pp. 29–44.
54. Edmonds, C. G. and Smith, R. D. (1990) Electrospray Ionization Mass Spectrometry. *Methods Enzymol.* **193,** 412–431.
55. Scoble, H. A. and Martin, S. A. (1990) Characterization of recombinant proteins. *Methods Enzymol.* **193,** 519–536.
56. Hillenkamp, F. and Karas, M. (1991) Matrix assisted laser desorption ionization mass spectrometry of biopolymers. *Anal Chem.* **63,** 1193A.
57. Jardine, I. (1990) Molecular weight analysis of proteins. *Methods Enzymol.* **193,** 441– 455.
58. Gordon, M. J., Huang, W., Pentoney, S. L., and Zare, R. N. (1988) Capillary Electrophoresis. *Science* **242,** 224–228.
59. Mazzeo, J. R. and Krull, I. S. (1991) Coated capillaries and additives for the separation of proteins by capillary zone electrophoresis and capillary isoelectric focusing. *BioTechniques* **10,** 638–645.
60. Hitchcock, A. and Thatcher, D. R. (1994) Protein folding in biotechnology, in *Protein Folding* (Pain, R. H., ed.), Oxford University Press, pp. 229–262.
61. Blundell, T. L. and Johnson, L. (1976) *Protein Crystallography.* Academic, New York.
62. Arakawa, T., Timasheff, S. N., Feher, G., Kam, Z., and McPherson, A. (1985) Theory of Protein Solubility. *Methods Enzymol.* **114,** 49–77.

63. Ries-Kautt, M. and Ducruix, A. (1992) Phase diagrams, in *Crystallization of Nucleic Acids and Proteins: A Practical Approach* (Ducruix, A. and Giege, R., eds.), IRL, Oxford, UK, pp. 219–239.
64. Mikol, V. and Giege, R. (1992) The physical chemistry of protein crystallization, in *Crystallization of Nucleic Acids and Proteins: A Practical Approach* (Ducruix, A. and Giege, R., eds.), IRL Press, Oxford University Press.
65. McPherson, A. (1976) The growth and preliminary investigation of protein and nucleic acid crystals for X-Ray diffraction analysis. *Methods Biochem. Anal.* **23,** 249–345.
66. McPherson, A. (1982) *Preparation and Analysis of Protein Crystals.* John Wiley, New York.
67. McPherson, A. (1990) Current approaches to macromolecular crystallisation. *Eur. J. Biochem.* **189,** 1–23.
68. McPherson, A. (1976) Crystallisation of Proteins from Polyethylene Glycol. *J. Biol. Chem.* **251,** 6300–6303.
69. Wood, S. P. (1990) Purification for crystallography, in *Protein Purifications: A Practical Approach* (Harris, E. L. V. and Angal, S., eds.), IRL, Oxford, UK, pp. 45–58.
70. Zeppezauer, M., Eklund, H., and Zeppezauer, E. (1968) Micro diffusion cells for the growth of single protein crystals by means of equilibrium dialysis. *Arch. Biochem. Biophys.* **126,** 564–572.
71. Cox, M. J. and Weber, P. C. (1987) Experiments with automated protein crystallisation. *J. Appl. Cryst.* **20,** 366–373.
72. Salemme, F. R. (1972) A free interface diffusion technique for the crystallization of proteins for X-Ray crystallography. *Arch. Biochem. Biophys.* **163,** 423–425.
73. Jancarik, J. and Kim, S. H. (1991) Sparse matrix sampling: a screening method for crystallization of proteins. *J. Appl. Cryst.* **24,** 409–411.
74. Carter, C. W., Jr. (1990) Protein and nucleic acid crystallization. *Methods, A Companion to Methods Enzymol.* **1,** 12.
75. Carter, C. W., Jr. (1992) Design of crystallization experiments and protocols, in *Crystallization of Nucleic Acids and Proteins: A Practical Approach* (Ducruix, A. and Giege, R., eds.), IRL, Oxford, UK, pp. 47–71.
76. Carter, C. W., Jr. INFAC, Department of Biochemistry and Biophysics, University of North Carolina at Chapel Hill, NC.
77. Fox, K. M. and Karplus, P. A. (1993) Crystallization of old yellow enzyme illustrates an effective strategy for increasing protein crystal size. *J. Mol. Biol.* **234,** 502–507.
78. McPherson, A., Koszelak, S., Axelrod, H., Day, J., Williams, R., Robinson, L., McGrath, M., and Cascio, D. (1986) An experiment regarding crystallisation of soluble proteins in the presence of β-octyl glucoside. *J. Biol. Chem.* **261,** 1969–1975.
79. Michel, H. (1983) Crystallising membrane proteins. *Trends Biochem. Sci.* **8,** 56–59.
80. Deisenhofer, J. Epp, O., Miki, K., Huber, R., and Michel, H. (1985) Structure of the protein subunits in the photosynthetic reaction centre of *Rhodopseudomonas viridis* at 3Å resolution. *Nature* **318,** 618–624.
81. Robert, M. C., Provost, K., and Lefaucheux, F. (1992) Crystallization in gels and related methods, in *Crystallization of Nucleic Acids and Proteins, A Practical Approach.* IRL, Oxford, UK, pp. 127–143.

82. Thaller, C., Weaver, L. H., Wilson, E., Karlsson, R., and Jansonius, J. N. (1981) Repeated seeding technique for growing large single crystals of proteins. *J. Mol. Biol.* **147,** 465–469.
83. Stura, E. A. and Wilson, I. A. (1992) Seeding techniques, in *Crystallization of Nucleic Acids and Proteins, A Practical Approach* (Ducruix, A. and Giege, R., eds.), IRL, Oxford, UK, pp. 99–126.
84. Malkin, A. J. and McPherson, A. (1994) Light-scattering investigations of nucleation processes and kinetics of crystallization in macromolecular systems. *Acta Cryst.* **D50,** 385–395.
85. Ferre'-d'Amare, A. and Burley, S. K. (1994) The use of dynamic light scattering to assess crystallisability of macromolecules and macromolecular assemblies. *Struct. Curr. Biol.* **2,** 357–359.

CHAPTER 3

Preliminary Characterization of Crystals

Sherin S. Abdel-Meguid, David Jeruzalmi, and Mark R. Sanderson

1. Introduction

1.1. General

To determine whether crystals of macromolecules are suitable for crystallographic studies, they must be fully characterized. Characterization is intended to answer many important questions about the crystals. Do they diffract X-rays? Do they diffract to a suitable resolution? Are they stable in the X-ray beam? What is their overall symmetry? What are the dimensions of their repeating unit? How many protein molecules are in that repeating unit? The answer to these questions is important to ensure a successful structure determination. There is nothing more frustrating than growing large beautiful crystals that do not diffract X-rays, diffract poorly, have a large repeating unit, or are unstable in the X-ray beam. Crystals that diffract to less than 3.5-Å resolution may be useless for answering most structural questions because of the inability to trace unambiguously the polypeptide chain and define clearly the side chains at such resolutions. Large repeating units (with any dimension >250 Å) and unstable crystals, on the other hand, complicate data acquisition from most laboratory equipment.

The initial stage of a structure determination starts with the correct determination of the space group. This initial determination of the space group, together with a crystal density determination, can give an insight into the symmetry of the macromolecular assembly that has been crystal-

From: *Methods in Molecular Biology, Vol. 56: Crystallographic Methods and Protocols*
Edited by: C. Jones, B. Mulloy, and M. Sanderson Humana Press Inc., Totowa, NJ

lized. For a crystal of a macromolecule, space group determination usually requires taking X-ray diffraction photographs using a Buerger precession camera *(1)*, which is designed to take photographs of an undistorted reciprocal lattice. For macromolecular crystals, the X-ray source for the diffraction work is preferentially a rotating anode generator, which delivers a more intense X-ray beam compared with sealed-tube generators, often employed in small molecule structure determination. **It should be stressed that X-ray equipment should only be used under the supervision of a trained operator and not used under any circumstances by an untrained beginner.** Space groups can also be determined *ab initio* by using an area detector (*see* Section 3.2.7.). For a discussion of space group determination of viruses, the reader is referred to Chapter 13.

Before discussing X-ray photography, it is necessary to cover basic aspects of crystal symmetry and discuss how symmetry elements lead to systematic absences in X-ray photographs.

1.2. Crystal Lattices and Symmetry

A crystal is made up of atoms or molecules arranged in a pattern that repeats periodically in three dimensions. The periodicity or regularity of the pattern is the most important feature of a crystal. The diffraction resolution of a crystal is directly proportional to the degree to which this pattern repeats at regularly spaced intervals with its atoms or molecules in the same orientation throughout the crystal. A crystal lattice is, thus, a geometrical construction in which points are defined with respect to the atoms or molecules, so that each point is in an equivalent environment throughout the lattice. A lattice may be thought of as being built up of unit cells, which are parallelepipeds, whose lengths are chosen to have the shortest possible dimensions and angles closest to 90° (Fig. 1). The atoms or molecules representing the unit cell are invariably related by crystal symmetry; crystals that do not exhibit symmetry are less common than those that do. The unique portion of the unit cell, i.e., the portion that is not related to other portions by crystal symmetry is called the asymmetric unit. The asymmetric unit, when acted on by the symmetry elements (Table 1 describes the symmetry elements found in crystals) of the crystal, generates the contents of the unit cell. The size and shape of the unit cell are described by a set of three lengths (a, b, and c) representing the edges of the cell and three angles (α, β, and γ) between these

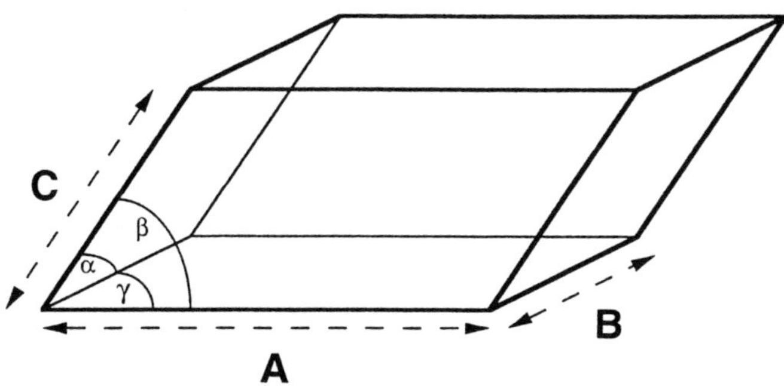

Fig. 1. A unit cell. The six values *a*, *b*, *c*, α, β, and γ specify its size and shape.

Table 1
Elements of Crystal Symmetry

Type of element	Description of element
Rotation axis	Counterclockwise rotation of 360°/*n* about axis, where *n* is 1, 2, 3, 4, or 6; a twofold axis involves rotation by 180°
Mirror plane	Reflection through a plane
Inversion center	All points inverted through a center of symmetry
Screw axis	Same as rotation axis, but followed by a translation of *p*/*n* along the rotation axis; *p* is an integer $<n$; a 3_1 screw axis involves rotation by 120° followed by translation of 1/3 of a unit cell
Glide plane	Combination of reflection through a plane followed by a translation of half the unit cell parallel to the plane; glide planes are not relevant in macromolecular crystallography as a result of the chirality of the biological building blocks

edges (Fig. 1). The angle α is between *b* and *c*, β is between *a* and *c*, and γ is between *a* and *b*. These six parameters are referred to as lattice constants.

Crystals are grouped into seven crystal systems based on their symmetry. Table 2 describes the limitation imposed on the lattice constants for each of the systems. For example, there are no restrictions for a triclinic cell, whereas a cubic cell must have equal cell lengths and 90° angles. A unit cell is often chosen to contain only one lattice point at each of its eight corners (Fig. 2). Such a cell is called primitive *(P)*. Sometimes, however, it is advantageous to select a unit cell that contains additional lattice points. A unit cell with one additional lattice point at the center of the cell is known as a body-centered (*I*; German for *Innenzentrierte*) cell, whereas that having an additional lattice point at the center of each of the

Table 2
The Seven Crystal Systems

Crystal system	Possible bravais lattices	Minimum symmetry of unit cell	Restrictions on unit cell parameters
Triclinic	*P*	No symmetry	$a \neq b \neq c$; $\alpha \neq \beta \neq \gamma$
Monoclinic	*P, C*	One twofold axis (parallel to *b*)	$a \neq b \neq c$; $\alpha = \gamma = 90°$; $\beta \neq 90°$
Orthorhombic	*P, C, I, F*	Three mutually perpendicular twofold axes	$a \neq b \neq c$; $\alpha = \beta = \gamma = 90°$
Tetragonal	*P, I*	One fourfold axis (parallel to *c*)	$a = b \neq c$; $\alpha = \beta = \gamma = 90°$
Trigonal/	*P*	One threefold axis (parallel to *c*)	$a = b \neq c$; $\alpha = \beta = 90°$; $\gamma = 120°$
rhombohedral	(or R)[a]		$a = b = c$; $\alpha = \beta = \gamma \neq 90°$
Hexagonal	*P*	One sixfold axis (parallel to *c*)	$a = b \neq c$; $\alpha = \beta = 90°$; $\gamma = 120°$
Cubic	*P, I, F*	Four threefold axes along the diagonal of the cube	$a = b = c$; $\alpha = \beta = \gamma = 90°$

[a]Rhombohedral is a subset of the trigonal system in which the unit cell can be chosen on either hexagonal or rhombohedral axes.

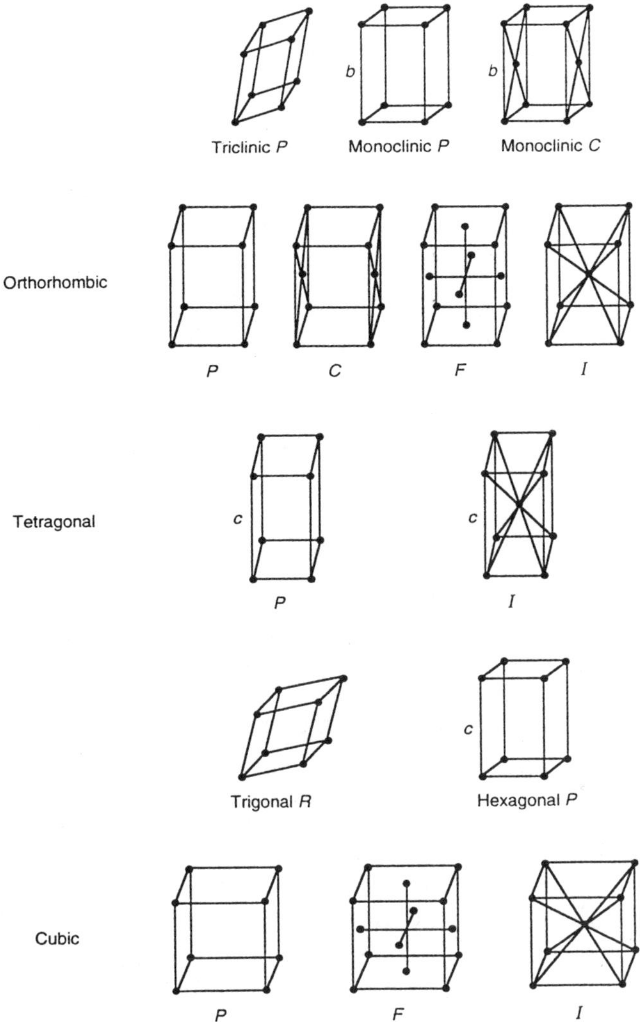

Fig. 2. The 14 Bravais lattices.

six cell faces is a face-centered *(F)* cell. A cell that contains an additional lattice point at the center of only one of the six faces (by definition the *ab* face) is referred to as a *C*-centered *(C)* cell. Note that the volume of *I* and *C* cells is double that of the *P* cell, whereas that of the *F* cell is quadruple. Centering, thus, expands the number of unique lattices to 14 (Table 2; Fig. 2). These 14 lattices are known as Bravais lattices; they were first described by Frankheimer and Bravais in the mid-19th century.

There are only 230 different combinations of symmetry elements in crystals; each of these is called a space group. Since biological molecules are inherently enantiomorphic (having only one hand, for example, proteins are made up of L-amino acids), of the 230 space groups only the 65 enantiomorphic ones are of relevance to macromolecular crystallographers. These 65 space groups are listed in Table 3.

So far we have only discussed crystal (crystallographic) symmetry. However, the macromolecules in the crystal may possess molecular symmetry that is not coincident with the crystallographic symmetry. In this case, the additional symmetry in the asymmetric unit is referred to as noncrystallographic symmetry. For example, icosahedral viruses have five-, three-, and twofold symmetry. Although it is possible for the two- and threefold axes of the virus molecules to be coincident with crystallographic axes, crystals of these viruses must at least have fivefold noncrystallographic symmetry in the asymmetric unit, since fivefold crystallographic symmetry does not exist. Noncrystallographic symmetry is often used to improve the phases necessary for structure determination through averaging of the electron density.

2. Equipment

2.1. Mounting of Crystals

Most of these tools are required for all crystal-mounting protocols. A few are necessary for some protocols, but not for others.

1. Dissecting microscope: This is used to visualize the crystal during mounting.
2. X-Ray capillaries (Lindemann tubes): They are thin-walled (0.001-mm thick) glass or quartz capillaries in which the crystal is mounted and sealed. The glass capillaries are more fragile than the quartz, but absorb less X-rays; thus, they are preferred for data acquisition. However, if crystals are only used for preliminary X-ray diffraction studies, then quartz is preferred because it is less fragile. Capillaries can be purchased in a variety of sizes ranging from 0.1 to 2.0 mm.
3. Diamond-tip pen: This is used for cutting quartz X-ray capillaries.
4. Long-nosed forceps: These are used for cutting glass X-ray capillaries.
5. Dental or bee's wax: This is used for sealing X-ray capillaries and attaching them to mounting pins.
6. Pasteur Pipets: These are used for transferring crystals.
7. Pasteur pipets with capillary tips: These are pipets that have their tips drawn in a flame to a fine diameter capable of fitting into an X-ray capillary. They are used to fill the X-ray capillaries with mother liquor or stabi-

Table 3
The 65 Enantiomorphic Space Groups Allowed for Protein and Nucleic Acid Crystals

Crystal system	Diffraction symmetry[a]	Space groups[b]
Triclinic	$\bar{1}$	P1
Monoclinic	$2/m$	P2, $P2_1$, C2
Orthorhombic	mmm	P222, $P222_1$, $P2_12_12$, $P2_12_12_1$, C222, $C222_1$, F222, [I222, $I2_12_12_1$]
Tetragonal	$4/m$	P4, ($P4_1$, $P4_3$), $P4_2$, I4, $I4_1$
	$4/mmm$	P422, ($P4_122$, $P4_322$), $P4_222$, $P42_12$, ($P4_12_12$, $P4_32_12$), $P4_22_12$, I422, $I4_122$
Trigonal	$\bar{3}$	P3, ($P3_1$, $P3_2$), R3
	$3m$	[P321, P312], [($P3_121$, $P3_221$), ($P3_112$, $P3_212$)], R32
Hexagonal	$6/m$	P6, ($P6_1$, $P6_5$), ($P6_2$, $P6_4$), $P6_3$
	$6/mmm$	P622, ($P6_122$, $P6_522$), ($P6_222$, $P6_422$), $P6_322$
Cubic	$m3$	P23, $P2_13$, F23, [I23, $I2_13$]
	$m3m$	P432, ($P4_132$, $P4_332$), $P4_222$, F432, $F4_132$, I432, $I4_143$

[a]An overbar = inversion axis; *m* = mirror plane.

[b]Space groups in parentheses and brackets cannot be distinguished from one another by inspection of diffraction patterns. All others can be uniquely assigned. Those in parentheses are enantiomorphs.

lizing solution, and to remove the solution after the crystal has been mounted.

8. Glass-fiber capillaries: These are used to remove excess mother liquor from around the crystal through capillary action. They must have a very small diameter to fit easily above the crystal in the X-ray capillary. They usually come from the tips of the flame-drawn Pasteur pipets.
9. Bunsen burner or low-temperature soldering iron: This is used for sealing glass X-ray capillaries and melting wax. The former is used in making Pasteur pipets with capillary ends.
10. Syringe: This is used to suck the crystal in the X-ray capillary. Often it is a 1-mL glass syringe.
11. Rubber tubing: A small piece of rubber tubing plugged at one end is often attached at the end of Pasteur pipets to permit finer control while transferring crystals. Also, a small piece of rubber tubing is used for connecting the syringe to the X-ray capillary.
12. Mounting pins: These are small metal tubes in which the X-ray capillary containing the crystal is inserted and fastened with wax. These pins are then fastened to a goniometer head.
13. Goniometer head: This is a device to which the X-ray capillary containing the crystal is fastened at one end, while the other end can be attached to the X-ray camera. It allows for alignment of the crystal such that it rotates perfectly about a central axis. Thus, when the crystal is aligned, it will not wobble in the X-ray beam on rotation. It also allows for orientation of the crystalline lattice. These alignments are accomplished through two perpendicular translations and two perpendicular arcs controlled by worm screws.
14. Plasticine: This is often used to hold the X-ray capillary while mounting the crystal and can be used to fasten the capillary to the goniometer head.
15. Filter paper: Finely cut strips of filter paper are used to remove the mother liquor from around the crystal before sealing. Also, a piece of wet filter paper may be wrapped around the X-ray capillary containing the crystal before sealing with melted wax; this should prevent heat transfer to the crystal.
16. Scissors or guillotine: This is used for cutting filter paper in fine strips.
17. Nylon fibers or paint brush hairs: These are used to manipulate the crystals and separate the one to be mounted from the rest.
18. Vaseline or high vacuum grease: This can be used instead of wax to seal the end of the X-ray capillary away from the goniometer head.
19. Scalpel: This is used for cutting crystals; it must have a very sharp blade. Cutting a crystal or portions of it must not be done except when it is absolutely necessary. It is a very delicate operation that could easily destroy the crystal.
20. Q-tips: These are used to apply the Vaseline.

2.2. Precession Photography

The main requirements are a precession camera mounted on an X-ray generator, X-ray films, and a dark room.

3. Methods

3.1. Preparation of Crystals

Crystals of macromolecules are extremely fragile; they must be handled with considerable care. Once crystals have reached their maximum size, they can often be left sealed in the container they grew in or can be transferred to a stabilizing solution until mounted. A stabilizing solution is one that maintains the integrity of crystals. If properly stored, crystals can survive for years without any loss in diffraction quality, particularly if 0.02% azide is present in the buffer to prevent bacterial growth. Most crystallization experiments produce many crystals, not all of which are suitable for X-ray diffraction. Some crystals are too small; others are not single. Thus, it is important to take the time to select suitable crystals.

3.1.1. Selection of a Crystal for Mounting

The best crystals must always be saved for high-resolution data acquisition. Such crystals should ideally be well shaped with sharp edges and well-defined faces. They must be single, free of satellite crystals, and of reasonable size. A crystal having the shape of a cube of 0.2–0.5 mm on an edge is ideal. However, if the crystal is only used for preliminary characterization, it does not necessarily have to be the best. A less ideal crystal can be used for that purpose; however, it must be single and free of cracks. At times, crystals grow in clumps from which a single crystal can be separated. This can be accomplished by a very gentle nudge with a nylon fiber or cutting with a sharp scalpel blade. It is absolutely necessary to have a steady hand and apply as little pressure as possible when cutting crystals.

Inspection of crystals is done under a dissecting microscope, with magnification around 40×. Crystals that are not easily seen under such a microscope are often too small for X-ray diffraction studies. The light used to illuminate the microscope must not heat the stage to ensure safety of the crystals. Unlike salt crystals, most crystals of macromolecules contain 50–60% solvent *(2)*, either hydrogen bonded to the macromolecule or free between its molecules in the lattice. Solvent is absolutely necessary for maintaining the integrity of the crystal lattice; thus, dehydration results in a rapid loss of diffraction.

3.1.2. Selection of a Stabilizing Buffer

Crystals can either be mounted directly from their mother liquor (*see* Section 3.1.4.) or by transferring to a stabilizing solution (*see* Section 3.1.5.). Often crystals are transferred to a stabilizing solution for storage. The stabilizing solution must be selected to mimic the mother liquor as much as possible, particularly in buffer conditions and ionic strength. It should result in minimal change in the environment in which the crystal grew. A good place to start is often the solution used to induce crystallization. This is the solution from the well, assuming that the vapor diffusion technique was used for crystallization; it is the solution used for dialysis if crystals were grown by equilibrium dialysis. If the macromolecule under study is available in ample supply, it should be included in the stabilizing buffer to maintain equilibrium conditions. Often, however, the macromolecule is scarce, and a higher concentration of the precipitant is needed to ensure that the crystals do not dissolve. The concentration of the precipitant should only be raised to the point at which the crystals are stable; concentrations that are too high may cause the crystals to crack or shatter. In practice, one prepares a number of solutions containing increasing concentrations of the precipitant to which test crystals, usually those that are not ideal for X-ray diffraction studies, are transferred and observed for several hours or overnight. The solution with lowest precipitant concentration that maintains the integrity of the crystal should be selected as the stabilizing buffer.

3.1.3. Mounting and Aligning of Crystals

Mounting crystals is a delicate operation that requires a steady hand and patience, particularly for crystals that are small or unusually fragile. Two techniques are used for mounting crystals depending on whether the crystal is mounted directly from its mother liquor or from a stabilizing solution. In any case, there is no right or wrong way for mounting crystals, and each individual develops his or her own technique, modifying it to suit the crystals under study.

3.1.4. Mounting Directly from the Mother Liquor

Break the sealed end of an X-ray capillary evenly. Quartz capillaries are relatively sturdy and can be easily broken after etching with a

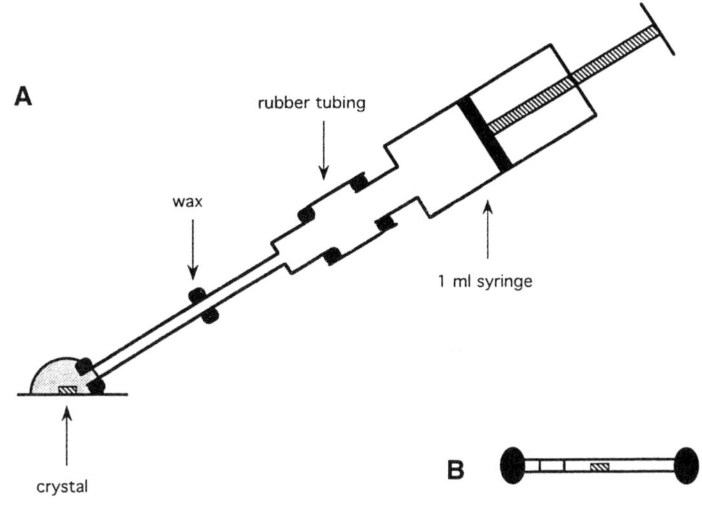

Fig. 3. Mounting of a crystal directly from the mother liquor.

diamond-tip pen. For glass capillaries, it is wise to apply a drop of wax to the capillary before snapping off the end with long-nosed forceps. The capillary is then washed by running a stream of distilled water through it, followed by ethanol to remove water traces, and finally dried with a stream of clean air. Two additional drops of wax are applied around the capillary, as can be seen in Fig. 3A. The top drop is needed to ensure a good seal when the capillary is attached to the syringe or Gilson pipet (the latter can give greater sensitivity when drawing up the crystal), and the other is needed to strengthen the point at which the capillary is broken before sealing. The capillary is attached to a 1-mL syringe through a small piece of rubber tubing (Fig. 3A). Operating under a dissecting microscope, the crystal is sucked directly into the capillary (Fig. 3A), it is positioned approx 1 cm away from the central drop of wax, the mother liquor is removed from around the crystal with a finely cut strip of filter paper, a drop of mother liquor is sucked in the capillary to prevent the crystal from drying, and the capillary is sealed with wax. Now it is safe to detach the capillary from the syringe and to seal the other end of the capillary with wax. This end can either be sealed directly or after removal of the unneeded portion of the capillary (Fig. 3B). While sealing the capillary with wax, it is wise to wrap the portion of the capillary containing the crystal with a wet piece of filter paper to prevent heat transfer to the

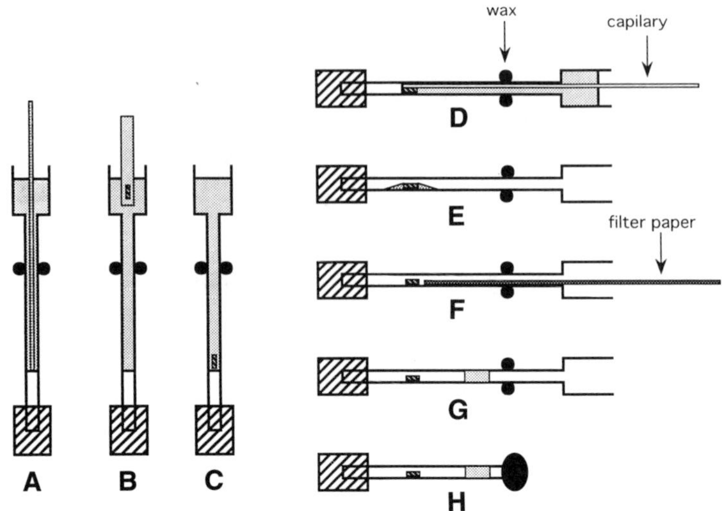

Fig. 4. Mounting of a crystal from a stabilizing solution.

crystal. If the crystals are extremely heat-sensitive, then it may be necessary to seal the tube with Vaseline, epoxy resin, or immersion oil. Finally, the capillary is attached to the goniometer head with plasticine or a mounting pin that inserts into the goniometer head.

3.1.5. Mounting from a Stabilizing Solution

Follow the same procedure described above to clean the X-ray capillary, and then seal the broken end with wax. Glass capillaries can be sealed by melting. The capillary is attached to a mounting pin with wax, and the pin is fastened to the goniometer head. Fill the capillary with the stabilizing solution, starting approx 1 cm above the mounting pin (Fig. 4A). This can be accomplished by inserting a Pasteur pipet with a capillary end containing the stabilizing solution all the way into the X-ray capillary and extruding the mother liquor from the bottom up. The crystal is then transferred with a Pasteur pipet to the mother liquor (Fig. 4B) and allowed to drop to the meniscus at the bottom of the capillary (Fig. 4C). Operating under a dissecting microscope, most of the stabilizing solution is removed using a Pasteur pipet with a capillary end (Fig. 4D). The remainder of the solution is removed by capillary action using a glass-fiber capillary inserted above the crystal (it must not touch the crystal) (Fig. 4E), followed by drying with a finely cut strip of filter paper (Fig. 4F). A drop of stabilizing solution is inserted in the X-ray capillary to prevent the crystal from drying (Fig. 4G), and the capillary is then sealed with Vaseline or high-vacuum grease (Fig. 4H).

Once the crystal is mounted and attached to the goniometer head, it may be made to rotate about a crystal axis by adjusting the translations (sledges) and arcs on the goniometer head. Generally, crystal axes are parallel or perpendicular to the faces. The arcs on the goniometer head can be adjusted by approximately ±20°. However, larger adjustments can be made by manually tilting a crystal fastened to the goniometer head with plasticine, followed by finer adjustments using the translations and arcs of the goniometer head.

3.1.6. Flash Freezing

The technique of flash freezing, which entails data collection at liquid nitrogen temperatures, was developed to allow data measurements from crystals that decay rapidly in the X-ray beam at room temperature. Once cooled to liquid nitrogen temperatures, crystals are in essence "immortalized." For example, this technique has allowed data measurements from one ribosome crystal over several interspersed sessions at the synchrotron, the crystal being returned to a liquid nitrogen storage vessel in the intervening periods *(3)*.

Using the technique of flash freezing, it is possible to mount protein crystals without using X-ray capillaries. The technique involves mounting the crystal on a nylon loop in the presence of a cryoprotectant and then freezing to –196°C in a stream of liquid nitrogen. The freezing has to be fast, so that the water in the protein lattice does not form ice crystals, but turns to amorphous ice.

In recent years, a variety of strategies has evolved by which the crystal may be positioned in the path of the X-ray beam. We shall not discuss strategies of crystal mounting that involve glass spatulas *(4)* or wire loops *(5)*, since the consensus is that mounting in a loop of thin nylon is the best technique for most protein crystals. Two different cryocooling techniques have evolved; one involves first cooling the crystal in a bath of liquid propane that has been cooled to liquid nitrogen temperatures and then inserting the crystal into the nitrogen stream. The second involves no intermediate cooled liquid propane step; it involves cooling the crystal directly to liquid nitrogen temperatures. The first technique was proposed by Parak et al. *(6)*; further development of crystal handling and cryocooling liquids by Jeruzalmi and Steitz *(7)* has allowed the technique to develop into a routine procedure for data collection in the Molecular

Table 4
Cryoprotectants Used in Flash Freezing *(9)*

Cryoprotectant	Concentration
Glycerol	10–18% (used with high-salt crystals)
MPD	15–25%
Sucrose	25%
Isopropanol	25% (only tested with synthetic peptides)
Jeffamine	5–20% (T-403 to T-600 Polyoxypropylenetriamine by Texaco)
Polyethylene 8000	20–25%
Polyethylene 6000	18%
Polyethylene 3000	20%

Biophysics and Biochemistry Laboratories at Yale University *(7)*. A description of the second technique as it is used at Harvard University can be found in ref. *8*. Other laboratories have developed their own in-house techniques for cryo-data measurement *(3,4,8)*.

Crystals are mounted in a small loop of nylon; the best material is American grade 200 suture nylon. An extensive survey by one of the authors (Jeruzalmi) of different nylon materials' opacity to X-rays found this suture material to be the best. A loop of nylon is glued to a mounting pin; the glue has to withstand cooling to liquid nitrogen temperatures. Loctite Prism surface insensitive gel adhesive no. 454 has been found to work well. The crystal is first placed in a small vessel containing a stabilizing solution that contains a cryoprotectant. Table 4 lists a variety of cryoprotectants that have been found by Cascio (Molecular Biology Institute, UCLA) to work well *(9)*. Earlier work by Rossmann and Petsko using cryoprotectants is found in refs. *10* and *11*. The loop is made slightly smaller than the crystal. The crystal is gently agitated to allow it to rise in solution, fall on the loop, and get scooped through the liquid surface (Fig. 5). Most crystals survive such treatment, although very fragile ones have to be mounted on glass spatulas. Using a propane plunger (Fig. 6) that has been precooled to liquid nitrogen temperatures (Fig. 7), the crystal is then transferred to a bath of propane suspended in a dewar of liquid nitrogen (Fig. 8) for about 5 min (if left longer, the propane may freeze). The cryovial of cooled liquid propane (at about –150°C), attached to the propane plunger, is then transferred to the liquid nitrogen stream at the goniometer (Fig. 9). Using tweezers and insulated gloves, the mounting pin and loop containing the crystal are transferred from the propane plunger to the goniometer

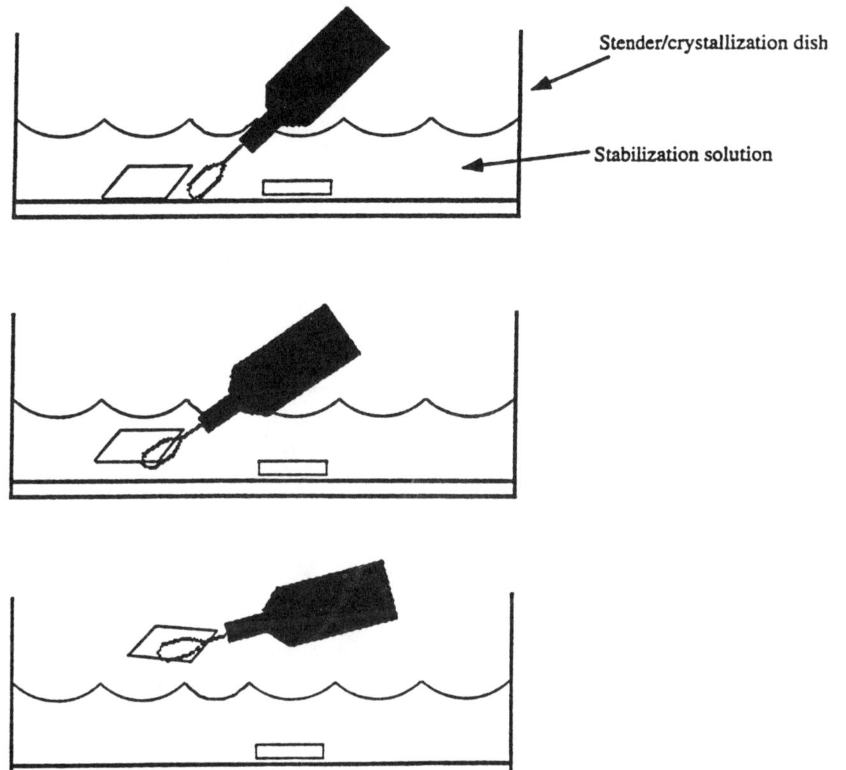

Fig. 5. Scooping up the crystal on a nylon loop.

head in a smooth action (Fig. 10). The crystal must be kept at liquid nitrogen temperatures in the cold stream while data acquisition is proceeding.

3.2. Preliminary X-Ray Data

As indicated above, the most common method for the determination of the crystal space group and lattice constants is that using a Buerger precession camera *(1)*. Recently, however, the increasing use of area detectors has made it possible to characterize the crystals directly from data measured on a detector. Here, we will primarily focus on characterization using the precession camera and briefly describe characterization using area detectors.

3.2.1. Optical Alignment of the Crystal

Once the crystal is attached to the goniometer head, it must be optically aligned using the microscope of the precession camera to ensure

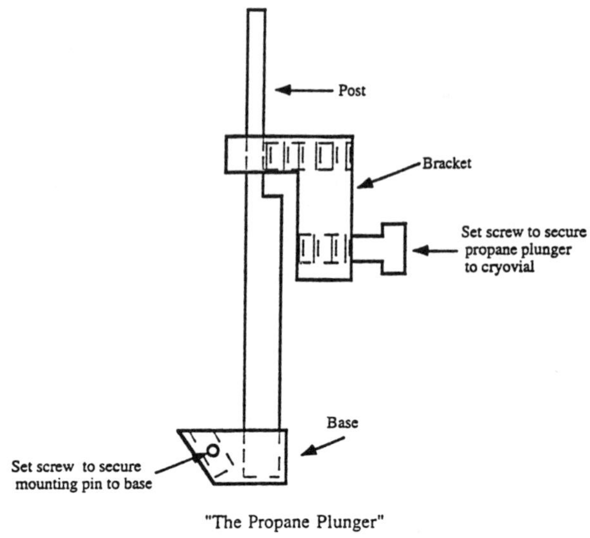

Fig. 6. Propane plunger.

that it always remains in the X-ray beam on rotation of the camera's spindle. This involves adjusting the sledges so that the crystal does not precess about the center of the crosshairs of the microscope. This is best done by iterative centering of the crystal at 0 and 180° on the spindle, and then at 90 and 270°. It is important that the crystal, not the X-ray capillary, remains centered in the X-ray beam on a 360° rotation of the spindle. A large amount of time spent taking setting photographs may be saved by adjusting the arcs and sledges to bring a crystal plane along the axis of rotation of the camera spindle as described above.

3.2.2. Photographic Alignment of the Crystal

The purpose of the photographic alignment is to position accurately one of the crystallographic axes parallel to the X-ray beam, and thus enhance one's ability to record the crystal reciprocal lattice on a piece of film. If one is satisfied that the crystal is correctly centered, load a piece of film in the film cassette. Position the film holder on the precession camera, and set the spindle to 0°. Make sure that the precession angle is also zero, the nickel filter is removed from behind the collimator, and the film is marked to specify a unique orientation. Record the crystal-to-film distance; it should be set at 75 or 100 mm, more commonly the latter. Remove the beamstop, and quickly open and close the shutter to record

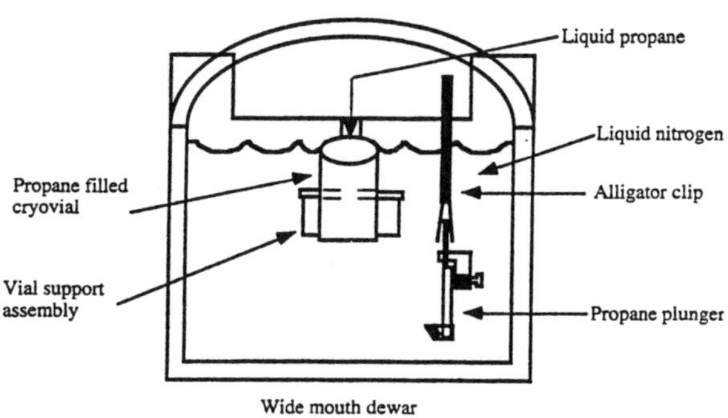

Fig. 7. Precooling the plunger in the nitrogen bath.

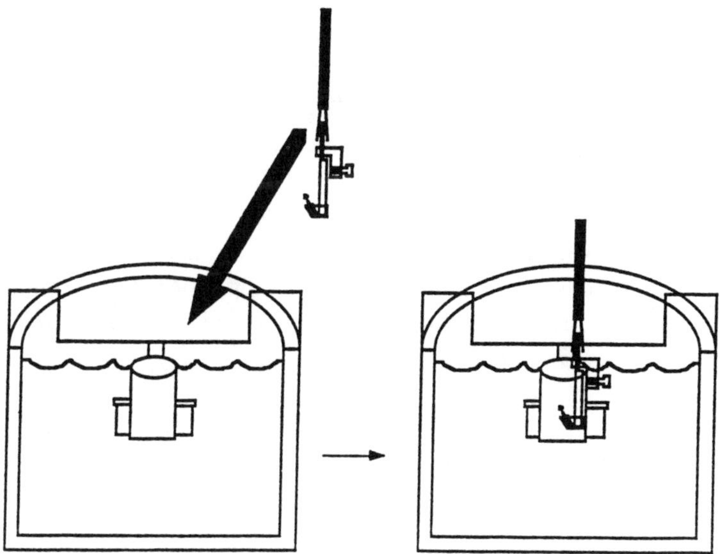

Fig. 8. Transferring the plunger to the propane bath.

the position of the direct beam. You do not have to remove the crystal for this operation. Be sure that the beamstop is back in its proper position, and expose the crystal for a few minutes with the X-ray generator power turned to maximum. Exposure time differs from one crystal to another and can only be determined empirically, but it usually ranges from 2 to 20 min.

Develop the film. You should observe a set of concentric circles (Fig. 11A), one of which (the zero-order circle) goes through the origin (position of the direct beam). This type of photograph is referred to as a "still" or a

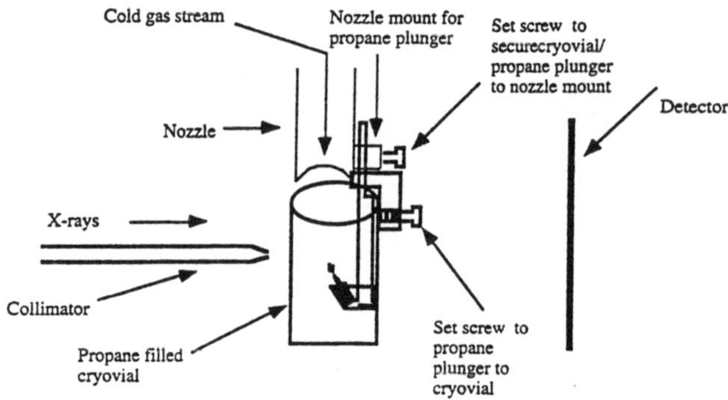

Fig. 9. Transfer of plunger and liquid propane bath to the nitrogen stream.

setting photograph. The next step is to calculate the vertical and horizontal corrections, which are required to bring the primary cone (zero-order circle) behind the beamstop. This is best done by having a set of concentric rings drawn onto a transparent sheet and overlaid on the film so that the center of the zero-order circle may be marked using a needle spike. The vertical correction is applied to the vertical arc of the goniometer head, while the horizontal correction is applied to the horizontal arc. Make sure that the correction is applied in the proper direction, particularly when applying the horizontal correction. The correction is calculated from the following formula:

$$\tan \varepsilon = (\Delta/F) \qquad (1)$$

where ε is the correction to be applied to the arc in degrees, Δ is the measure horizontal or vertical displacement off the film in millimeter, and F is the crystal-to-film distance in millimeter.

The photographic alignment described above is usually repeated two more times. Rotate the spindle 90° away from its current position, take a "still" photograph, develop it, and make the appropriate corrections. Turn the spindle back by 90° toward its original position and repeat. The alignment is complete, however, when the zero-order circle is no longer seen on the film, i.e., it is completely superimposed on the direct beam position (Fig. 11B). This may require additional setting photographs as described above. However, if the remaining corrections are about 1°, it is best to use small-angle precession photographs (*see* Section 3.2.3.1.) to complete the alignment.

A few additional "still" photographs may be useful. For example, if one knows from the crystal morphology that it may belong to either a trigonal or a

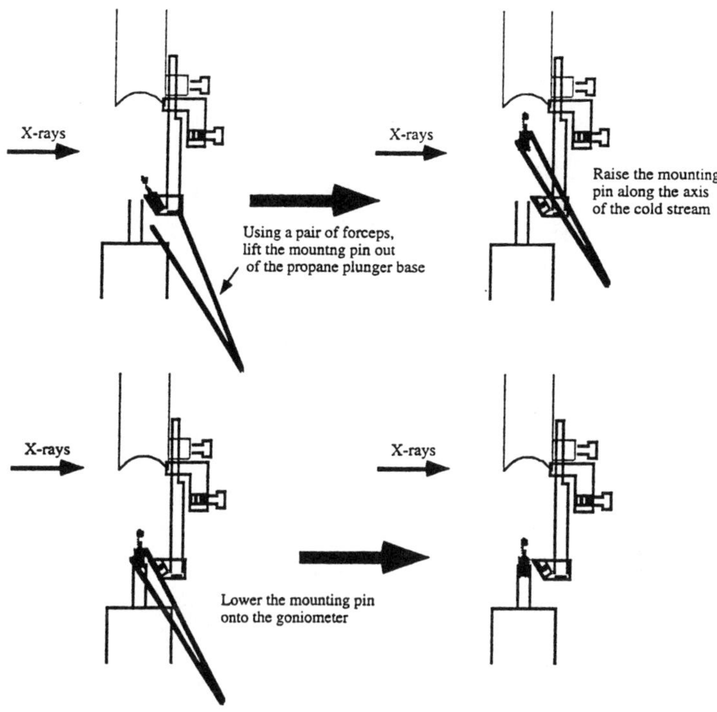

Fig. 10. Transfer of the loop to the goniometer for data measurement.

hexagonal space group, and that the crystal is mounted with its c^* axis along the spindle axis, "still" photographs every 60° should show a repeating zone.

3.2.3. Precession Photography

Although in the early days of macromolecular crystallography precession photography was used to measure data, its role is now confined to the determination of lattice constants and space groups, and to the search for heavy-atom derivatives. Even the latter role is being quickly replaced. Now, it is more common to measure data rapidly from crystals of potential heavy-atom derivatives and compare them to the native data. Thus, here we will focus on precession photography's role in preliminary characterization of crystals.

3.2.3.1. Screenless (Small-Angle Precession)

For crystals with very large unit cells, such as viruses, small-angle precession photographs (4°) are all that are needed to determine the space

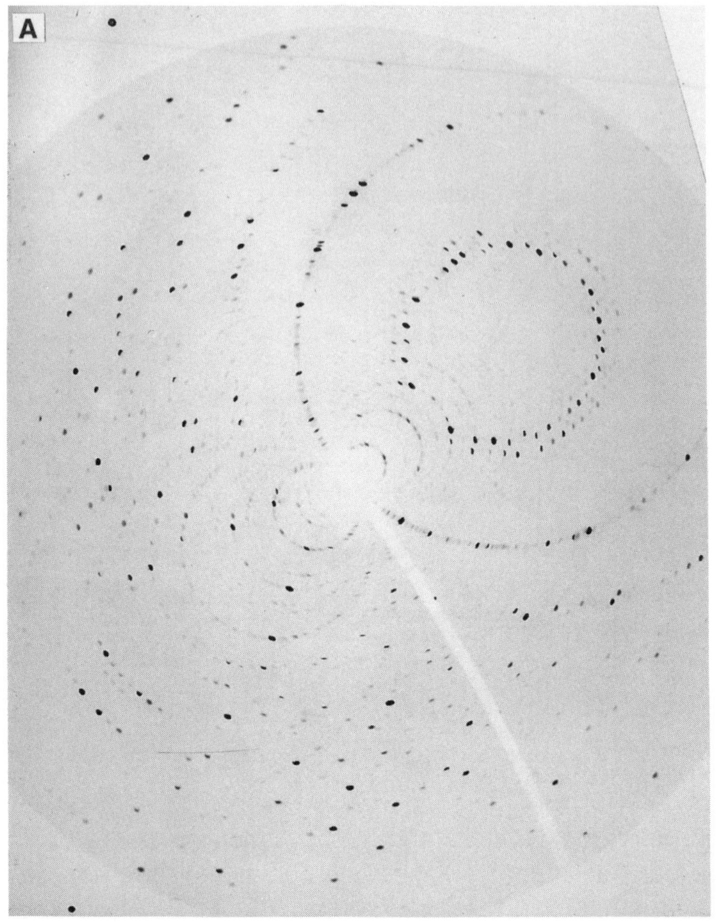

Fig. 11. **(A)** Alignment photograph. A misaligned "still" photograph.

group. In such cases, a considerable amount of information is present on each precession photograph to allow for an unambiguous determination of the space group. However, for crystals of most proteins and nucleic acids, small-angle precession photographs are used for fine crystal alignments of about 1° or less.

Load a piece of film in the film cassette. It is not necessary to use a full piece of film; a third of a full sheet is sufficient. Position the film holder on the precession camera, and set the spindle to either of the two final positions for crystal alignment using "still" photographs. Set the preces-

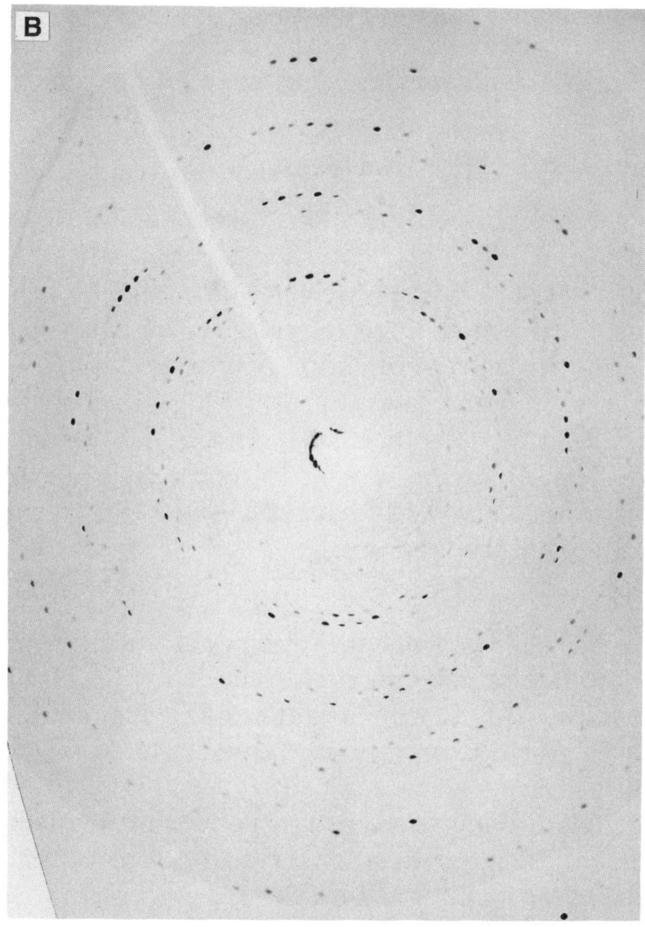

Fig. 11. **(B)** Alignment photograph. An aligned "still" photograph.

sion angle (μ) to 2–3°, and switch the power of the precession motor on. Again, make sure that the nickel filter is removed from behind the collimator, and expose the crystal for approximately four times as long as was needed for a "still" photograph, with the X-ray generator power turned to maximum. Develop the film. If the crystal is well aligned, the pattern on the film will be symmetric. This should be particularly evident in the inner circle, the zero layer. If the pattern is not symmetric, additional corrections are required. Measure the distance from the center of the diffraction (direct beam position) to the edge of the diffraction in the two vertical directions (left and right) and calculate the difference (Δ). Similarly, calculate the difference in the horizontal direction. The corrections that should be applied to the vertical arc and to the horizontal arc of the goniometer head can each be calculated from the following equation:

$$\tan \varepsilon = (\Delta/4F) \quad (2)$$

where ε is the correction to be applied to the arc in degrees, Δ is the measured horizontal or vertical displacement off the film in millimeter, and F is the crystal-to-film distance in millimeters. This equation gives a reasonable approximation of the correction when μ and ε are small.

Note that the screenless precession photographs show in addition to the zero layer, other higher layers. Thus, if the photograph does not show clear demarcation between the zero layer and the first layer as can be seen in Fig. 12A, it would be wise to repeat the photograph using a smaller precession angle. Only when there is a clear demarcation is it possible to align the crystal properly.

3.2.3.2. Screened

After it has been determined that the crystal is well aligned, with one of its crystallographic axes parallel to the X-ray beam, one may wish to take additional photographs that should unambiguously assign the space group. These usually are large-angle, screened precession photographs.

The first step in taking a screened precession photograph is to set and align the screen. The screen position (crystal-to-screen distance) depends on the radius of the annulus and the desired precession angle. For a nth layer photograph, it can be calculated from the following formula:

$$S_n = r_S \cot \cos^{-1}(\cos\mu - nd^*) \quad (3)$$

where s_n is the screen position for the nth reciprocal layer, r_S is the annulus radius, μ is the precession angle, and n corresponds to the nth upper level of a reciprocal spacing d^*. Thus, for a zero-layer photograph, the formula for the screen setting can be reduced to:

$$S_O = r_S \cot \mu \quad (4)$$

Upper-layer photographs also require movement of the film cassette toward the crystal so that the center of the film follows the processing movement of the center of the reciprocal lattice layer. The distance moved is Fnd^*.

Now that the screen is properly positioned, it must be aligned. To do that, set the spindle to either of the two final positions for crystal alignment; this must be done with the precession angle set to 0°. Remove the beamstop, and ensure that the direct beam and the center of the screen annulus coincide. This can be accomplished by marking the center of the screen with fluorescent material and observing its fluorescence on exposure to X-rays. Make sure that the beamstop and the nickel filter are now into their proper positions. Load a full piece of film in the film cassette, and position the film holder on the precession camera. Set the precession angle (μ) to the desired angle, but do not start the precession motor. Take a "still" photograph at that spindle position. Move the spindle axis by 90, 180, and 270°, and take three additional "still" photographs on the same piece of film. Finally, develop the film. If the screen is well aligned, one should observe four full zero layer circles in the shape of a cloverleaf.

Once the screen is well aligned, load a full piece of film in the film holder, position it on the camera, turn the precession motor on, and expose the crystal overnight. Develop the film; it should appear similar to that in Fig. 12B.

3.2.4. Calculation of Resolution

Resolution can be easily determined from "still" photographs by measuring the distance (r) from the center of diffraction (direct beam position) to the furthest spot on the film, and applying Bragg's equation:

$$d = \lambda/2\sin[(1/2)\tan^{-1}(r/F)] \qquad (5)$$

where d is the resolution in Å, λ is the radiation wavelength in Å (1.5417 Å for Cu Kα), r is the distance measured off the film from the center to the diffraction edge in millimeters, and F is the crystal-to-film distance in millimeters. Resolution measured this way only represents the upper limit assuming that the crystal-to-film distance allows for the highest resolution spots to be recorded on the film. Data to that resolution may be limited and may represent only a small percentage of the theoretically possible intensities. Note that smaller crystal-to-film distances allow higher resolution data to be recorded on a particular size film.

3.2.5. Determination of Lattice Constants

From an aligned "still" photograph, one can obtain a good estimate of the lattice constant. Measure the radius (r) of the nth circle, and calculate the lattice constant (d) from the following formulas:

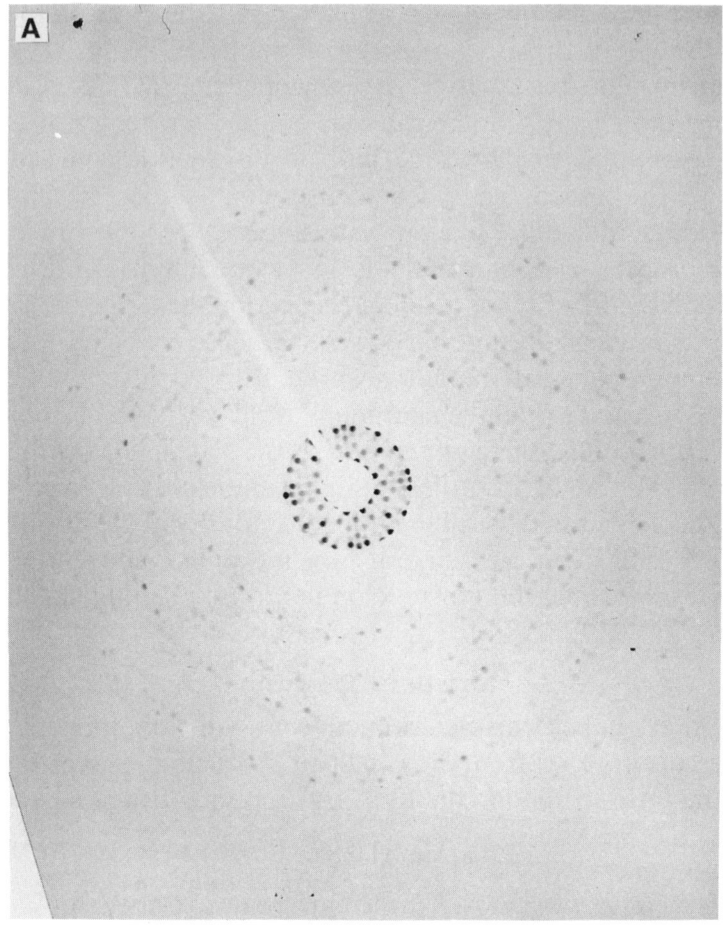

Fig. 12. **(A)** A screenless precession photograph for lysozyme.

$$d = (1/d^*) = [n\lambda/(1 - \cos \nu)]$$

$$\tan \nu = (r/F) \tag{6}$$

where d^* is the reciprocal lattice spacing, λ is the radiation wavelength in Å (1.5417 Å for Cu Kα), n is the measured ring order, and F is the crystal-to-film distance. Repeat the measurement for a number of rings, and take the average.

More accurate lattice constants determination comes from high angle precession photographs, where the spacing between consecutive rows of

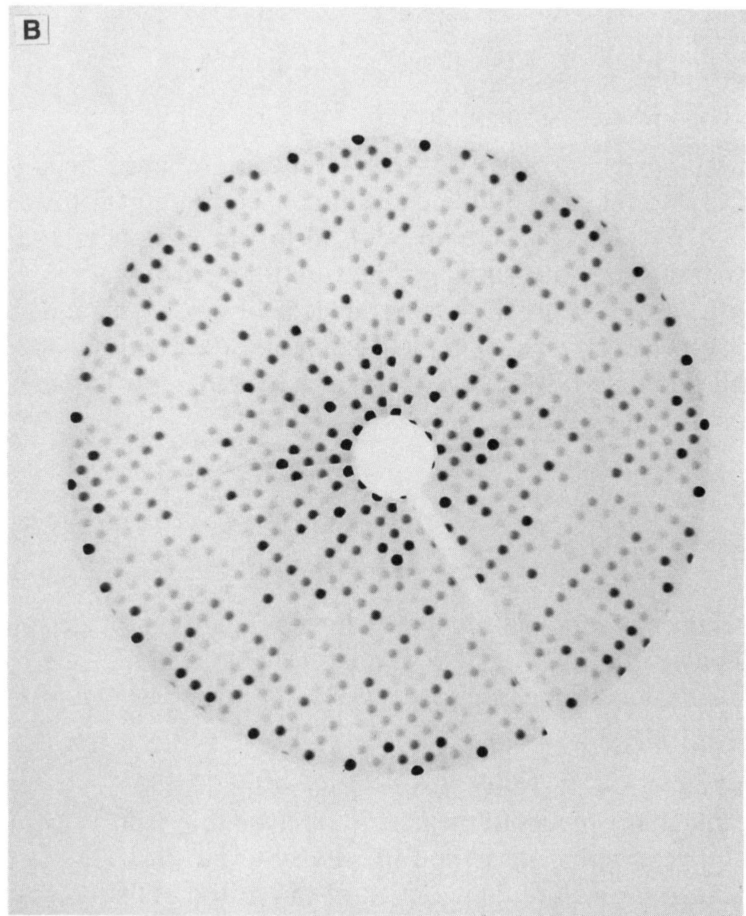

Fig. 12. **(B)** A 9 ½° screened precession photograph for lysozyme.

spots is inversely proportional to the length of the axis. Measure the distance between as many spots in a row as are fully recorded, and calculate the average distance between any two consecutive spots. Note that in case of centering or the presence of other systematic absences, spots may be missing. Thus, one must account for absences resulting from centering and must not measure distances along axial rows; these are the rows that cross the direct beam position. In addition to the distance between consecutive rows, one must measure the angle between rows. This information can be used to calculate the lattice constants (a, b, or c) from the following formulas:

$$a = (\sin \alpha^*/va^*) = (\lambda F \sin \alpha^*/vx)$$
$$b = (\sin \beta^*/vb^*) = (\lambda F \sin \beta^*/vx)$$
$$c = (\sin \gamma^*/vc^*) = (\lambda F \sin \gamma^*/vx) \quad (7)$$

where $v = (1 - \cos^2 \alpha^* - \cos^2 \beta^* - \cos^2 \gamma^* + 2 \cos \alpha^* \cos \beta^* \cos \gamma^*)^{1/2}$, a^*, b^*, and c^* are reciprocal lattice spacing, λ is the radiation wavelength (1.5417 Å for Cu Kα), F is the crystal-to-film distance, and x is the interplanar distance measured off the film for a particular axis, and also where α^*, β^*, and γ^* are the angles between b^* and c^*, a^* and c^*, and a^* and b^*, respectively. Thus, in the case of an orthorhombic space group where all three angles are 90°, the above formulas for lattice constants calculations can be reduced to:

$$a = (1/a^*) = (\lambda F/\mathrm{x})$$
$$b = (1/b^*) = (\lambda F/\mathrm{x})$$
$$c = (1/c^*) = (\lambda F/\mathrm{x}) \quad (8)$$

Note that conventions have been adopted regarding the designation of lattice constant parameters (*see* Table 2). For example, the angle not equal to 90° of a monoclinic space group is β, whereas that of a hexagonal space group is γ.

3.2.6. Determination of Space Group

Now that many pieces of the puzzle are available as information in the various photographs, it is time to solve the puzzle. This is done by inspection of the nature and content of the different "still" and precession photographs. For example, if a particular pattern on a "still" or a precession photograph repeats every 60°, it is reasonable to assume that the crystal belongs to either a trigonal of a hexagonal space group, with its c^* axis (in this case c^* and c coincide) along the spindle axis. Finally, however, identification of the crystal class and space group comes primarily from analyses of the diffraction symmetry recorded on the various precession photographs. It is important to note that zero-level diffraction photographs will always contain an inversion center. Thus, on a zero-level photograph, one cannot distinguish between a twofold axis in the plane of the photograph and a mirror plane.

The first step in a space group determination is to determine the crystal class (Table 2). For example, a tetragonal space group must have $a = b$ and $\alpha = \beta = \gamma = 90°$. Once the crystal class has been determined, the presence or absence of centering must be investigated by indexing the spots on the different precession photographs, and looking for systematic absences representative of a particular centering (*see* the International Tables, vol. 1 or A for systematic absences). Indexing is accomplished by recognizing that the direct beam position represents the index *0, 0, 0*. Note that a zero-level photograph by definition has one of its three indices *(h, k, l)* as zero. Therefore, a zero-level photograph can be either *h, k, 0*; *h, 0, l*; or *0, k, l*. By indexing the various precession photographs the systematic absences become apparent. Systematic absences along axial rows represent screw axes. For example, the absence of every other spot along a particular axial row indicates a 2_1 screw along that axis. Further inspection of the precession photographs should reveal the presence or absence of rotation axes and mirror planes.

The information found in the various precession photographs should allow for an unambiguous determination of the space group. For crystals of biological macromolecules, the space group must be one of those listed in Table 3. If it is not possible to determine unambiguously the space group, additional photographs are needed. Note that it is not possible from photographs to distinguish between space groups that differ only in the directionality of their screw axes, such as $P6_1$ and $P6_5$.

3.2.7. Space Group Determination from Area Detector Data

Space groups may be determined directly from native data collected on an area detector, without having to use precession photography. In fact, a number of laboratories have now parted with their precession cameras in order to mount additional image plate detectors on the freed up X-ray ports of their rotating anodes. A number of programs are available that enable the symmetry of the crystal to be determined (a Bravais lattice assignment), but the software that may be used depends on the type of area detector that was used to measure the data. At the time of writing Kabsch's XDS computer program, containing the algorithm IDXREF capable of calculating vectors between reflections with low indices and building up full data indexing *(12–14)*, can be used on data from both the Siemens (Xentronics) multiwire detector and the Hamburg MAR image plate. Table 5 shows the successful assignment of a space group for data from a quinone reductase crystal measured on a Siemens detector. The correct Bravais-type is mP in this case; this is given by the lowest value of the quality of fit (12.5). A Bravais-type mP corresponds to possible space groups P2 and $P2_1$. Once the Bravais lattice is assigned, a full space group

Table 5A

Assignment of Space Group for Data from a Quinone Reductase Crystal Using Dr. W. Kabsch's Program XDS

Lattice-character	Bravais-lattice	Quality of fit	Unit cell constants a(Å)	b(Å)	c(Å)	alpha(deg.)	beta(deg.)	gamma(deg.)	Reindexing card						
1	cF	474.5	186.5	189.2	186.5	110.6	100.8	116.1	1	-1	1	0	1	1	-1
2	hR	228.8	143.7	154.4	191.0	97.8	86.7	115.5	1	-1	0	0	-1	0	1
3	cP	245.6	99.4	107.0	118.9	89.6	90.4	91.8	1	0	0	0	0	-1	0
4	cI	411.8	155.5	143.7	159.4	61.0	56.2	64.7	-1	0	-1	0	-1	1	0
5	hR	241.3	148.3	155.5	186.5	99.3	86.4	116.8	1	1	0	0	-1	0	-1
6	tI	999.0	160.5	155.5	148.3	63.2	59.9	54.1	0	1	1	0	1	0	1
7	tI	999.0	154.4	143.7	160.5	61.2	56.2	64.5	1	0	-1	0	1	-1	0
8	oI	999.0	143.7	155.5	159.4	56.2	61.0	64.7	1	-1	0	0	1	0	1
9	hR	812.4	99.4	143.7	385.0	91.6	105.1	131.9	1	0	0	0	-1	1	0
10	mC	79.5	148.3	143.7	118.9	90.0	90.5	94.2	1	1	0	0	1	-1	0
11	tP	108.9	99.4	107.0	118.9	90.4	89.6	91.8	-1	0	0	0	0	1	0
12	hP	324.7	99.4	107.0	118.9	90.4	89.6	91.8	-1	0	0	0	0	1	0
13	oC	91.8	143.7	148.3	118.9	90.5	90.0	85.8	-1	1	0	0	1	1	0
15	tI	600.4	99.4	107.0	277.8	68.4	69.4	91.8	-1	0	0	0	0	1	0
16	oF	999.0	148.3	143.7	281.4	92.2	122.3	85.8	-1	-1	0	0	1	-1	0
14	mC	91.8	143.7	148.3	118.9	90.5	90.0	85.8	-1	1	0	0	1	1	0
17	mC	999.0	148.3	143.7	155.5	64.7	116.8	85.8	-1	-1	0	0	1	-1	0
18	tI	744.4	159.4	186.5	99.4	59.3	90.9	95.7	0	-1	1	0	1	-1	-1
19	oI	623.7	99.4	159.4	186.5	84.3	59.3	89.1	-1	0	0	0	0	-1	1
20	mC	149.3	160.5	159.4	99.4	90.9	91.5	96.0	0	1	1	0	0	1	-1
21	tP	165.9	107.0	118.9	99.4	89.6	91.8	90.4	0	1	0	0	0	0	-1
22	hP	447.0	107.0	118.9	99.4	89.6	91.8	90.4	0	1	0	0	0	0	-1
23	oC	161.7	159.4	160.5	99.4	88.5	90.9	84.0	0	1	-1	0	0	-1	-1
24	hR	999.0	262.0	159.4	99.4	89.1	69.5	101.6	1	-2	-1	0	0	1	-1
25	mC	153.6	160.5	159.4	99.4	89.1	91.5	84.0	0	-1	-1	0	0	1	-1
26	oF	608.1	99.4	233.0	257.1	81.0	112.3	113.4	1	0	0	0	-1	2	0
27	mC	487.3	233.0	99.4	159.4	89.1	127.2	66.6	-1	2	0	0	-1	0	0
28	mC	250.2	99.4	257.1	107.0	90.4	91.8	67.7	-1	0	0	0	-1	0	2
29	mC	235.1	99.4	233.0	118.9	89.8	90.4	66.6	1	0	0	0	1	-2	0
30	mC	289.8	107.0	260.1	99.4	90.4	91.8	66.1	0	1	0	0	0	1	-2
31	aP	0.0	99.4	107.0	118.9	89.8	89.6	88.2	1	0	0	0	0	1	0
32	oP	29.6	99.4	107.0	118.9	90.4	89.6	91.8	-1	0	0	0	0	1	0
40	oC	310.7	107.0	260.1	99.4	89.6	91.8	113.9	0	-1	0	0	0	1	-2
35	mP	25.3	107.0	99.4	118.9	89.6	90.4	91.8	0	-1	0	0	1	0	0
36	oC	271.5	99.4	257.1	107.0	89.6	91.8	112.3	1	0	0	0	-1	0	2
33	mP	25.5	99.4	107.0	118.9	90.4	89.6	91.8	-1	0	0	0	0	1	0
38	oC	245.4	99.4	233.0	118.9	89.6	89.6	113.4	1	0	0	0	-1	2	0
34	mP	12.5	99.4	118.9	107.0	90.4	91.8	89.6	1	0	0	0	0	0	1

42	oI	560.7	99.4	107.0	277.8	111.6	110.6	91.8	1	0	0	0	0	-1	0
41	mC	306.6	260.1	107.0	99.4	91.8	90.4	66.1	0	-1	2	0	0	-1	0
37	mC	267.3	257.1	99.4	107.0	91.8	90.4	67.7	1	0	-2	0	1	0	0
39	mC	241.1	233.0	99.4	118.9	89.6	90.2	66.6	1	-2	0	0	1	0	0
43	mI	999.0	143.7	277.8	107.0	111.6	136.3	58.9	-1	1	0	0	-1	1	-2
44	aP	4.1	99.4	107.0	118.9	90.4	89.6	91.8	-1	0	0	0	0	1	0

Table 5B
Possible Space Groups for Protein Crystals for Each Bravais Type

Bravais-type	Possible space-groups for protein crystals
aP	[1,P1]
mP	[3,P2] [4,P2(1)]
mC,mI	[5,C2]
oP	[16,P222] [17,P222(1)] [18,P2(1)2(1)2] [19,P2(1)2(1)2(1)]
oC	[21,C222] [20,C222(1)]
oF	{22,F222]
oI	[23,I222] [24,I2(1)2(1)2(1)]
tP	[75,P4] [76,P4(1)] [77,P4(2)] [78,P4(3)] [89,P422] [90,P42(1)2]
	[91,P4(1)22 [92,P4(1)2(1)2] [93,P4(2)22] [94,P4(2)2(1)2]
	[95,P4(3)22 [96,P4(3)2(1)2]
tI	[79,I4] [80,I4(1)] [97,I422] [98,I4(1)22]
hP	[143,P3] [144,P3(1)] [145,P3(2)] [149,P312] [150,P321] [151,P3(1)12]
	[152,P3(1)21] [153,P3(2)12] [154,P3(2)21] [168,P6] [169,P6(1)]
	[170,P6(5)] [171,P6(2)] [172,P6(4)] [173,P6(3)] [177,P622]
	[178,P6(1)22] [179,P6(5)22] [180,P6(2)22] [181,P6(4)22] [182,P6(3)22]
hR	[146,R3] [155,R32]
cP	[195,P23] [198,P2(1)3 [207,P432] [208,P4(2)32] [212,P4(3)32]

can be worked out by inspecting the systematic absences when the data are indexed into the lowest symmetry (in this case P2). In this example, the space group was assigned to $P2_1$ based on the absence of every other reflection along k ($2n + 1$ are absent).

For data measured from a Rigaku R-axis II image plate, the space group assignment may be made using the "blind" option within the R-axis Rigaku software written by T. Higashi; the data are then analyzed for systematic absences. Otwinowski's computer program DENZO *(15)* may also be used; Table 6 gives an example of the space group assignment for data from herpes simplex virus type 1 thymidine kinase crystals. A lattice type of *C*-centered orthorhombic was given a quality of fit index of 1.08%, since this was the highest lattice symmetry with the lowest quality of fit. Here again, indexed reflections were checked for systematic absences, and the correct space group was found to be $C222_1$. DENZO uses Minor's peak searching and displaying computer program XDISPLAYF.

For symmetry assignment of crystal data measured on the Hamburg MAR detector (both the large, 30-cm and the small, 18-cm plates), the computer programs XDS, DENZO, or MOSFLM *(16)* that have incorporated Kabsch's IDXREF *(12–14)* may be used. For data measured on an Enraf-Nonius Fast detector, space group determination may be made using Pflugrath's computer program MADNES *(17)*, which also incorporates Kabsch's IDXREF routine *(12–14)*. MADNES has been further developed to handle data from the R-axis, MAR, and Siemens detectors. The most recent version of DENZO also handles data from the MacScience detectors.

3.3. Density Measurement

The number of molecules in the asymmetric unit may be determined by calculating the ratio of the volume of the asymmetric unit and the molecular weight of protein contained in it. This ratio is referred to as the V_m and is expressed as Å^3/Dalton. In 1968, Matthews showed the V_m for most proteins ranges from 1.8 to 3.0 Å^3/Dalton *(2)*. When the number of molecules in the asymmetric unit is ambiguous from the calculation of the V_m, then the determination of the crystal density may provide an indication of the correct number of molecules in the asymmetric unit.

Density determination is initiated by making a density gradient solution. The solution is made by mixing two solutions of different densities like water-saturated xylene and carbon tetrachloride *(18–20)*, or low- and high-density solutions of Ficoll *(21)*. The solution is calibrated by intro-

Table 6
Space Group Assignment for Data from Herpes Simplex Virus Type 1 Thymidine Kinase Crystals Using Dr. Z. Otwinowski's Program DENZO

	Fit	Unit cell constants a(Å)	b(Å)	c(Å)	alpha(deg.)	beta(deg.)	gamma(deg.)
Primitive cubic	28.58%	91.68	91.68	91.68	90.00	90.00	90.00
		80.74	83.08	108.6	90.32	90.67	88.48
I centred cubic	32.71%	129.72	129.72	129.72	90.00	90.00	90.00
		117.38	134.57	136.36	130.10	115.65	65.53
F centred cubic	22.05%	158.1	158.1	158.1	90.00	90.00	90.00
		158.95	157.99	157.35	92.93	116.62	118.74
Primitive rhombohedral	28.58%	91.68	91.68	91.68	89.25	89.25	89.25
		80.74	108.60	83.08	89.68	88.48	89.33
Primitive hexagonal	12.70%	81.92	81.92	108.60	90.00	90.00	120.00
		80.74	83.08	108.60	90.32	89.33	91.52
Primitive tetragonal	1.74%	81.92	81.92	108.60	90.00	90.00	90.00
		80.74	83.08	108.60	90.32	90.67	88.48
I centred tetragonal	12.51%	81.92	81.92	245.03	90.00	90.00	90.00
		80.74	83.08	245.03	109.59	71.93	88.48
Primitive orthorhombic	0.84%	80.74	83.08	108.60	90.00	90.00	90.00
		80.74	83.08	108.60	90.32	90.67	88.48
C centred orthorhombic	1.08%	114.31	117.38	108.60	90.00	90.00	90.00
		114.31	117.38	108.60	90.69	90.23	91.64
I centred orthorhombic	15.99%	80.74	136.36	157.35	90.00	90.00	90.00
		80.74	136.36	157.35	76.61	60.59	89.60
F centred orthorhombic	11.55%	80.74	182.81	230.84	90.00	90.00	90.00
		80.74	182.81	230.84	98.40	109.81	65.32
Primitive monoclinic	0.45%	80.74	108.60	83.08	90.00	91.52	90.00
		80.74	108.60	83.08	89.68	91.52	90.67
C centred monoclinic	1.00%	117.38	114.31	108.60	90.00	90.69	90.00
		117.38	114.31	108.60	89.77	90.69	88.36
Primitive triclinic	0.00%	80.74	83.08	108.60	89.68	89.33	88.48
Autoindex unit cell		80.71	83.05	108.59	90.00	90.00	90.00
crystal rotx, roty, rotz		-56.059	12.484	-3.994			
Volume of the primitive cell		728215					
Autoindex Xbeam, Ybeam		97.44	100.52				

ducing droplets of solutions of known density covering the range expected for the protein crystal. A crystal is then introduced into the gradient, and its density measured relative to calibration droplets.

References

1. Buerger, M. J. (1964) *The Precession Method.* John Wiley, New York.
2. Matthews, B. (1968) Solvent content of protein crystals. *J. Mol. Biol.* **33,** 491–497.
3. Hope, H., Frolow, F., von Bohlen, K., Makowski, I, Kratky, C., Halfon, Y., Danz, H., Webster, P., Bartels, K. S., Wittmann, H. G., and Yonath, A. (1989) Cryocrystallography of ribosomal particles. *Acta Cryst.* **B45,** 190–199.
4. Hope, H. (1988) Cryocrystallography of biological macromolecules: A generally applicable method. *Acta Cryst.* **B44,** 22–26.
5. Teng, T.-Y. (1990) Mounting of crystals for macromolecular crystallography in a free standing thin film. *J. Appl. Cryst.* **23,** 387–391.
6. Parak, F., Moessbauer, R. L., Hoppe, H., Thomanek, U. F., and Bade, D. (1976) Phase-determination of a diffraction peak of a myoglobin single crystal by nuclear λ resonance scattering. *J. Phys. Colloq. (Paris)* **C6 37,** 703–706.
7. Jeruzalmi, D. and Steitz, T. A., manuscript in preparation.
8. Gablin, S. J. and Rogers, D. W. (1993) Data collection and processing. *Proceedings of the CCP4 Study Weekend* (Sawyer, L., Issacs, N., and Bailey, S., eds.), Daresbury Laboratories, Warrington, UK, pp. 28–32.
9. Cascio, D., Molecular Structure Corporation Users Meeting, Houston, 1994.
10. Haas, D. J. and Rossmann, M. G. (1970) Crystallographic studies on lactate dehydrogenase at –75° C. *Acta Cryst.* **B26,** 998–1004.
11. Petsko, G. A. (1975) Protein crystallography at sub-zero temperatures using cryo-protective moter liquors for protein crystals. *J. Mol. Biol.* **96,** 381–392.
12. Kabsch, W. (1988) Automatic indexing of rotation diffraction patterns. *J. Appl. Cryst.* **21,** 67–71.
13. Kabsch, W. (1988) Evaluation of single-crystal X-ray diffraction data from a position-sensitive detector. *J. Appl. Cryst.* **21,** 916–924.
14. Kabsch, W. (1993) Data collection and processing. *Proceedings of the CCP4 Study Weekend* (Sawyer, L., Issacs, N., and Bailey, S., eds.), Daresbury Laboratories, Warrington, UK, pp. 63–70.
15. Otwinowski, Z. (1993) Data collection and processing. *Proceedings of the CCP4 Study Weekend* (Sawyer, L., Issacs, N., and Bailey, S., eds.), Daresbury Laboratories, Warrington, UK, pp. 56–62.
16. Leslie, A. (1993) Data collection and processing. *Proceedings of the CCP4 Study Weekend* (Sawyer, L., Issacs, N., and Bailey, S., eds.), Daresbury Laboratories, Warrington, UK, pp. 44–51.
17. Pflugrath, J. W. (1993) Data collection and processing. *Proceedings of the CCP4 Study Weekend* (Sawyer, L., Issacs, N., and Bailey, S., eds.), Daresbury Laboratories, Warrington, UK, pp. 52–55.
18. Low, B. W. and Richards, F. M. (1952) The use of the gradient tube for the determination of crystal densities. *J. Am. Chem. Soc.* **74,** 1660.
19. Low, B. W. and Richards, F. M. (1952) Determination of protein crystal densities. *Nature* **170,** 412.
20. Matthews, B. W. (1985) Determination of protein molecular weight, hydration, and packing from crystal density. *Methods Enzymol.* **114,** 176–187.
21. Westbrook, E. M. (1985) Crystal density measurements using aqueous ficoll solutions. *Methods Enzymol.* **114,** 187–196.

CHAPTER 4

Modern Methods for Rapid X-Ray Diffraction Data Collection from Crystals of Macromolecules

Elspeth F. Garman

1. Introduction

During the last 8 years, there has been a revolution in X-ray crystallographic data-collection technology, resulting in an enormous increase in data-acquisition rates and in the range of macromolecules that can be investigated in most laboratories. Provided that the macromolecule forms crystals of reasonable size (minimum of 100 μm in the largest dimension for in-house experiments) and quality, the next step in the determination of its three-dimensional structure is to collect X-ray diffraction data. Some fairly sophisticated equipment is required for this: an X-ray generator, an X-ray detector, and a system of stepping motors ("a goniometer") with translational slides and rotational arcs above them ("goniometer head") on which the crystal in a glass or quartz capillary tube is usually held using plasticene and then aligned in the X-ray beam. This chapter will outline the principles of the most commonly used laboratory equipment, and give the basic steps involved in data collection and processing.

Methods of detecting diffracted X-rays are moving away from diffractometers, and from the photographic film used on oscillation and precession cameras, to techniques that give immediate digital output of a two-dimensional area of the diffraction pattern, and that do not require wet developing and subsequent digital scanning. These so-called area

From: *Methods in Molecular Biology, Vol. 56: Crystallographic Methods and Protocols*
Edited by: C. Jones, B. Mulloy, and M. Sanderson Humana Press Inc., Totowa, NJ

detectors have several advantages, not least of which is that in human terms, the data take much less time to collect and process, so that much more data can be acquired and easily handled. Compared to conventional diffractometers employing scintillation counters, which detect a single reflection at a time, area detectors are much more time-efficient, providing the *x-y* coordinates of the incident X-rays over a large part of the diffraction pattern at once. For example, a data set complete to 2.4 Å (1 Å = 10^{-10}m) on a protein, such as T-state phosphorylase *b* (cell 128 Å × 128 Å × 116 Å, space group $P4_32_12$), can be collected in 20 h with an area detector as opposed to the many weeks it once took with a diffractometer. Since many protein crystals have a limited lifetime in the X-ray beam (*see* Section 4.2.), rapid data collection can be an enormous advantage.

The general requirements for a useful area detector are several: that the sensitivity to X-rays across the face should be uniform, or at least well known; that the spatial distortion should be well defined to ensure reliable calibration; that the spatial resolution, active area, and dynamic range (range of intensities that can be recorded) be adequate; that the response to X-rays is well parameterized; and that a large proportion of incident X-rays are recorded (i.e., a high "detective quantum efficiency" [DQE], defined as the ratio of the square of the output signal-to-noise ratio over the square of the input signal-to-noise ratio). An excellent general review of area detectors, and a summary of recent developments can be found in refs. *1* and *2*.

The area detectors currently commercially available utilize three different physical principles to detect the X-ray photons. In the first, the X-ray ionizes a heavy gas (xenon), and the ion pairs so produced are accelerated toward a plane of closely spaced thin wires held at high voltage, ionizing more xenon on the way (*see* Fig. 1). This results in multiplication (by a factor of about 10^5) of the original signal, so that it is then large enough to be detected electronically. Signals taken from two wire planes in the chamber are processed and the *x-y* coordinate of the incident X-ray derived. The San Diego *(3–6)* and Siemens (originally called Xentronics, then Nicolet, and now Siemens) *(7,8)* area detectors both use this principle of operation, and are generally called multiwire proportional counters (MWPCs). The devices are single photon counters, with a global count rate maximum of between 20 and 35 kHz (note that the improved Siemens detector available in late 1993, the Hi-Star, should operate at up to 100 kHz). Above this count rate, the dead time (propor-

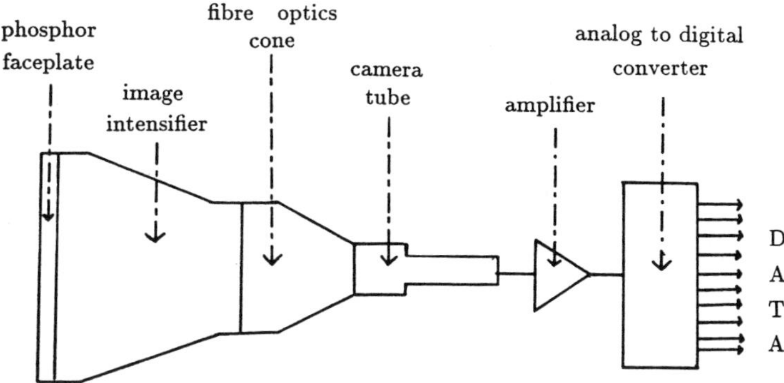

Fig. 1. The principle of operation of the multiwire area detector. Xenon gas is ionized by the incident 8-keV X-ray, and the resultant ions move under the influence of the applied electric field, ionizing more gas on the way.

tion of time for which the system cannot detect new X-rays arriving because it is occupied encoding previous X-rays) becomes unacceptably large.

In the second detection method, the X-rays hit a phosphor-covered fiberoptics screen in which they excite visible wavelength fluorescence. These light photons then enter an image intensifier. Within the image intensifier, the light signal is amplified by being converted into electrons, and the image is read out by a commercially available television scanning system. The x-y coordinates and intensities are accumulated in an image store, interfaced to a computer. These detectors are integrating devices like film, rather than single photon counters, and can thus operate at much higher count rates. The “FAST” *(9,10)* is such a device, and a schematic of it is shown in Fig. 2.

In the final method, employed in the “image plate” detector, a barium halide phosphor of BaFBr doped with europium (Eu^{2+}) is excited into metastable electron states (Eu^{3+}) by incident X-rays. These states, which have half-lives of about 10 h, can be stimulated into emitting violet luminescence ($\lambda \approx 390$ nm) by a red He-Ne laser ($\lambda \approx 633$ nm). Since the intensity of the emitted violet light is proportional to the number of absorbed X-rays, the image can be read out by scanning the plate with a fine laser beam and detecting the emitted light with a photomultiplier system. The residual image can then be erased from the plate by exposure to bright visible (optimally yellow) light and the same plate exposed again. Figure 3 illustrates the basic principle of the plate. It is again an integrating device

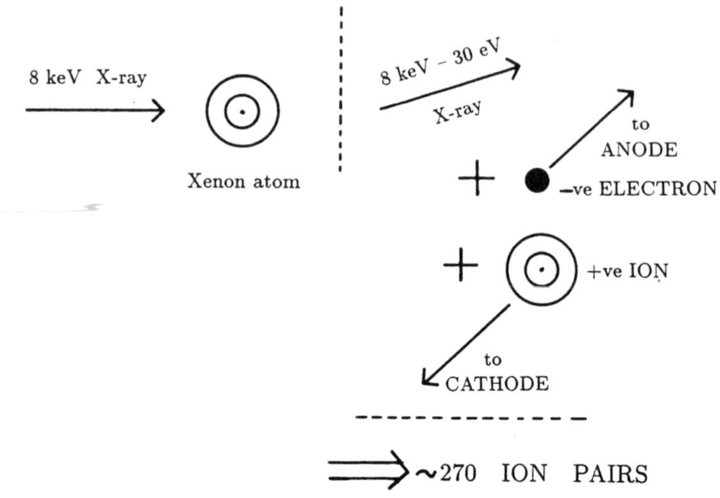

Fig. 2. Schematic of the FAST television area detector.

and is used similarly to conventional photographic film, although it has several important advantages: the plates can be much larger, there is no intrinsic chemical "fog" on the image plate (on which the noise is <3 X-ray photons/100 μm^2 *(2)*, compared with 1000/100 μm^2 for film), and the dynamic range is in principle 10^5 *(11)* whereas for film it is theoretically 10^5, but in practice only about 200, since the bottom two orders of magnitude are lost in the fog. Additionally, the DQE of image plates is >70% over the whole dynamic range and for a large range of incident wavelength, whereas for film, the DQE for 8 keV X-rays ranges from 5% at 10^2 photons/100 μm^2 to over 60% at 10^4 photons/100 μm^2, again because of the fog, which contributes dominantly to the output noise component of the DQE ratio at low signal intensity, making weak reflections harder to measure accurately on film. However, as far as the experimenter on the ground is concerned, the single most important feature of image plates is that, unlike film, the data collection and digitization can be automated *(12)*. The spatial resolution of film is excellent (20 μm; a limit imposed by the available densitometers: intrinsically 1 μm), and this is now its only advantage over image plates (100–150 μm). In most cases, the image plate resolution is quite adequate, and the crystal-to-plate distances can be reasonably short, enabling a large amount of data to be collected at once. At larger distances, data from crystals with very

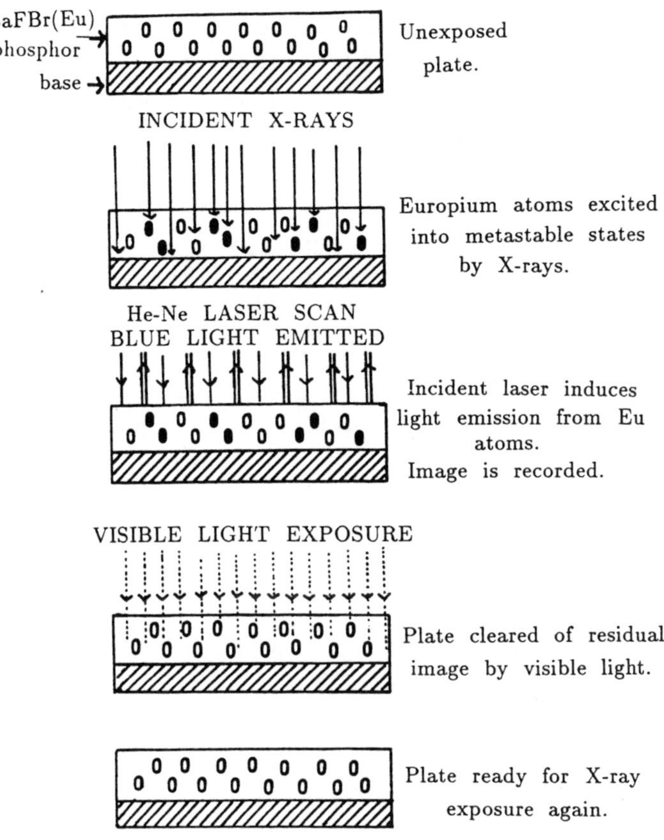

Fig. 3. The principle of operation of image plate phosphor.

big cells can also be obtained. Commercially available image plate systems include the Mar Research (Norderstedt, Germany) *(13)* and the Rigaku (Tokyo, Japan) R-axis II *(14)* scanners.

Within the next few years it is likely that a charged coupled device (CCD) detector will become available for protein crystallography, but these are still at the development stage *(15)*.

Modern X-ray generators are both more powerful in terms of the number of X-ray photons/unit area/second ("brilliance") and much more reliable than their predecessors.

Most protein crystallographic laboratories now have at least one rotating anode generator (common types are the Elliot [now Enraf, Delft, The Netherlands] GX18, 20, 21, and FR571, the Rigaku RU200 and RU300, and the Mac Science [Tokyo, Japan] MX-18). The anode, usually a drum of copper, rotates at high speed (e.g., 6000 rpm for the RU200) while being bombarded by a beam of focused electrons coming from a heated tungsten filament. The whole assembly is enclosed in a high vacuum (10^{-5} Pa). A beam of copper X-rays is produced, and one wavelength, λ, can then be selected (e.g., for $Cu_{K\alpha}$, $\lambda = 1.542$ Å) using a graphite crystal ("monochromator"), or the beam can be focused with mirrors. The rotation of the anode makes available a much larger surface area of copper, so that heat induced by power dissipation is less concentrated and the copper does not melt. Thus, more power can be applied to the filament and a significantly higher (greater than a factor of 5 *[16]*) X-ray flux obtained than from a conventional sealed-tube generator. Recently, a rotating anode generator with a direct drive motor has become available (Mac Science SRA M18XHF), as opposed to the pulley on the anode shaft and an external motor, which are usually used.

With all the detectors mentioned above, data are usually collected using the oscillation method *(17)*. The crystal is oscillated over a small angle (0.1–1.5°) about an axis perpendicular to the X-ray beam, for anything between 1 and 40 min, this "frame" is recorded, and the contiguous same small angle then collected as the next frame, until the desired range of angle has been covered. The necessary range for a complete data set is determined by the crystal symmetry, or space group, and the geometry of the experiment g (i.e., detector swing angle 2θ, and crystal orientation relative to the beam and detector).

Data-collection strategies with area detectors differ in three substantial ways from those used with diffractometers. First, the crystal orientation does not have to be known before data collection commences, saving time and optimizing the use of crystal lifetime. Second, the speed of analysis of each frame is extremely rapid, and can proceed as fast as the data are collected. Third, the high degree of automation now possible with area detectors allows the oscillation range ("frame width") to be much smaller (0.1–0.3°) than that practicable with film (typically between 0.3 and 4.0°), where the penalty of small frame widths is an enormous number of films to develop and digitize. Depending on the mosaic spread of the crystal and the properties of the incident X-ray beam, each diffraction spot usually spreads over only between 0.1 and 0.5° of the rotation range. For a 0.2° frame from an area detector, the collection can thus be from a spot for most of the angular range covered. However, for a 1° oscillation, collection will be from a spot over only about 0.2° of the angular range and from background alone for the remaining 0.8°. Thus, much better signal-to-noise ratios are obtained with small oscillation width frames, so that good-quality weak data can be collected. This advantage, coupled with the improvement in X-ray sources and with advances in area detector technology, enables weakly diffracting crystals and crystals with larger unit cells to be examined in the laboratory rather than at a synchrotron.

It should be noted that the previous discussion does not really apply to data collection with an image plate, where owing to the rather long readout time per image (between 2 and 8 min), larger oscillation (0.75–1.5°) frames are collected to minimize the total time taken for a data set. However, owing to the inherent advantages over film already outlined and other considerations, such as reducing the number of partial reflections measured per image, minimizing the background by using smaller oscillation widths becomes a secondary factor, although data quality could be improved by doing so. Thus, compared to the other detection methods described here, the main disadvantage of image plates is the long readout times for each image, since this is the primary reason for choosing larger oscillation widths.

Although data-collection methods using a synchrotron will not be covered here, it should be stressed that the use of synchrotron radiation is an increasingly important and vital tool in macromolecular crystallography. This is clear from ref. *18*, which is a review of worldwide protein crystallographic synchrotron beam lines and their characteristics. The Laue technique of using intense white synchrotron radiation (i.e., multiwavelength: $\Delta\lambda/\lambda \approx 10$ as compared to ≈ 0.002 for monochromatic) to obtain many orders of diffraction simultaneously, enables an entire data set to be collected on a small number of images. Owing to the greater incident X-ray flux (>1000×), it is an extremely rapid method *(19,20)* of data acquisition; for example, 40,000 reflections can be collected on one 800-ms exposure. However, due to the paucity of low-resolution data that can be extracted from Laue images, data analysis can be rather problematic.

2. Data Collection Equipment

This section briefly describes the mode of operation and use of five currently commercially available area detectors. The main characteristics of the different detectors are summarized in Table 1. A protocol for setting up a data collection on each one is presented. The basic steps are the same,

Table 1
Characteristics of Commercially Available Area Detectors in Common Use

Detector	Size	Pixel size	Number of pixels	DQE at λ of $Cu_{K\alpha}$	Max. global count rate	Dynamic range
Siemens	11.5-cm diameter	200 × 200 μm^2	512 × 512	78%	35 kHz	N/A
San Diego	30 × 30 cm^2	2 × 1 mm^2	144 × 256	50%	20 kHz	N/A
FAST	48 × 64 mm^2	90 × 130 μm^2	512 × 512	60%	1000 kHz	N/A
Big-Mar IP	30-cm diameter	150 × 150 μm^2	2000 × 2000	>70%	N/A	~10^4–10^5
Small-Mar IP	18-cm diameter	150 × 150 μm^2	1186 × 1186	>70%	N/A	~10^4–10^5
R-axis II IP	20 × 20 cm^2	102 × 105 μm^2 204 × 210 μm^2	1900 × 1900 950 × 950	>70%	N/A	~10^4–10^5

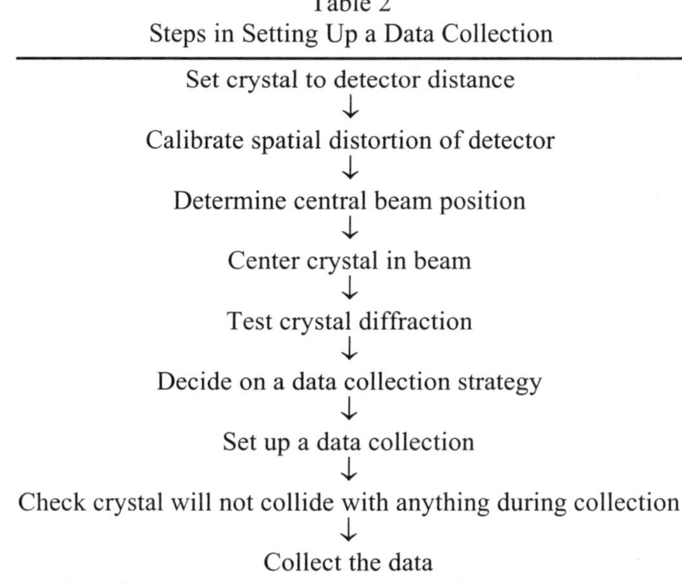

Table 2
Steps in Setting Up a Data Collection

Set crystal to detector distance
↓
Calibrate spatial distortion of detector
↓
Determine central beam position
↓
Center crystal in beam
↓
Test crystal diffraction
↓
Decide on a data collection strategy
↓
Set up a data collection
↓
Check crystal will not collide with anything during collection
↓
Collect the data

with variation of detail for different detectors. These broad steps are summarized in Table 2. However, it is attention to detail in data collection that optimizes the quality of the resulting data, and can thus make the difference between success and failure at a later stage in the structure determination.

There are other "one-off" detectors that have been developed in various laboratories *(21–24)*, but these will not be discussed here. The protein crystallographer usually does not have a choice of instrument, and must use the one available locally, or collaborate with another laboratory. It must be emphasized that these detectors should not be used without adequate training from the person in charge of the equipment and that inexperienced experimenters should adhere to protocols developed locally from experience. All the detectors described below have the capability of producing consistently good-quality data, but only if used correctly and carefully. Some ways of improving general data quality utilized in the author's laboratory are outlined at the end of this section.

2.1. Multiwire Area Detectors

There are two commercially available multiwire area detectors commonly used in the laboratory that are photon counters utilizing the first of the detection principles (*see* Fig. 1) outlined in Section 1: the Siemens

(Madison, WI) area detector and the San Diego Multiwire Diffractometer (SDMW). They are both basically MWPCs, and had their origins in detectors invented for particle detection in high-energy nuclear physics experiments. Their response has been tuned to detect $Cu_{K\alpha}$ X-rays, and neither is sensitive for use in experiments with shorter wavelength radiation.

2.1.1. Siemens Area Detector

This detector was developed by Burns in collaboration with a group at Harvard University *(7,8)*. It consists of a sealed chamber filled with a xenon gas mixture at 4 bar pressure, in which there are three wire planes comprising front and back cathode grids (wires perpendicular to each other) at ground potential with a wire every 0.3 mm, and a central anode grid with a 1.0-mm wire spacing and held at a potential of about 5 kV. The detector uses a system of capacitative readout to derive positional information for the incident X-rays. The active area is 11.5 cm in diameter, defined by a 1-mm thick concave (r = 24 cm) beryllium entrance window (N.B. **poisonous and fragile**), and the positional resolution is 200 μm in both the x and y directions. A 512 × 512 array of counting bins ("pixels") is used, each pixel representing approx 200 × 200 μm^2. The output of the detector is the x-y coordinate of the incident X-ray, and the appropriate pixel is then incremented by one count in the computer memory. Seventy-eight percent of 8-keV X-rays from $Cu_{K\alpha}$ penetrate the beryllium window, and then 100% are absorbed and detected, giving an overall maximum DQE of 78%. The uniformity of response over the entire detector face should be ±1% (manufacturer's specification), so no global correction should be necessary for nonuniform response. There is no electronic noise or background giving false counts, so this detector is particularly suitable for collection of weak data.

The X-100 (1985–1989) and X-1000 (1989–1993) models have the same detector chamber, but different control computers. The Hi-Star (1993–present) is also of the same design, but improved electronics allow it to be used in a 1024 × 1024 pixel (100 × 100 μm^2) mode, and it has an advertised count rate capability of up to 100 kHz without significant dead time losses. It is also said to have improved long-term stability, which should obviate the need for recalibration unless the crystal-to-detector distance is changed.

2.1.1.1. Protocol for Setting Up a Data Collection

1. Set crystal-to-detector distance. As a rule of thumb, which ensures that the diffraction spots will be resolved, use a_{max}/Q cm (assuming $Cu_{K\alpha}$ radia-

tion: Q is in Å cm^{-1}), where a_{max} is the longest unit cell dimension of the crystal (in Å), and Q is determined by the X-ray beam divergence and by the software package to be used for data reduction (e.g., $Q = 8$ for the XENGEN package with a monochromated beam and wide collimation [0.7 mm]). There are four commonly available software packages *(25–29)* for processing Siemens detector data (*see* Section 3.), and they have different capabilities for resolving close spots. The diffracted spot size depends critically on the properties of the beam (*see* Section 2.4., step 4) and if there is a low beam divergence, it may be possible to place the detector as close as $a_{max}/12$. Local practice should be followed. Clearly, the nearer the detector to the crystal, the higher the resolution of the data that can be collected with one setting of the swing angle. (*See* Section 4.1. for how to calculate the maximum resolution obtainable.) If there is no previous determination of the unit cell, the detector distance should be set according to the size of the protein: the larger the protein, the longer the distance required for finding the cell dimensions. Set the swing angle (angle between main beam and detector center), 2θ, to zero.

2. A radioactive source of ^{55}Fe X-rays (λ = 2.05 Å, E=5.9 keV) is provided for calibration purposes. Set the detector bias knob to the locally determined optimum (usually around 9.00 on the 10 turn potentiometer) for these X-rays.
3. Perform the “flood field” correction by placing the ^{55}Fe source at the crystal position and counting until about 10^7 X-ray photons have been detected. The necessary time will obviously depend on the crystal-to-detector distance and the strength of the source, e.g., 15 min at 10 cm for a 100 μCi source. The flood field corrects for any spatial irregularities in sensitivity owing to imperfections in the wire spacing of the grids. It provides an on-line photon-by-photon local rebinning of counts, which is invisible to the user after the correction has been loaded into the computer memory. Although it is not usually necessary, the uniformity of response over the detector face can now be checked by collecting an image with the source, and irregularities (owing, for instance, to broken wires or blown preamplifiers) will be evident and should be reported to the person in charge of the detector.
4. Perform the “brass plate” correction by screwing a plate, with an accurately drilled array of holes in it, to the detector face, and counting for double the time used for the flood field (criterion for adequate statistical accuracy is that there should be at least 50 counts in the most populated pixel). This provides a pixels-to-centimeters (and vice versa) calibration, and also a long-range correction for the geometric distortion of the detector, including the parallax introduced by projecting a spherical diffraction

pattern onto a flat grid of wires. The image should be processed straight away, and the errors between the observed and fitted pixel positions checked. If they are large (>0.5 pixels for the root mean square [r.m.s.] error in *x* and *y* [XENGEN *{26}*], or >0.18 overall r.m.s. error [XDS *{27,28}*]), the detector bias may need adjustment. The optimum bias increases over time as the detector gas gradually leaks out, and should be checked every 8 wk or so by experienced personnel. Spots at the edge of the diffraction pattern will also appear to be smeared radially outward if the bias is not optimum.

This step and step 3 should be followed every time the detector is moved to a new distance. If the distance is kept constant, the above calibrations should be repeated at least every 2 wk to maintain the quality of the high spatial resolution, and to avoid introducing unnecessary uncertainties and errors into the data. It is not advisable to use the brass plate image from a previous run, since the detector may not be exactly in the same place this time!

5. Change the detector bias to the locally determined value for $Cu_{K\alpha}$ X-rays (E = 8 keV), as a rule 0.14 lower than for ^{55}Fe X-rays. Note that if this is not done, a large (up to 65%) sensitivity variation over the detector face will be introduced for the $Cu_{K\alpha}$ radiation that will distort the values of the diffracted intensities and result in unusable data.
6. Place a Perspex (Plexiglas) or aluminum absorber in front of the detector to attenuate the beam of X-rays from the generator, and measure the main beam position at low generator power (e.g., 30 kV, 30 mA for a 3.5-cm perspex absorber). One software package (*25*; BUDDHA) requires an image containing the main beam, which should be taken and stored at this stage, at the swing angle (2θ) at which data are to be collected. Alternatively, with the backstop in place, a piece of amorphous polyethylene in the beam path will give diffraction rings, and the center of these rings can be determined with a suitable computer program.

   Note that in the newer Siemens PC-controlled systems (X-1000-type detector system), an oriented test crystal of ammonium tartrate is provided and can be used to determine the main beam position, obviating the need to remove the backstop at all. The flood field and brass plate calibrations are then performed at a detector swing angle of around 40° to avoid casting a backstop shadow.
7. Position the backstop, and ensure it intercepts the main beam. The main beam can cause **permanent** damage to the detector by leaving a localized dead area at the main beam position. This area will not detect X-rays again until the detector is dismantled and the grids cleaned. Remove the attenuator and check again.

8. Center the crystal. Most Siemens detectors are mounted on three axis goniometers in which the angle between ω and ϕ, χ, is fixed at 45° (Fig. 4A). A few detectors are mounted on four circle goniometers with variable χ (Fig. 4B). The new Hi-star has a fixed χ of 54.74°.
9. Check the strength of diffraction to enable the necessary exposure time per oscillation range for one frame ($\Delta\omega$) and the generator power to be decided. The counting rate should be below 25 kHz to keep the dead time, and thus the proportion of lost counts, well below 10%. This is especially important for data collection from smaller unit cells (<50Å) (e.g., for crystals of DNA fragments) since the low-resolution reflections are very intense, and the dead time temporarily increases when oscillating over them, as well as giving local saturation and dead time effects. This results in the measured low-resolution intensities (>8Å) being depressed relative to the higher-resolution data, for which the dead time fraction was less. For such a data collection, the power of the generator should be decreased and the exposure time per frame increased. (The data rate can be higher on the newer X-1000 system than on the old X-100 PCS-controlled system owing to decreased event processing time.) Squaring the "lates/totals" ratio gives a rough minimum estimate of the fraction of counts lost because of dead time.
10. Start collecting a data set. Typically, $\Delta\omega$ is in the 0.1–0.3° range. The ω span required for a full data set will depend critically on the space group, swing angle, and crystal orientation. Various interactive computer programs exist (LATTICEPATCH *[30]*, RSPACE *[31]*, and ASTRO *[32]*) to enable optimization of data-collection strategy, but experience with the particular crystal system is often the best guide. It is unnecessary to know the orientation of the crystal or to find it before proceeding to take data, since this will be determined by the software ("autoindexing"). However, the crystal can be aligned if required, using an available option in the data-collecting software. Several useful diagnostic numbers ("overbytes" [X-100 only], total counts, and "lates/totals" [X-100 only]) are output for each frame when viewed on the display, and these can indicate crystal decay and general data quality.
11. Data frames are transferred, usually automatically as collection proceeds, to a different computer for processing. Software exists in most laboratories for viewing frames, and adding them together to simulate film oscillation data collection and inspect for multiple crystal lattices. These should be used during data collection, so that detector time is not wasted on substandard crystals.

2.1.2. SDMD

This detector, developed and marketed by Hamlin in collaboration with Xuong *(3–6)* is also a gas-filled (72% xenon, 28% dry hydrocarbon free air) MWPC, but is held at a pressure of 1 bar. It has a flat active area of 28.8 × 28.8 cm^2, and a positional resolution of 0.5 mm horizontally and 2 mm vertically. Thus, to resolve diffraction patterns, the crystal-to-detector distances, *S*, must be large (e.g., S = 52 cm for 100 Å cell for $Cu_{K\alpha}$ radiation). Also multiple settings of swing angle are required to obtain complete high-resolution data, and commercially, the detectors are supplied in pairs to facilitate this. An advantage of the long crystal-to-detector distances is improved signal-to-noise ratios over closer settings, since the background decreases as a function of $1/S^2$ (owing to decrease in the solid angle subtended), whereas the signal decreases only as $1/D^2$ (where *D* is the distance from the X-ray source to the detector). Since $D > S$, $1/D^2$ changes less rapidly than $1/S^2$, and the signal/noise improves with larger *S*.

The large detector entrance window (30 × 30 cm^2) is constructed from thin (0.75-mm) beryllium (N.B. **fragile and poisonous**) and forms the front cathode. The anode consists of a grid of 144 horizontal wires 2.13 mm apart (12/in.) held at around 2.7 kV. The back cathode grid has 288 vertical wires, at half the spacing of the anode ones. Delayline readout from the anode and back cathode provide the *X* and *Y* coordinates, respectively, where *X* is defined as the downward vertical coordinate, and *Y* the horizontal one, with the origin at the top left-hand corner of the detector as seen by the X-ray beam. The X-ray counts are stored in a 144 × 256 pixel array with 1 pixel representing 2 × 1 mm^2 ($X \times Y$). The detector has an overall maximum DQE of 50%; because of the lower gas pressure and the smaller depth (11 mm between cathodes), this is less than for the Siemens detector. The sensitivity variation over the face should be ±0.5%, so no correction need be made for it. Since a count rate of 20 kHz/detector gives 10% dead time, this is an upper limit for the data collection rate.

Data are commonly collected using a variation of the conventional oscillation technique called the "short-range rotation method" *(33)* in which exposures are taken while the crystal is rotating about the ω axis (χ and ϕ fixed) over between 0.07 and 0.2°, dependent on the mosaic spread of the crystal. The instrument is equipped with a four circle goniometer (*see* Fig. 4B).

Software is provided with the detector to enable precession photography to be mimicked *(34)* by the data collection, for use in determination of unknown space groups.

2.1.2.1. Protocol for Setting Up a Data Collection

1. Decide on the crystal to detector distance: use approx $a_{max}/2$ cm, where a is your maximum cell dimension (Å). Note that if your cell has one long axis and two short ones, it may be possible to have the detector closer by orientating the crystal so the 0.5-mm resolution rather than 2-mm resolution is in the plane with closest spaced spots. Since the crystal-to-detector distances are large, the X-ray path is routinely enclosed in a helium-filled cone to minimize attenuation of the diffracted X-rays, which is 1%/cm in air. Start filling the helium cone for the required distance well **before** data collection to ensure that it is properly flushed of air (e.g., it takes 3–4 h to fill a 50-cm long box, and about 8 h for a 120-cm box.)

   Note that no calibrations or checks using an X-ray source are usually necessary. These are done approximately once every six months by experienced personnel, or after each filament change and subsequent beam alignment on the rotating anode. It is thus unnecessary to remove the backstop (N.B. the main beam **permanently** damages the detector), since the direct beam position is known, and the alignment is adjusted so that it does not change with detector-to-crystal distance.
2. Set the crystal-to-detector distance.
3. Set the swing angle of the two detectors. (*See* Section 4.1. for how to calculate the maximum resolution obtainable.) The recommended mode of operation is to position the detectors symmetrically around $\theta = 0°$, so that they both collect the same resolution data, providing more equivalent reflections for scaling and making a complete data set easier to obtain. Take care not to drive the detectors into each other.
4. Center the crystal in the beam.
5. Check that the diffraction quality is acceptable, and profile a few reflections to ensure that their widths are <10 frames, which is the maximum allowed by the processing software.
6. Collect three sets of 20 frames each, at radically different crystal orientations (e.g., [a] $\omega = 30°$, $\phi = 0°$, and $\chi = 0°$, [b] $\omega = -30°$, $\phi = 0°$, and $\chi = 0°$, [c] $\omega = 0°$, $\phi = 90°$, and $\chi = 0°$), and the software will calculate the orientation of the crystal. The refined parameters are the detector distance, the main beam position, and the cell orientation, but not the detector swing angle or tilt.
7. Check, by taking a few stills at different orientations, that the predicted spot positions coincide with the observed diffraction pattern.
8. Work out the data-collection strategy *(33)*, and start data collection. The crystal is positioned (to within 0.2°) at an orientation chosen by the experimenter to collect as much data as possible. Since each detector is 30 × 30 cm^2, the two detectors side by side make a "superdetector" 60 cm wide and

30 cm high, which gives an angular acceptance that is always more in the horizontal than in the vertical plane. The usual data-collection strategy with a four-circle goniometer is to take a set of frames by sweeping in ω, and then to repeat the sweep with χ adjusted by the vertical angular acceptance of the detector for that crystal-to-detector distance. The sweeps in ω at different χ are repeated until 90° of data in the vertical plane has been collected. Note that the total range of ω for the collection should be below 65° to avoid a crash of the goniometer with the helium cone on one side or with the collimator on the other. There are limit switches to prevent collisions with the collimator, but not with the cone, the nose of which can be guillotined by the goniometer.

9. The data are normally processed (i.e., the spot intensities extracted; *see* Section 3.) on the fly by the data-collection computer, and the frames need not be stored: hence, the requirement first to orient the crystal, and know its space group and cell dimensions. The spot intensities list is then transferred to another computer for further processing. However, the frames can be stored for reprocessing and investigation of the space group if necessary.

2.2. FAST Area Detector

The fast scanning area sensitive TV detector diffractometer (FAST), developed by Arndt and collaborators *(9,10)*, and marketed by Enraf-Nonius, uses the second detection method described in Section 1. (*see* Fig. 2). The active face of the detector consists of a 48 × 64 mm Gd_2O_2S:Tb phosphor screen deposited on a flat fiberoptics disk that is coupled to a 4:1 demagnifying image intensifier. This in turn is connected, via a reducing fiberoptics cone, to a silicon-intensifier-target (SIT) television tube (photocathode at –9 kV). The output is proportional to the incident X-ray flux: the device integrates, like film. The television tube is scanned 25 times/s and the video output digitized into a mass accumulation store with 512 × 512 elements. This store can then be read out in whole or part for processing.

The whole detector assembly is enclosed in a light-tight box with thin **fragile** (**not** to be touched) black paper over the front screen. Radioactive sealed tritium sources (so-called beta-lights) at three corners of the input surface provide light patches so the detector gain can be monitored and controlled. The box interior is kept at a constant temperature (around 9.5°C), regulated by a Peltier cooling device. A stable operating temperature (±0.1°) is **vital** to keep the dark current (signal when no X-rays are incident) constant: the current increases exponentially with absolute temperature. The current is monitored during data collection, since it constitutes a major and unfortunately variable source of background noise. The detector response is also affected by changing magnetic fields.

The "picture element" (pixel) size of the FAST is approx 90 × 130 μm². The positional resolution is defined in terms of the Point Spread Function (PSF) *(9)*, and a diffracted spot will normally cover 3 × 3 pixels, spots being resolved if they are separated by 5 pixels, although it is usual to take data with about 10 pixels in between spots. Recently, a new and improved image intensifier has been developed with an output window that matches the input size of the SIT tube, so that the fiberoptic-reducing cone is no longer necessary *(35)*. This change has eliminated the long tails on the spot profiles, thus reducing the background under neighboring reflections and the overall noise level. This gives improved signal/noise ratios compared to the older (pre-1991) model and enables shorter crystal-to-detector distances to be used. The uniformity of response across the screen should be ±1.5% after the nonuniformity correction mentioned in the next section has been made. The DQE of the FAST is around 60%, and the maximum global count rate can be $>10^6$ Hz, a feature that enables it to be used on synchrotrons *(16)*. The FAST can be swung in 2θ around ±45°, or optionally from –70 to +20°.

2.2.1. Protocol for Setting Up a Data Collection

1. Two predetermined corrections are applied during data collection: for the spatial distortion introduced by the photon detection chain (mainly the SIT tube), and for the pixel-by-pixel nonuniformity of response over the detector face. For these, the detector, with and without a precisely drilled plate screwed on the front, is flooded with generator X-rays. The positions and relative intensities of the three light sources (beta-lights) are also required, and a background image with no X-ray beam has to be collected. The corrections are found to be very stable (given a constant temperature), and are usually performed only about four times a year by an experienced experimenter. Note that ideally, and especially if the instrument does not have antimagnetic rails, each crystal-to-detector distance requires a different spatial distortion curve (derived from the image taken with the drilled plate), so check that there is one for your distance stored on the computer. If not, one must be collected, preferably at the 2θ to be used for data collection, since the magnetic field of the generator varies over the table top, and so the position of the detector on it will affect the calibration. Some FASTs have mu-metal shields to protect them from local magnetic fields.
2. Set the crystal-to-detector distance to $a_{max}/1.0$ mm for $Cu_{K\alpha}$ radiation (for the improved FAST mentioned above, this can be $a_{max}/1.5$ mm), and set

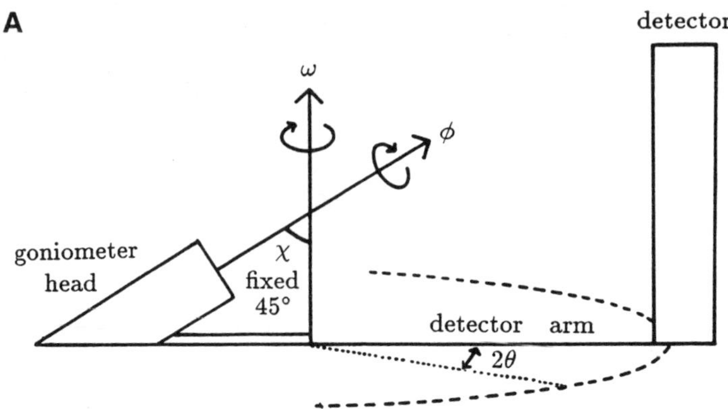

Fig. 4A. Three-axis goniometer.

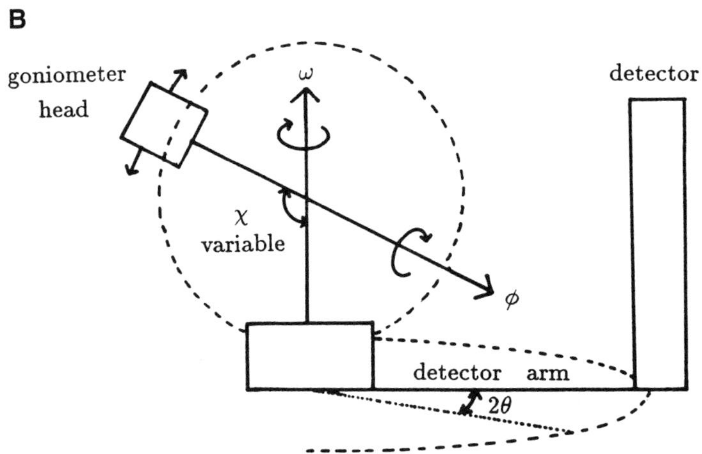

Fig. 4B. Four-axis goniometer.

the swing angle (N.B. maximum range ±45° for detector with 50-cm arm, ±30° for detectors fitted with 100-cm arm) to obtain the required resolution (*see* Section 4.1) using the pocket terminal that controls the hardware. The FAST is supplied with a κ geometry goniostat (*see* Fig. 4C) to allow for maximum flexibility in orienting the crystal.

3. Remove the beam stop and measure the direct beam position at low generator power (e.g., 10 kV, 2 mA) with an attenuator of Perspex (Plexiglas) or brass in front of the detector. Allowing the main beam on the detector is **not** recommended. A semitransparent beamstop can be used, obviating the

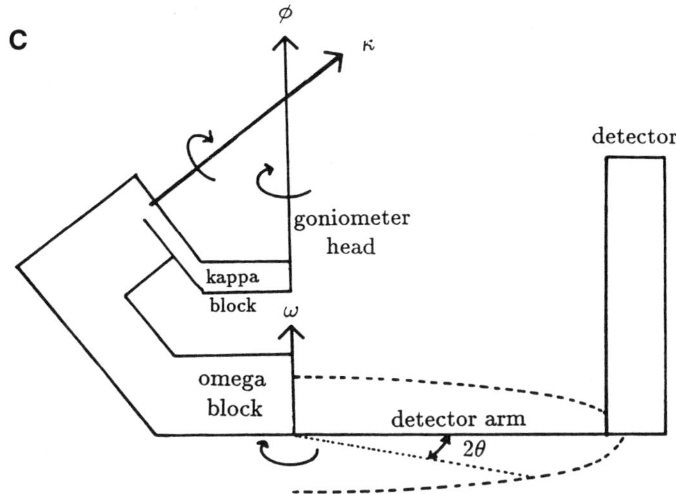

Fig. 4C. κ geometry goniometer.

need for its removal to determine the main beam position. The software package, MADNES *(36)*, provides control of the on line experimental parameters as well as the programs for off-line data analysis. It defines the vertical detector coordinate as Y_{ms}, and the horizontal as Z_{ms}. MADNES allows for local definition of the other data-collection coordinates, so it is advisable to find out what they are.

4. Replace the beamstop, and check it intercepts the beam.
5. Use an optical goniometer to center the crystal prior to mounting it on the goniometer. Drive the detector to 0.0° and to about 120 mm back, to prevent accidents to the black paper front. Slide the telescope forward, and align the crystal. The pocket terminal will register "CRASH" when the telescope is moved, but can be overridden by depressing the red button on the left-hand side of the telescope rail. Slide the telescope back out of the way, and return the detector to the data-taking position.
6. Oscillate the crystal, usually over 0.1–0.2°, and check for diffraction at at least two different orientations. Estimate the dimensions of a box that would contain one reflection, the "shoebox," measured in pixels in *Y* and *Z*, and in "images" (i.e., frames) about the rotation axis. Odd (not even) numbers must be chosen.
7. Set thc datum—the zero position of the goniostat for the data collection—and check that the goniometer will not collide with anything in the planned angular range of the data. Start collecting data without determining the crystal orientation, and save all the images on disk to start processing after

a few hours. Alternatively, if the crystal cell and space group are already known, orient the crystal and process the data on line, so that the complete image does not have to be stored on disk (although it is advisable also to save them for off-line analysis). To orient the crystal, collect two 4° sections of data at starting ϕ values about 90° apart. Around 50 reflections are required for the software to find and refine an orientation matrix. An exposure with no X-ray beam ("a dark image") of the same duration as an image should be measured about twice every 100 images to monitor and update the dark current background.

2.3. Imaging Plates

The use of image plates for macromolecular crystallographic data collection has become well established in the last four years, and many laboratories now have at least one. The physical principle involved (*see* Fig. 3) has been outlined in Section 1., and the technological challenge has been to develop a fast enough scanning system with the required reproducibility in positional accuracy. There are now several plate scanners available commercially, for instance, the Rigaku R-axis II *(14)*, the Mar Research scanner *(13)*, and the Mac Science DIP series, and various of these have been installed at synchrotrons around the world *(18)*. There is also an image plate facility at the Photon Factory in Japan that uses Weissenberg geometry *(37,38)*. For home laboratory data collection, the R-axis II and Mar Research scanners are most widely used. Neither require calibration by the user before data collection or any corrections for variation of sensitivity across the plate (although this is at least 1–2%), so in principle they are straightforward to use.

The image plates are constructed from a 150-μm thick layer of BaFBr:Eu^{2+} mounted on flexible plastic sheet, and are manufactured by the Fuji Photo Co. Ltd. (Tokyo, Japan). They can be reused many times without deterioration. Since the X-ray-induced images are erased by white light, the image plate must be sealed in a light-tight box with a front window that is transparent to X-rays. The scanner is required to deliver an HeNe laser beam to stimulate photoluminescence from the metastable phosphor and to collect, with good positional accuracy, as much of this light as is practicable. The light is detected in a conventional high-quantum efficiency photomultiplier tube, which converts it to electrons and thus to an analog signal that can then be digitized and stored on computer disk.

The design of the scanner optics will not be described in detail here, and is slightly different for the R-axis and Mar scanners. They basically consist of a moving combined reading head containing the fiberoptic light guide, photomultiplier, and the focusing and scanning optics for the laser beam. Thus, a beam from the fixed laser can impinge on the whole area of the plate, and the light guide can collect the photo-stimulated luminescence for input to the photomultiplier.

Data are collected by the oscillation method *(17)*, with ranges of 0.75–2.0°/image, as for film data collection. This is wider than for electronic area detectors because of the plate readout time, which becomes a significant proportion of the collection time if smaller-range, shorter time interval images are collected. Multiple oscillations are performed within one exposure to average out errors induced by beam fluctuations and the natural decay of the image plate luminescent centers (10% in the first few minutes).

Knowledge of the direct beam position to an accuracy of at least half the spot separation is **vital** for indexing the crystal lattice. It can easily be determined by exposing a piece of wax placed at the crystal position, and analyzing the observed rings, or by using a semitransparent beamstop (except on the older Mar scanners, which do not scan right to the center of the image).

2.3.1. Mar Research Scanner

This is an 18-cm (small-Mar) or 30-cm (big-Mar) diameter imaging plate, which employs a spiral readout method, whereby the image plate rotates at an increasing rate under the scanning optics, which themselves move inward along a radius at a constant speed. After readout, the data are transformed into Cartesian coordinates with a pixel size of 150 × 150 μm^2. It has a single plate that is exposed, read, and then erased before collection of the next image is started. The time between exposures is 2 min for the small-Mar and 4 min for the big-Mar. The front window is made of darkened plastic and is fairly robust, but should not be touched. The scanner is supplied with a one-circle goniometer that has a horizontal ϕ axis and no 2θ arm. Since the minimum crystal-to-detector distance is 66 mm (usual maximum is 400 mm), the resolution limit of data that can be collected using $Cu_{K\alpha}$ radiation is 1.7 Å with the small-Mar and 1.4 Å with the big-Mar.

The Mar is equipped with two sets of manually adjustable slits near the crystal position (second set about 6 cm upstream) with an ionization gage downstream of each. Since the slit assembly can be moved by remotely controlled motors, the beam intensity can be conveniently optimized by maximizing the ion gage currents. There is an additional shutter between the slits and the crystal that is computer-controlled and synchronized with the data collection, and is closed during plate readout and erasure: the main generator shutter remains open all the time. The main X-ray beam is stopped on a metal beamstop mounted on a horizontal mylar strip. This is attached to a holder, situated just downstream of the crystal, and the crystal-to-holder distance can be varied for easier access to the crystal position.

The crystal is centered with the aid of a television camera that views the crystal perpendicular to the X-ray beam from below. The data collection is controlled from a menu-driven computer program. The scanner has two erase lamps, and data collection will not stop automatically if one of them blows. In the laboratory, this is a rare occurrence, but at a synchrotron where exposure times are much shorter and the lamps are thus used more often, their lifetime is correspondingly shorter. Thus, care must be taken to monitor the top-level computer-control window, which gives a warning message if a lamp has blown. The newer Mar scanners read out the image all the way to the center of the plate, rather than leaving a hole in the middle of the image. This facility is useful for measuring the attenuated main beam position.

2.3.1.1. Protocol for Setting Up a Data Collection

1. Set the slits (two horizontal and two vertical). You may have to change these when you see your crystal diffraction, but a reasonable start is $0.25 \times 0.25\ mm^2$.
2. Before putting the crystal on the machine:
   a. Move the detector back to at least 200 mm under computer control using "set distance" in menu to give safe access for goniometer head adjustment.
   b. Gently pull the backstop supports two or three clicks away from the crystal position to give better access.
      Note that this step and step a should be carried out **whenever** access to the crystal position is required.
   c. Check that ϕ is at 0° on the goniometer, so that the locating lug for the goniometer head is upward. If not, set it to 0° using the "set phi" menu command.
      Note: The "big click" of the goniometer should be at 0°. If it is not, confusion will result, especially if the crystal + head is removed and then attempts are made to replace it in the same orientation to take more data. There are also "small clicks" at 90, 180, and 270°.

3. Mount the crystal on the machine:
   a. Screw the crystal + goniometer head onto the goniometer, and line the crystal up on television crosswires, being careful that the goniometer head, capillary, or goniometer head arcs do not hit anything when rotated.
      Note: The beam may not be lined up on the crosswires, so check where it is with the last user. Then ensure that the crystal is in this spot at a ϕ of 90° **less** than where you want to start taking data, and 90° **less** than the ϕ where you want to finish data collection; e.g., for a data collection between $\phi = 0$ and 90°, line the crystal up at 270 and 0°. This is because of the position of the camera relative to the crystal.
   b. Gently return the backstop nearer crystal.
   c. Check that the goniometer will not hit anything during your data collection by rotating it by hand carefully over your chosen range, and set ϕ back to 0° by hand.
4. Set the required image plate distance (*see* Section 4.1.) using "set distance" in menu. You may decide to change this when you see your crystal diffraction and spot separation.
5. Optimize the beam flux. This should be done at the distance used for data collection, because moving the detector may distort the frame that holds the slits. Select ion gage 1 ("align" menu command enables selection) with gain 10, and put its motor speed to medium (labeled –). With the X-ray shutter open (but the Mar menu-driven shutter still closed), press the "vertical" and "horizontal" switches on the left to optimize the reading shown on the unit. Now select ionization gage 2 and the use switches on the right to optimize the position of the downstream slits.
6. Check that the crystal diffracts by setting up for either a still and/or an oscillation exposure, specifying the file name, ϕ overlap, increment between successive images, rotation range for the whole data collection, sequence (mixture of stills and oscillations required: the sequence will be repeated until the specified rotation range has been covered, n times, where n = rotn. range/increment/no. oscillation images in one sequence), number of oscillations per image (should be at least 1/5 min of exposure to average out beam fluctuations and the exponential decay of the image plate excitation with time), and the exposure time. If the plate has not been read out for over an hour, erase it before commencing collection to clear it of cosmic radiation and of previously collected diffraction patterns that "seep" to the surface with time.
7. Align the crystal orientation (if desired), and collect data. After taking two stills 90° apart, you may want to move your crystal (*see* 2a and b) on the goniometer head arcs, or change your starting ϕ to optimize the completeness of the data collected. Take care to check that there will be no collisions

during collection. You may also decide to alter the slit settings and/or detector distance at this point. Set up your data collection as above for oscillations.

8. Warnings:
   a. Use "end" to stop a series of exposures after the current one: do not use "stop," since this often causes the collection system to hang, and **never** click on "stop" when taking stills.
   b. The data frames are large, and data collection will stop if the disk becomes full, so check that there is enough room for all the data before leaving the experiment.

2.3.2. R-Axis II Scanner

The Rigaku Automated X-ray Imaging System (R-axis II) is a dual-plate system marketed by the Molecular Structure Corporation (MSC) and made by Rigaku. The back-to-back parallel 20-cm^2 plates are rotated about a vertical axis, one being exposed while the other is read out. The readout is performed by a linear motor-driven optical head, which moves in x (constant fast-speed motor) and y (slower stepper motor) in a plane parallel to the plane of the plate. The plates can be read out in either a coarse (204 × 210 μm^2 pixel size, 950 × 950 pixels, 4-min readout time) or a fine (102 × 105 μm^2 pixel size, 1900 × 1900 pixels, 8-min readout time) scanning mode. The exchange and erasure of the plates takes 45 s, the erasure taking place at 90° to the exposure and readout positions. Thus, if exposures for a fine mode data collection are 8 min or longer, 45 s is the only loss in data-collection time. Lower exposure times (e.g., at a synchrotron) result in much more lost time.

The R-axis is supplied with a one-axis goniometer consisting of a vertical ϕ axis. A telescope at 45° to the collimator enables the crystal to be centered. The front entrance window is made of black paper and is fragile. The plate can be positioned between 55 and 200 mm away from the crystal, giving a maximum resolution of 1.5 Å (with $Cu_{K\alpha}$ radiation) for data collected. However, a manual 2θ stage can be purchased, which extends the crystal to detector range to between 53 and 450 mm, allowing $2\theta = 45°$ from 61 to a 450 mm ($2\theta_{max} = 101°$ at 61 mm, i.e., 1-Å data), and smaller swings between 53 and 61 mm.

Other optional additions include a system of mirror optics that significantly increase the X-ray flux and decrease the beam divergence compared to using a monochromator, an adjustable-length (85–450 mm), helium-filled path to cut down air absorption of the diffracted X-rays, and a 45° bracket for the goniometer head, making a pseudo three-circle goniometer, which enables blind region data to be obtained. This bracket significantly enhances the completeness of the data that can be collected and is highly recommended. The R-axis II is controlled by a menu-driven computer menu.

2.3.2.1. Protocol for Setting Up a Data Collection.

1. Before putting the crystal on the machine:
   a. Move the detector back to at least 200 mm to give safe access for goniometer head adjustment, since the detector face is made of paper and is easily damaged. Use the manual lead screw at the back of the plate box to adjust the distance.
   b. Place a strong lamp or fiberoptic light pointing at the crystal position to make crystal centering less problematic.
2. Screw the mounted crystal + goniometer head onto the goniometer, taking care not to knock the capillary on the beamstop. Center the crystal in the telescope crosshairs, and lock the translational adjustment with the square section key **before** finally centering the crystal, since the action of locking it will cause some sideways translation. Lock the rotation with the same key. Great care must be taken not to knock inadvertently the monochromator angle adjustment screw, which is rather near the left-hand side of the telescope.
3. Set the crystal-to-detector distance to a_{max}/Q, where a is your maximum cell dimension (A°) and where Q is in the range of $12 \leq Q \leq 18$ for monochromator optics, but follow local practice.
4. Measure the main beam position by swinging the backstop and placing an attenuator in front of the detector. There are built-in, menu-controlled attenuators provided. **Replace** the backstop.
5. Set 2θ, if available, to collect data of the required resolution range (*see* Section 4.1.). The air supply to the air pads must be turned on first. After moving to the desired 2θ, the air should be shut off and the 2θ value checked again.
6. Check the crystal diffraction, by taking a set of at least three stills, but preferably four or five, spaced throughout the range of ϕ contemplated for the data collection. In addition, take stills at 0 and 90°. For instance: 0, 30, 45, 60, 90° would be reasonable values for an orthorhombic system, or 0, 10, 20, 30, 90° for a trigonal system.
7. Autoindex these stills by updating the CRYSTAL.DAT file and using the software. Note that the file CRYSTAL.DAT must be updated for your crystal before this is run, and if the mosaicity is not known in advance, a generous value for the mosaicity should be used (0.5° for unfrozen samples, 1.0° for frozen ones). Some information must be input for each field, even if the parameters are just a best guess at this stage.

8. The crystal orientation having been determined, an R-axis data simulation program should now be run to determine the optimal oscillation angle for the distance being used, and, if a 2θ arm is installed, 2θ. Either use the program supplied or an equivalent program (e.g., RSPACE *[31]*) to test various data-collection strategies for redundancy, for completeness, and for efficient use of the overall collection time.
9. A trial oscillation exposure should then be taken to check for diffraction intensity, overlaps, and so forth. If there are overlaps or the spots are not resolved, either the oscillation angle will have to be reduced and/or the detector moved further away from the crystal. If the distance is altered, check the data-collection strategy again.
10. Set up a data collection. Having simulated the collection, the strategy should now be clear. Since the dead time is either 4 or 8 min (depending on pixel size), if possible, use a combination of exposure rate and oscillation angle such that the exposure of each frame will take at least this long. In general, a value of 10 min (for large crystals with small cell dimensions, 0.5-mm collimator) to 30 min of exposure/degree ϕ is reasonable. The final data-collection parameters can be checked using the menu command "Do," and the plate can be erased and ϕ set to 0° at this point. The software will then automatically check that there is enough disk space available for all the images and will not allow the collection to commence until there is.

    Remember to wait at least 10 min at the end of the data collection for the final readout to finish before beginning another experiment on the R-axis.
11. As mentioned above, a 45° mount enables more complete data to be collected without remounting the crystal, since without it, a single rotation in ϕ will leave at a minimum a large polar segment of data missing (the "blind region").

2.4. General Practical Considerations

The quality of data obtained, and therefore their usefulness, can be significantly improved by a number of seemingly small measures taken when planning and setting up the collection. Some factors affecting general data quality are given below: most can be applied in any laboratory. It must be emphasized that the single most important action that the experimenter can take is to monitor the images as they are collected and to try to identify problems as they occur, rather than waiting for the automatic data collection to finish and then discovering during processing that the data are suspect.

1. Minimize the collimator-to-crystal distance and the crystal-to-backstop distance, since this cuts down background air scatter significantly (e.g.,

reduction of 60% when the collimator-to-crystal and the crystal-to-backstop distances were reduced from 11 to 4 mm and 27 to 7 mm, respectively).

2. If using an electronic area detector, optimize the oscillation range by collecting a rocking curve over a reflection before data collection. This is usually determined by the beam geometry, but if the crystals are poor, the mosaic spread may be large. There is no point in taking 0.1° frames for a crystal with mosaic spread of 0.5°. The optimum oscillation width is a third of the full width at half the maximum (FWHM) intensity of the rocking width of the reflections.
3. An advantage of knowing the orientation prior to data collection is that it is easier to ensure collection of a complete data set or to collect reflections that have an anomalous difference on the same image.
4. The beam divergence critically affects the spot size at the detector, which in turn determines the crystal-to-detector distance, since spots must be resolved. The divergence is governed by the focus used in the generator cup, whether mirrors or monochromators are employed, and the slit or collimator settings. Mirrors in general give more parallel beams than a monochromator. The normal compressed graphite monochromators have a large mosaic spread, and a good area on it has to be found, whereas the single crystal monochromators are much better.
5. The crystal should be centered on the telescope crosswires, which should point at the intersection of all the axes of the goniometer. The beam should already be aligned to hit this point. If there is any suspicion that this is not so, move the crystal around to optimize the diffraction. In many systems, the telescope can be unwittingly knocked out of alignment. Internal misalignment between the axes is rare, but the beam alignment can drift off its optimum settings.
6. It is very useful to establish a standard for the X-ray flux at several collimation (if changed) and power settings for your local data-collection equipment. When a count rate monitor is available (e.g., Siemens detector and SDMD), an amorphous scatterer (such as four thicknesses of magnetic tape mounted at the crystal position) can be used to check the beam flux, and a table compiled against generator power and crystal-to-detector distance. (For the Mar scanner, beam optimization is straightforward: *see* Section 2.3.1.1., step 5.) Any doubts over the beam intensity can then easily be checked by remounting the scatterer. This check is always worthwhile, since it ensures that the data are as strong as possible.
7. Check the collimator and backstop for plasticene or other foreign bodies. They can seriously affect the data quality.
8. An absorption curve can be collected by measuring the attenuation of the direct beam, using a crystal that is larger than the cross-section of the beam

for the whole oscillation range. However, modern local scaling programs compensate well for the variation in absorption, so the correction applied using the correction curve usually makes little difference to the final data (around 0.2% on R_{merge} *(1)* [*see* Section 3.1., step 10]), although obviously it makes it more accurate. If a transparent beamstop is used, the data can be scaled to the beam.

9. For crystal-to-detector distances greater than about 18 cm, it is advantageous to insert a helium-filled tube or cone with thin front and back mylar windows to enclose the diffracted X-ray path. This minimizes absorption of the diffracted X-rays, since 10 cm of air absorb 11% of 8-keV X-rays, whereas there is negligible absorption in helium. The exact distance at which a helium cone becomes advantageous depends on the thickness of mylar used for the entrance and exit windows, in which the X-rays scatter.
10. Raise and lower the X-ray generator power slowly (if there is already some power on, not faster than 10 kV, 10 mA/5 min, but if there is no power on yet, use half this speed), thus minimizing excursions in the vacuum from outgassing, and thermal stress on the filament owing to sudden temperature changes. Careful powering up and down will significantly prolong filament lifetime (average 2500 h in the author's laboratory) and thus save time lost to filament changes. Note that a higher flux is obtained for the same power by using as high a voltage as possible. For instance, 60 kV, 50 mA give approx 25% more 8 keV photons than 40 kV, 75 mA.

3. Data Processing

For each area detector described above, one or more software packages have been developed. These reduce the raw data to a list of reflection indices and corresponding measured amplitudes or intensities. The commonly available program suites are:

1. Siemens area detector:
   a. BUDDHA *(25)*.
   b. XENGEN *(7,26)*.
   c. XDS *(27,28)*.
   d. SAINT *(29)*.
2. San Diego area detector:
   a. "UCSD software" *(39)*.
   b. MADNES *(36,40)*.
3. FAST: MADNES.
4. Image plates:
   a. CCP4 software package (MOSFLM program) modified for use with the Mar and R-axis II scanners *(41)*.
   b. WEIS software for the Photon Factory Weissenberg geometry scanner *(42)*.

Table 3
Steps in Data Processing

Corrections for nonuniformity of detector
↓
Calibrate spatial distortion: pixels to cm
↓
Determine active and inactive pixels
↓
Identify intense well-defined spots over a limited rotation range
↓
Autoindexing of spots to determine orientation of cell
↓
Prediction of reflection positions
↓
Integration of the reflection intensities
↓
Reduction of reflections to unique list
↓
Postprocessing: scaling
↓
Output of statistics
↓
Postrefinement

c. PROCESS (now BLIND), the Rigaku/MSC software for R-axis II scanner *(43)*.
d. PROFILE software for the R-axis II scanner and for film *(44,45)*.
e. Hokkaido University/MAC Science software for the DIP100 scanner *(46)*.
f. DENZO for all scanners (including Weissenberg) and for photographic film *(47)*.
g. XDS adapted for the Mar scanner.

5. Additionally, there is an EEC-funded collaborative effort to produce a general device independent package based on MADNES *(48)*.

The details of each package will not be described here: they utilize different algorithms and varying methods to achieve the same end. The basic steps required for data reduction are outlined below and are summarized in Table 3. However, since not all the packages named above can do all the steps in data reduction, the individual program manuals and protocols should be consulted before using the programs. Depending on the type of detector, some items are performed on-line during data collection (*see* Section 2.).

3.1. Steps in Data Reduction

1. Correction on a pixel-by-pixel basis for any nonuniformity of response over the detector face.
2. Calibration of the detector using a specially collected image: pixels to cm.
3. Determination of the active and inactive pixels (e.g., to identify and flag pixels in the backstop shadow).
4. Identification of bright, well-defined spots over a range of the rotation angle for use in finding the orientation of the crystal.
5. Autoindexing of spots to determine the orientation matrix of the cell. In the method developed by Kabsch *(49)* and independently by Howard *(50)*, a search is performed of difference vectors between reciprocal lattice points, after the vectors have been histogrammed, filtered, and refined. Both the orientation matrix and unknown cell dimensions can be found by this powerful technique. Difficulty in autoindexing is sometimes experienced with centered space groups, but otherwise, it fails only for multiple lattices or split crystals, or for data from very poor quality crystals. As already mentioned, it is **very** important to have a reliable central beam position, since this is vital for the data to be correctly indexed. Once an initial estimate of the orientation has been made, the matrix, the cell parameters and the detector parameters can all be refined by linear and nonlinear minimization methods. These can be re-refined during integration (step 7) to track, for instance, small movements of the crystal owing to either slippage or vibration.

   Note that several software packages (XENGEN, XDS, DENZO) now contain algorithms to autoindex data from crystals where there is **no** prior knowledge of cell and space group. This makes data analysis considerably easier.
6. Prediction of expected reflection positions from refined parameters and space group.
7. Integration of the reflection intensities. There are two methods for this: summation (e.g., BUDDHA, XENGEN, UCSD software, MADNES, DENZO) and profile fitting (e.g., XENGEN, XDS, MADNES, SAINT, MOSFLM, WEIS, DENZO).
   a. Summation: Definition of volume ("shoebox") containing an individual spot. Summation of counts in all the pixels in the region of the spot as defined by the shoebox, and subtraction of the background, determined by examining surrounding pixels from contiguous frames on one or both sides of the rotation axis.
   b. Profile fitting *(27,44,51)* of the reflection in two or three dimensions on a background. An empirically derived model reflection shape is scaled to the data and then integrated. This assumes that the reflection shape is independent of intensity. The observed profiles vary over the detector

face, so several model profiles are usually used, dependent on where on the face the reflection was detected. If large oscillation widths are collected (>0.4°), for instance on an image plate, two dimensional (*x*-*y*) profile fitting is appropriate, whereas for the smaller widths used on electronic area detectors (multiwire detectors and the FAST), the reflection profile is sampled at enough points to fit it in three dimensions.

8. Reduction of the measurements list to a unique and ordered reflection list for the specified space group.
9. "Post processing:" Scaling of the data to take account of the Lorentz factor, air and crystal absorption owing to asymmetry in the crystal shape, and crystal decay during data collection. Reference *52* gives valuable further details on the best strategies for scaling.
10. Output of final data statistics and list of indices with measured amplitudes or intensities, and associated statistical errors. The usual measure of data precision quoted is the merging R value on intensity, $R_{merge}(I)$ = defined as:

$$R_{merge}(I) = \frac{\Sigma_h \Sigma_i \mid (I_h - I_{hi}) \mid}{\Sigma_h \Sigma_i I_{hi}} \times 100 \quad (1)$$

where I_h is the weighted mean measured intensity of the observations I_{hi} in which the intensities of the symmetry-related reflections, which should be the same, are compared. $R_{merge}(I)$ gives an estimate of their disagreement.

11. Postrefinement: The unit cell, detector parameters, and central beam position can be re-refined using all or some of the reflections collected over the whole data set, and more accurate values obtained than in step 5. If necessary, the data can then be reintegrated using these new values (step 6 onward).

No mention has been made in this chapter of the collection and use of anomalous data from area detectors, which is an extensively and successfully employed method of adding to the phase information for solving the structure of a particular macromolecule. Some methods have been developed (e.g., *53*) to minimize systematic errors in measuring small intensity differences (anomalous differences) between symmetry related reflections by ensuring that they are detected on the same part of the detector, but these will not be covered here.

4. Notes

4.1. Calculation of the Maximum Resolution Obtainable

To calculate the maximum resolution, d_{max} (in Å), obtainable at the edge of a detector of dimension L set at distance *S* from the crystal. (*S*

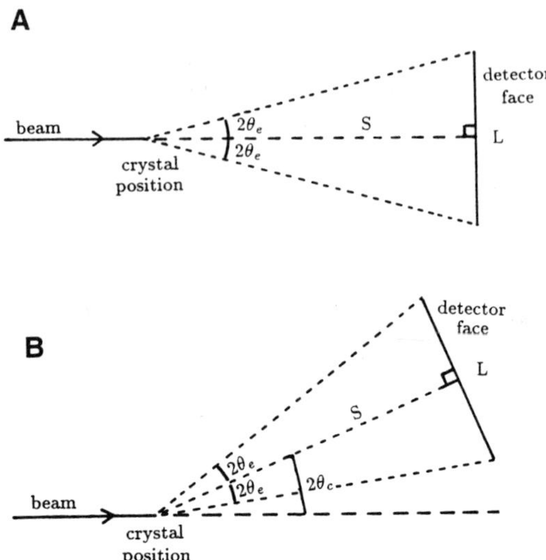

Fig. 5. Geometry for calculating the resolution of data collected.

is determined by the beam divergence and the spatial resolution of the detector.)

Case 1. For swing angle $2\theta = 0.0°$. Refer to Fig. 5A.

a. Calculate 2θ at the detector edge, $2\theta_e$:

$$2\theta_e = \arctan(L/2S) \quad (2)$$

b. Use Bragg's Law:

$$n\lambda = 2d\sin\theta \quad (3)$$

where for $Cu_{K\alpha}$ $\lambda = 1.542$ Å, so that:

$$d_{max} = \lambda/2\sin\theta_{max} \quad (4)$$

where $\theta_{max} = \theta_e$ as determined in a.

Case 2. For swing angle $2\theta \neq 0.0°$. Refer to Fig. 5B.

a. If the center of the detector is set at $2\theta_c$, then $\theta_{max} = \theta_c + \theta_e$ with θ_e calculated using Eq. (2):

$$d_{max} = \lambda/2\sin(\theta_c + \theta_e) \quad (5)$$

b. The resolution limit at the other edge of the detector, d_o, will be:

$$d_o = \lambda/2\sin(\theta_c - \theta_e) \quad (6)$$

If $\theta_c > \theta_e$, data to a higher number of Å than d_o will not be collected.

If $\theta_c < \theta_e$, data to a higher number of Å than d_o (i.e., so-called low-resolution data) will be collected.

4.2. Flash Freezing of Macromolecular Crystals to Prolong Crystal Lifetime

Some protein and virus crystals are extremely susceptible to radiation damage and only diffract for a few minutes in the X-ray beam, so that data collection presents a very difficult problem. A technique for such cases, successfully used for some time by small molecule crystallographers and now being increasingly tried on macromolecular crystals, is that of shock cooling the crystal to a temperature between 80 and 100 K. The crystal is plunged suddenly into a stream of cold gaseous nitrogen (boiling point of liquid nitrogen = 78 K) and kept in this stream during data collection. This cooling has been observed to increase crystal lifetime significantly, in some cases making it effectively infinite. If the cooling is fast enough, the disordered water and solvent in the crystal freeze to an amorphous glass and do not contribute to the diffraction pattern.

Several variations have been developed in the crystal treatment methods prior to cooling and in the mounting arrangements (e.g., *55–58*). The most successful and generally used technique for protein crystals is to soak the crystals in a cryoprotectant *(55)* and then suspend them in a small-diameter loop of thin wire or fiber in a free-standing thin film of liquid *(58)*. The crystal is held in the loop by surface tension and is frozen immediately. The function of the cryoprotectant is to ensure that an amorphous glass is formed on freezing. A procedure for this technique is given briefly below and is illustrated in Fig. 6. Some useful additional practical details can be found in ref. *59*.

The first step is to find a suitable cryoprotectant for the crystal; possibilities include ethylene glycol, glycerol, PEG 400, glucose, and MPD, at concentrations between 15 and 25% made up with your buffer. There are several different ways of introducing the crystal into the cryoprotectant: dialysis, cocrystallization with cryoprotectant in the buffer, sequential transfer into increasing concentrations (e.g., 4 min each in 10, then 20, and finally 30% glycerol), or direct transfer into the final concentration of cryoprotectant for a few min. All these methods may have to be tried for success. It is very important to **minimize** the handling of the crystals, since excess manipulation often causes an increase in mosaic spread.

The loops should be matched to the size of the crystal, since then the crystal is easier to locate when frozen. They can be made of any thin

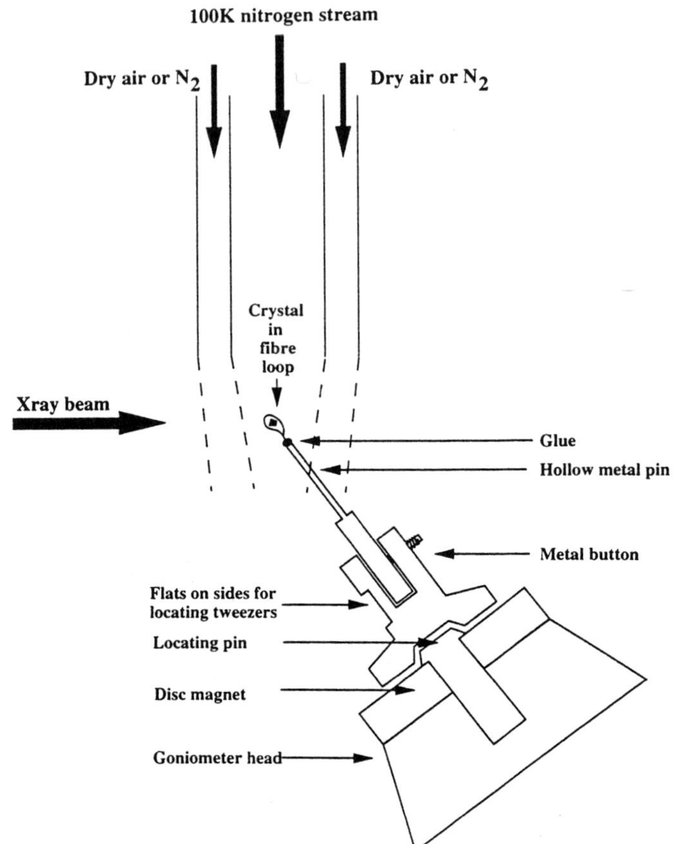

Fig. 6. The arrangement of a crystal on a goniometer head for low-temperature data collection.

(20–50 μm thick) fiber, for instance a strand from multistrand fishing line, angora wool, glass, or rayon. A loop of the required diameter can be made by tying the fiber around the blunt end of a drill bit, and glueing the knot into a hollow metal pin that fits into a metal button. This metal button is then held onto the goniometer head by a small disk magnet which has a beveled locating pin (base diameter about 4 mm) in the center. This arrangement (*see* Fig. 6) greatly assists in the speedy transfer of crystals.

A drop of the chosen cryoprotectant-buffer mixture alone, placed in the loop with a low-volume pipet, should first be flash frozen to check that it forms a transparent glass. A drop, rather than a film, should be used, since this better simulates the presence of a crystal. It is also worth testing the diffraction of a crystal in the cryoprotected buffer at room temperature, since it is important to determine if the addition of cryoprotectant or the flash freezing itself is the cause of unsuccessful freezing.

Note that experience shows that smaller crystals (0.1–0.4 mm in largest dimension) freeze better, since the volume is smaller and so the heat transfer is faster.

4.2.1. Protocol for Flash Freezing a Macromolecular Crystal

1. A stream of gaseous nitrogen at a stable (±2°C) temperature between 80 and 100 K is directed centrally onto the crystal position. Most laboratories now have a method of providing this stream: common commercially available devices are made by Enraf-Nonius (FR 5585H), Siemens (LT2), and Oxford Cryosystems: Cryostream, Oxford, UK. Some training in their operation is required before using them for an experiment. Most devices also provide for a concentric dry nitrogen or dry air stream around the cold nitrogen to isolate the cold gas from the warm humid air of the room. The flowrate of the dry stream is critical in minimizing turbulence at the boundary with the cold nitrogen, and thus preventing ice formation. Icing problems will also be minimized if the whole experiment is enclosed (e.g., in a radiation enclosure with fans turned off, or in a large plastic bag) so that there is little or no air turbulence.
2. A piece of light metal foil (e.g., aluminum cooking foil) should be placed over the goniometer to protect it from the nitrogen stream. The points of contact with the goniometer (and therefore the cold transfer) are minimized by crumpling the foil up first. The foil is light enough to move with the goniometer and not to interfere with it as it rotates.
3. The modified goniometer head is screwed onto the goniometer.
4. The loop of fiber on the pin, already attached to the metal button, is centered in the telescope **before** the crystal is mounted in it.
5. Under the microscope, the crystal is loosened from the sides of the well by disturbing the liquid around it, and the loop immersed so that one edge is beneath the crystal. The loop, with its plane perpendicular to the surface, is then gently pulled vertically upward to minimize the thickness of the film of liquid. The crystal is thus now trapped in the loop. Alternatively, a pipet with a 10-μL tip can be used both for transferring the crystal between cryoprotectant soaks and also for dropping it onto the loop held horizontally. Excess liquid can then be conveniently pipeted from the loop, although this can be difficult if the density of the crystal is very similar to that of the cryoprotected buffer, so that it does not sink to the bottom of the liquid.
6. A piece of cardboard is held over the nitrogen stream (most conveniently by a second person) while the loop and button are transferred swiftly using

tweezers (preferably self-closing ones) onto the goniometer head. The cardboard is then rapidly removed to shock cool the crystal. The crystal should already be approximately centered (step 4), but this is now checked. The faster the steps 5 and 6 can be performed (should be a matter of seconds), the better is the chance that the crystal will survive.

7. Data can now be collected. Great care should be taken to maintain a constant temperature at the crystal (±2.0°C) during the experiment, especially when the nitrogen supply dewar has to be replenished.
8. The crystal should not be removed from the stream until all the required data have been collected, since it may disintegrate when brought back up to room temperature. If more data are required later, or if the frozen crystal is to be transported to a synchrotron, it can be stored in liquid nitrogen for many months.

The flash cooling of protein crystals is, as yet, an empirical experimental technique for which there are no general rules regarding treatment of crystals, the optimum cryoprotectant, controlling the mosaic spread (which often becomes unacceptably large), and ensuring isomorphism with the crystal at room temperature. However, if a drop of the cryoprotected buffer freezes amorphously and the crystal is stable in this buffer, failure of the technique is almost certainly the result of operator error.

Three recent examples of successful cases from the author's experience illustrate the empirical nature of the technique.

1. Cocrystallization. Crystals normally grown by vapor diffusion against 18% PEG 4000. Equilibration was tried against 80% of 18% PEG 4000 and 20% glycerol, and the crystals grew and flash froze well. Prior to this, the experimenter had tried direct transfer of crystals into glucose, sucrose, trihalose, glycerol, MPD, and PEG, all unsuccessfully (concentrations of 10–25% tried).
2. Crystals grown in 2*M* ammonium sulfate. The author tried sequential transfer into 5, 10, 15, and 20% glycerol with no success. Transferring them directly into 20% glycerol for 1–4 min worked well.
3. Crystals grown in BES buffer: Freezing was tried with no cryoprotectant (to see the effects), then with sequential transfer to 10, 20, and 30% glycerol, and then directly into 30% glycerol, where diffraction was seen in addition to a huge increase in mosaic spread. Then crystal was put straight into 50% glycerol for 5 min, with success.

Use of the flash-freezing technique ("cryocrystallography") significantly improves the quality of X-ray data that can be obtained. A complete data set is collected from only one crystal, since on a laboratory X-ray source there is no radiation damage at 100 K. Cryoprotectant conditions can be optimized to minimize the mosaic spread of the crystal and to maximize the resolution of the data *(60)*. Also, as high resolution reflections do not decay with time, they can be measured with better accuracy. The end product is a more detailed macromolecular model, which in turn gives higher quality biological information.

As well as prolonging crystal lifetime in the X-ray beam, cryocrystallography is also of value for investigating reaction rates of macromolecules with various compounds and for substrate binding studies, since the reaction rates usually slow down significantly at low temperature, and data can be collected on crystals and substrates that would turn over too fast at room temperature.

5. Conclusions

Rapid X-ray diffraction data collection from crystals of macromolecules is becoming routine and widespread, as the number of protein crystallography laboratories worldwide expands, and as the range of proteins available in the milligram amounts required for crystallization becomes extended. Even though the data collection is now so automated, obtaining the best possible data from a given crystal is still an experimental challenge. An experimenter requires a basic understanding of the technologies involved to be capable of diagnosing problems *(54)* and optimizing the results.

Acknowledgments

I would like to thank Pat Baker, John Barnett, Peter Collins, Andrew Leslie, Paul McLaughlin, Peter Moody, Zbyszek Otwinowski, David Stuart, John Tate, Alan Wonacott, and especially Peter David for useful discussions during the preparation of this chapter.

References

1. Arndt, U. W. (1986) X-ray Position-sensitive detectors. *J. Appl. Cryst.* **19,** 145–163.
2. Pflugrath, J. W. (1992) Developments in X-ray detectors. *Curr. Opinion Structural Biol.* **2,** 811–815.
3. Cork, C., Hamlin, R., Vernon, W., and Xuong, Ng. H. (1975) A Xenon-filled multiwire area detector for X-ray diffraction. *Acta Cryst.* **A31,** 702,703.
4. Hamlin, R., Cork, C., Howard, A., Nielsen, C., Vernon, W., Matthews, D., and Xuong, Ng. H. (1981) Characteristics of a flat multiwire area detector for protein crystallography. *J. Appl. Cryst.* **14,** 85–93.
5. Hamlin, R. (1985) Multiwire area X-ray diffractometers, in *Methods in Enzymology*, vol. 114 (Wyckoff, H. W., Hirs, C. H. W., and Timasheff, S. N., eds.), Academic, Orlando, FL, pp. 416–451.
6. Xuong, Ng. H., Sullivan, D., Nielsen, C., and Hamlin, R. (1985) Use of the multiwire area detector diffractometer as a national resource for protein crystallography. *Acta Cryst.* **B41,** 267–269.

7. Durbin, R. M., Burns, R., Moulai, J., Metcalf, P., Freymann, D., Blum, M., Anderson, J. E., Harrison, S. C., and Wiley, D. C. (1986) Protein, DNA, and virus crystallography with a focused imaging proportional counter. *Science* **232,** 1127–1132.
8. Howard, A. J., Gilliland, G. L., Finzel, B. C., and Poulos, T. L., Ohlendorf, D. H., and Salemme, F. R. (1987) The use of an imaging proportional counter in macromolecular crystallography. *J. Appl. Cryst.* **20,** 383–387.
9. Arndt, U. W. (1982) X-ray television area detectors. *Nucl. Instrum. Methods* **201,** 13–20.
10. Arndt, U. W., and Thomas, D. J. (1982) High-speed single crystal television X-ray diffractometer (hardware). *Nucl. Instrum. Methods* **201,** 21–25.
11. Miyahara, J., Takahashi, K., Amemiya, Y., Kamiya, N., and Satow, Y. (1986) A new type of X-ray area detector utilising laser stimulated luminescence. *Nucl. Instrum. Methods* **A246,** 572–578.
12. Amemiya, Y. and Miyahara, J. (1988) Imaging plate illuminates many fields. *Nature* **336,** 89,90.
13. Hendrix, J. and Lentfer, A. (1988) An imaging plate scanner. *EMBL Research Reports.* 170,171.
14. Sato, M., Yamamoto, M., Imada, K., Katsube, Y., Tanaka, N., and Higashi, T. (1992) A high-speed data-collection system for large-unit-cell crystals using an imaging plate as a detector. *J. Appl. Cryst.* **25,** 348–357.
15. Strauss, M. G., Westbrook, E. M., Naday, I., Coleman, T. A., Westbrook, M. L., Travis, D. J., Sweet, R. M., Pflugrath, J. W., and Stanton, M. (1991) Large aperture CCD X-ray detector for protein crystallography using a fibreoptic taper, in *Charged Coupled Devices and Solid State Optical Sensors II.* SPIE **1447,** pp. 12–27.
16. Arndt, U. W. and Gilmore, D. J. (1979) X-ray television area detectors for macromolecular structural studies with synchrotron radiation sources. *J. Appl. Cryst.* **12,** 1–9.
17. Wonacott, A. J. (1977) Geometry of the rotation method, in *The Rotation Method in Crystallography* (Arndt, U. W. and Wonacott, A. J., eds.), North-Holland, Amsterdam, pp. 77–103.
18. Ealick, S. E. and Walter, R. L. (1993) Synchrotron beamlines for macromolecular crystallography. *Curr. Opinion in Structural Biol.* **3,** 725–736.
19. Hadju, J. and Johnson, L. N. (1990) Progress with Laue diffraction studies on protein and virus crystals. *Biochemistry* **29,** 1669–1678.
20. Clifton, I. J., Fulop, V., Hadfield, A., Nordlund, P., Andersson, I., and Hadju, J. (1991) Macromolecular structure, function and dynamics by fast crystallography with synchrotron radiation. *Nucl. Instrum. Methods* **A303,** 476–487.
21. Sobottka, S. E., Chandross, R. J., Cornick, G. C., Kretsinger, R. H., and Rains, R. G. (1990) Design and performance of the multiwire area X-ray diffractometer at the University of Virginia. *J. Appl. Cryst.* **23,** 199–208.
22. Baru, S. E., Proviz, G. I., Savinov, G. A., Sidorov, V. A., Khabakhpashev, A. G., Shekhtman, L. I., Shuvalov, B. N., and Yasenev, M. V. (1983) Two-coordinate X-ray detector. *Nucl. Instrum. Methods* **208,** 445–447.
23. Gruner, S. M., Milch, J. R., and Reynolds, G. T. (1982) Slow-scan silicon-intensified target-TV X-ray detector for quantitive recording of weak X-ray images. *Rev. Sci. Instrum.* **53(11),** 1770–1778.

24. Kahn, R., Fourme, R., Bosshard, R., and Saintage, V. (1986) An area-detector diffractometer for the collection of high resolution and multiwavelength anomalous diffraction data in macromolecular crystallography. *Nucl. Instrum. Methods* **A246,** 596–603.
25. Blum, M., Metcalf, P., Harrison, S. C., and Wiley, D. C. (1987) A System for collection and on-line integration of X-ray diffraction data from a multiwire area detector. *J. Appl. Cryst.* **20,** 235–242.
26. Howard, A. (1993) XENGEN Version 2.1 1993. Unpublished.
27. Kabsch, W. (1988) Evaluation of single-crystal X-ray diffraction data from a position-sensitive detector. *J. Appl. Cryst.* **21,** 916–924.
28. Kabsch, W. (1993) Automatic processing of rotation diffraction data from crystals of initially unknown cell constants. *J. Appl. Cryst.* **26,** 795–800.
29. SAINT (1993) Siemens Area Detection Integration Software. Unpublished.
30. Klinger, A. L. and Kretsinger, R. H. (1989) LATTICEPATCH—an interactive graphics program to design data measurement strategies for area detectors. *J. Appl. Cryst.* **22,** 287–293.
31. Harris, M. R., Fitzgibbon, M., and Hage, F. (1989) RSPACE- a reciprocal-space modelling tool. *J. Appl. Cryst.* **22,** 624–627.
32. ASTRO (1992) Area detector strategy organiser. Siemens, unpublished.
33. Xuong, N. H., Nielsen, C., Hamlin, R., and Anderson, D. (1985) Strategy for data collection from protein crystals using a multiwire counter area detector diffractometer. *J. Appl. Cryst.* **18,** 342–350.
34. Edwards, S. L., Nielsen, C., and Xuong, Ng. H. (1988) Screened precession method for area detectors. *Acta Cryst.* **B44,** 183–187.
35. Shierbeek, A. and Parlevliet, D. (1991) New developments of an X-ray television detector. *Nucl. Instrum. Methods* **A310,** 571–575.
36. Messerschmidt, A. and Pflugrath, J. W. (1987) Crystal orientation and X-ray pattern prediction routines for area-detector diffractometer systems in macromolecular crystallography. *J. Appl. Cryst.* **20,** 306–315.
37. Sakabe, N. (1991) X-ray diffraction data collection system for modern protein crystallography with a Weissenberg camera and an image plate using synchrotron radiation. *Nucl. Instrum. Methods Phys. Res.* **(A) 303,** 448–463.
38. Stuart, D. I. and Jones, E. Y. (1993) Weissenberg data collection for macromolecular crystallography. *Curr. Opinion in Structural Biol.* **3,** 737–740.
39. Howard, A. J., Nielsen, C., and Xuong, Ng. H. (1985) Software for a diffractometer with multiwire area detector, in *Methods in Enzymology*, vol. 114 (Wyckoff, H. W., Hirs, C. H. W.,Timasheff, S. N., eds.), Academic, Orlando, FL, pp. 452–471.
40. Thomas, D. J. (1989) Calibrating an area-detector diffractometer: Imaging geometry. *Proc. R. Soc. Lond.* **A425,** 129–167.
41. Leslie, A. G. W. (1992) Recent changes to the MOSFLM package for processing film and image plate data. CCP4 and ESF-EACMB Newsletter on Protein Crystallography, Number 26.
42. Higashi, T. (1989) The processing of diffraction data taken on a screenless weissenberg camera for macromolecular crystallography. *J. Appl. Cryst.* **22,** 9–18.
43. Higashi, T. (1990) Auto-indexing of oscillation images. *J. Appl. Cryst.* **23,** 253–257.

44. Rossmann, M. G. (1979) Processing oscillation diffraction data for very large unit cells with an automatic convolution technique and profile fitting. *J. Appl. Cryst.* **12,** 225–238.
45. Kim, S. (1989) Auto-Indexing oscillation photographs. *J. Appl. Cryst.* **22,** 53–60.
46. Tanaka, I., Yao, M., Suzuki, M., Hikichi, K., Matsumoto, T., Kozasa, M., and Katayama, C. (1990) An automatic diffraction data collection system with an imaging plate. *J. Appl. Cryst.* **23,** 334–339.
47. Otwinowski, Z. (1993) Oscillation data reduction program, in *Data Collection and Processing* (Sawyer, L., Isaacs, N., and Bailey, S., eds.), SERC Daresbury Laboratory, Warrington, UK, DL/SC1/ R34, pp. 56–62.
48. Bricogne, G. (1987) The EEC cooperative programming workshop on position-sensitive detector software, in *Computational Aspects of Protein Crystal Data Analysis* (Helliwell, J. R., Machin, P. A., and Papiz, M. Z., eds.), SERC Daresbury Laboratory, Warrington, UK, DL/SC1/R25, pp. 120–135.
49. Kabsch, W. (1988) Automatic indexing of rotation diffraction patterns. *J. Appl. Cryst.* **21,** 67–71.
50. Howard, A. (1986) Autoindexing, in *Proceedings of the EEC Cooperative Workshop on Position-Sensitive Detector Software (Phases I and II)*, LURE, Paris, May 26–June 7, 1986, pp. 89–94.
51. Diamond, R. (1969) Profile analysis in single crystal diffractometry. *Acta Cryst.* **A25,** 43–55.
52. Evans, P. R. (1993) Data reduction, in *Data Collection and Processing* (Sawyer, L., Isaacs, N., and Bailey, S., eds.), SERC Daresbury Laboratory, Warrington, UK, DL/SC1/R34, pp. 28–32.
53. Derewenda, Z. and Helliwell, J. R. (1989) Calibration tests and use of a Nicolet/ Xentronics imaging proportional chamber mounted on a conventional source for protein crystallography. *J. Appl. Cryst.* **22,** 123–137.
54. Garman, E. F. (1993) Problematic data sets: give up or persist? in *Data Collection and Processing* (Sawyer. L., Isaacs, N., and Bailey, S., eds.), SERC Daresbury Laboratory, Warrington, UK, DL/SC1/ R34, pp. 28–32.
55. Petsko, G. A. (1975) Protein crystallography at sub-zero temperatures: Cryoprotective mother liquors for protein crystals. *J. Mol. Biol.* **96,** 381–392.
56. Dewan, J. C. and Tilton, R. F. (1987) Greatly reduced radiation damage in ribonuclease crystals mounted on glass fibers. *J. Appl. Cryst.* **20,** 130–132.
57. Hope, H. (1988) Cryocrystallography of biological macromolecules: a generally applicable method. *Acta Cryst.* **B44,** 22–26.
58. Teng, T.-Y. (1990) Mounting of crystals for macromolecular crystallography in a free-standing thin film. *J. Appl. Cryst.* **23,** 387–391.
59. Gamblin, S. J. and Rodgers, D. W. (1993) Some practical details of data collection at 100K, in *Data Collection and Processing* (Sawyer, L., Isaacs, N., and Bailey, S., eds.), SERC Daresbury Laboratory, Warrington, UK, DL/SC1/R34, pp. 28–32.
60. Mitchell, E. P. and Garman, E. F. (1994) Flash freezing of protein crystals: investigation of mosaic spread and diffraction limit with variation of cryoprotectant concentration. *J. Appl. Cryst.* **27,** 1070–1074.

CHAPTER 5

Use of Multiple-Wavelength Anomalous Diffraction Measurements in *Ab Initio* Phase Determination for Macromolecular Structures

H. M. Krishna Murthy

1. Introduction

The determination of three-dimensional structures of macromolecules in the crystalline state depends on the acquisition and processing of diffraction data generated by the interaction of X-rays with the crystals of the macromolecule under investigation. In general, there are two points at which the process languishes because of lack of necessary material or data. The first is the initial crystallization of the macromolecule; although work is being done to understand the principles involved *(1)*, the crystallization of a macromolecule usually depends heavily on empirical reasoning *(2)*. The second bottleneck is acquisition of phases. Each diffraction maximum has associated with it both an amplitude and a phase. The phases are not directly measurable in a typical diffraction experiment and must therefore be estimated by indirect means. By far the most widely used method for the derivation of phases is that of Multiple Isomorphous Replacement (MIR). The MIR method depends on the making of isomorphous derivative crystals of the macromolecule under investigation, and estimation of phases from the differences between the amplitudes of native and derivative diffraction maxima *(3)*. At least two isomorphous derivatives are required for reliable phase determination, although in favorable cases, anomalous diffraction information from a derivative may be used as a second pseudoderivative. Although anoma-

From: *Methods in Molecular Biology, Vol. 56: Crystallographic Methods and Protocols*
Edited by: C. Jones, B. Mulloy, and M. Sanderson Humana Press Inc., Totowa, NJ

lous diffraction information has historically been used very successfully *(3)* in this secondary role, a more direct role for it has long been advocated *(4)*. Recent theoretical analyses *(5)*, development of requisite instrumentation *(6,7)*, and advances in data-acquisition and processing strategies *(7)* have enabled an unambiguous experimental demonstration of the use of anomalous diffraction measurements in direct phasing of macromolecular diffraction patterns. Such direct phase determination using anomalous diffraction information involves the measurement of diffraction data at several different wavelengths, and is termed Multiple-wavelength Anomalous Diffraction (MAD) phasing.

2. Anomalous Diffraction

X-rays are diffracted by the electrons in atoms of the specimen being studied. The structure factor for each reflection, h, is the vector sum of the contribution to that reflection of diffracted amplitude from every atom in the unit cell. The intensity of diffraction of X-rays from each type of atom is determined by a quantity f, called the scattering factor, which is proportional to the number of electrons in the atom, and is inversely proportional to the angle of diffraction 2θ. A schematic, representative illustration of a classical atom is given in Fig. 1, as an electron orbiting a positively charged nucleus. Such a situation confers on the electron a natural frequency of oscillation, ω *(8)*. When the wavelength (λ_1) of incident radiation is distant from that corresponding to that of the natural frequency (Fig. 1A), the diffraction is termed normal, and f is a real, positive quantity. If, however, the wavelength (λ_2) of the incident X-radiation is close to a natural frequency, then the electron executes resonant vibrations (Fig. 1B), and diffraction occurs from such a resonating entity. In the latter case, the scattering factor f is a complex quantity, comprised of f' (which is real), called the dispersive component, and if'', a complex entity, called the Bijvoet component. Such resonant diffraction is termed anomalous diffraction. The magnitudes f' and f'' vary as a function of the wavelength of incident radiation. A plot of the f' and f'' values measured from crystals of *Clostridium acidi-urici* ferredoxin *(9)* is shown in Fig. 2. The figure illustrates variation of dispersive and Bijvoet signals as a function of X-ray wavelength, and also depicts the extreme values assumed by these quantities at discrete points of the wavelength spectrum.

The effect of anomalous diffraction on the measured structure amplitudes is represented in Fig. 3. Vectors drawn as solid lines above the real axis in Fig. 3 depict the structure factor for a particular reflection, from a

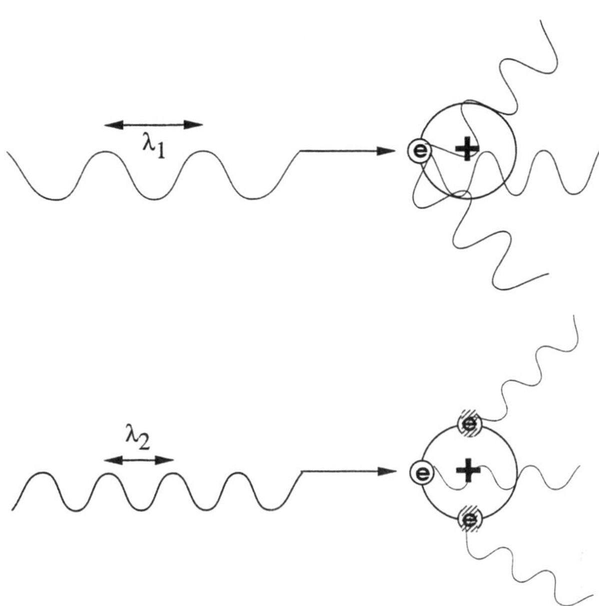

Fig. 1. Schematic illustration of anomalous scattering. An electron (e) is shown orbiting around a positively charged nucleus (+), the entire system representing a hypothetical, classical atom. **(A)** Diffraction of X-radiation with a wavelength (λ_1) far away from that of the natural frequency of oscillation of the electron: normal diffraction. **(B)** Diffraction of X-radiation with a wavelength (λ_2) close to the natural frequency of oscillation of the electron: anomalous diffraction.

unit cell composed of a large number of normal diffractors (such as carbon, nitrogen, and oxygen atoms) and a single anomalous diffractor–perhaps a metal ion. The sum of diffraction by all the normal diffractors is represented by the vector F^{+}_{0}, the normal diffraction vector from the anomalous diffractor by F^{+}_{A}, and the two components of anomalous diffraction by vectors f' and f''. The resultant structure factor vector is denoted by F^{+}. A similar construction for the F^{-} vector, representing the structure factor of the Friedel mate of F^{+} (the *-h* reflection), is also shown below the real axis in Fig. 3. It is clear that the two vectors F^{+} and F^{-}, which would be equal in magnitude in the absence of the f'' vector, become unequal in its presence. This nonequality of the amplitudes of Friedel mates is owing to the fact that at any given wavelength, although magnitudes of f'' vectors are equal for Friedel mates, phase of the f'' vector is always $\pi/2$ ahead of all other diffracted vectors *(10)*. Thus, the

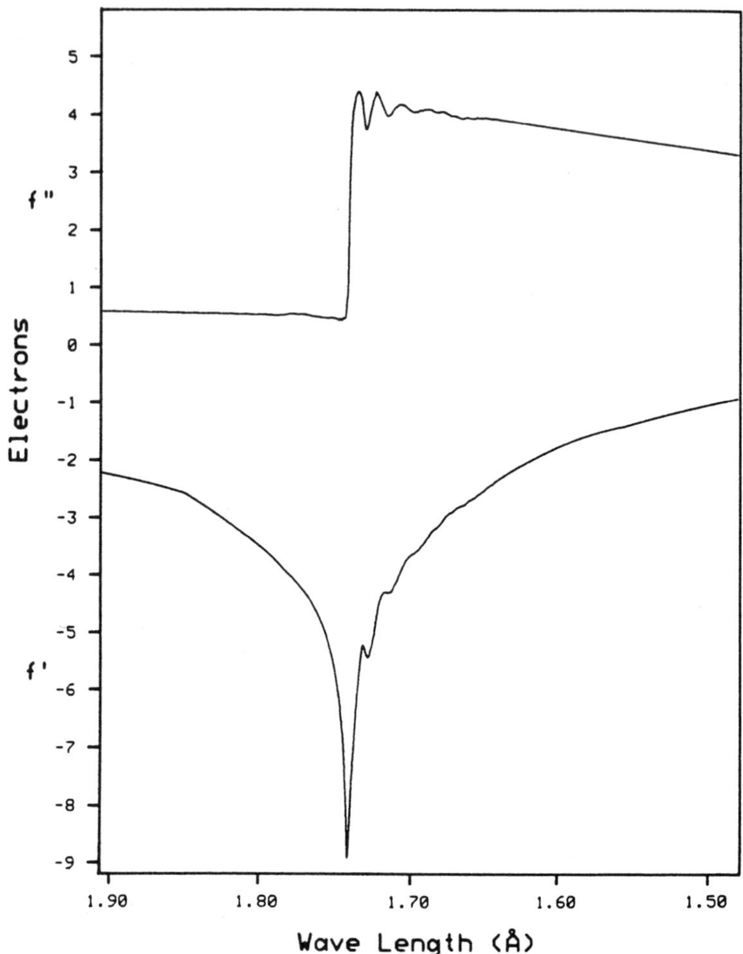

Fig. 2. Experimental determination of f' and f''. The top curve shows the scaled and fitted absorption spectrum of *Clostridium acidi-urici* ferredoxin crystals. Background correction was applied to the measured absorption spectrum by fitting a suitable polynomial to values away from the edge region. The spectrum was then scaled to the theoretically computed *(34)* spectrum at points away from the vicinity of the absorption edge. The curve shown is a composite of the scaled, theoretically computed values before and after the edge region with the experimentally determined edge region spliced in at the appropriate position. The lower curve shows the Kramers-Kronig transformation *(44)* of the upper curve. The absorption spectrum used to produce these curves was measured with the polarization vector parallel to the *a* axis. Reprinted with permission from ref. *9*.

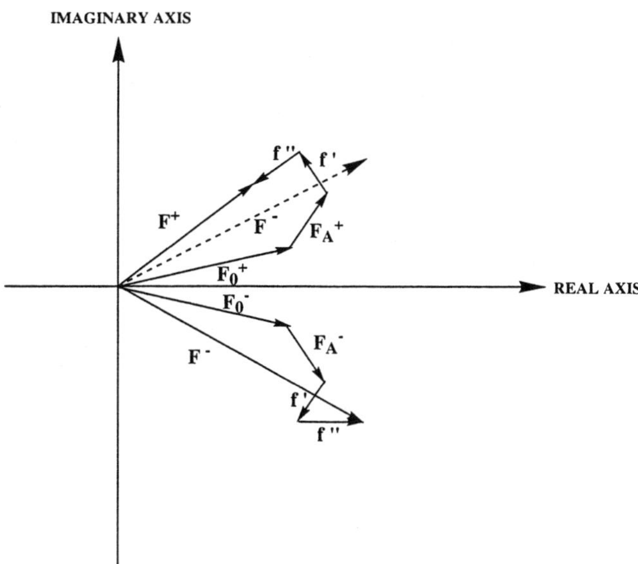

Fig. 3. Nonequality of Friedel mates. The vectors drawn as solid lines above the real axis represent the structure factor of the positive Friedel mate. F_0^+ represents the diffraction owing to all the normally diffracting atoms, F_A^+ the normal diffraction owing to all the anomalously diffracting atoms, and f' and f'' the dispersive and Bijvoet components of anomalous diffraction, respectively. The total structure factor vector for the Friedel positive reflection is represented by F^+. A similar construction for the total structure factor vector (F^-) of the negative Friedel mate is shown below the real axis. To facilitate the comparison of the lengths of the F^+ and F^- vectors, the F^- vector has been reflected across the real axis and shown by dashed lines on the same side of the real axis as F^+.

Bijvoet signal can be derived by measuring differences in intensities of Friedel mates at a suitably chosen wavelength. As is also obvious from Fig. 3, the magnitude of f' is the same for Friedel mates. The f' signal can, however, be utilized by measuring intensities of a given reflection at a suitably chosen pair of wavelengths and determining the difference between those measurements.

3. The MAD Method

3.1. Theoretical Background

The MAD method has been very succintly reviewed *(11,12)*, and its theoretical underpinnings comprehensively detailed in several places

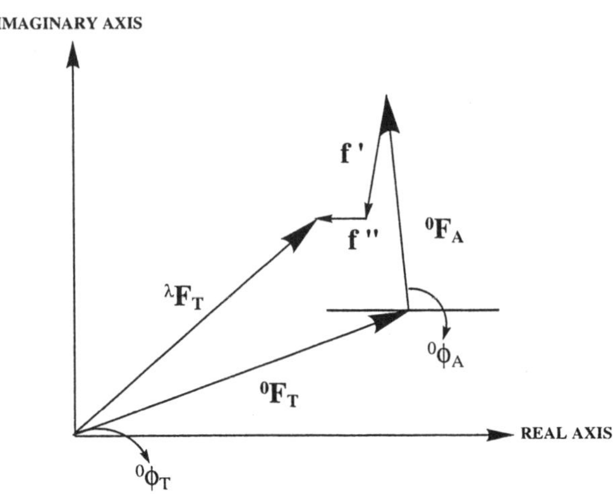

Fig. 4. Geometric representation of quantities in Eq. 2. As in Fig. 3, lengths of vectors denote amplitudes and their offset from the real axis represents phase angles. *See* Section 2.1. for a detailed description.

(5,13,14). Only a brief description of the relevant background will be given here. The structure factor for a particular reflection, h, is written as:

$$F(h) = \Sigma f_j \exp(2\pi i h\cdot x_j - B_j s^2) \quad (1)$$

to represent the fact that it depends on the positions x_j, the thermal motions Bj, and the scattering factor f_j of the atoms in the unit cell, as well as on the scattering angle s (= sin θ/λ). In the presence of anomalous diffraction, however, f_j is a complex quantity, and the equation will have to be rewritten to take this circumstance into account. Thus, the wavelength-dependent structure factor may be written:

$$^{\lambda}F_T(h) = {}^0F_T(h) + \Sigma\,[(f'_k/f^0_k) + (if''_k/f^0_k)]\;{}^0F_{Ak}(h) \quad (2)$$

where $^{\lambda}F_T(h)$ denotes the structure factor at wavelength λ, $^0F_T(h)$ the structure factor including contributions from all normal diffractors, and $^0F_{Ak}(h)$ the normal contribution from all anomalous diffractors; f'_k and f''_k are the dispersive and Bijvoet components of anomalous diffraction, respectively, at wavelength λ, for the kth kind of anomalous diffractor. f'_k and f''_k enter into Eq. (2) as fractions of the normal scattering factor f^0. A geometric representation of amplitudes and phase angles involved is given in Fig. 4 in vector form. Assuming the presence of only a single kind of anomalous diffractor, and denoting phases of $^0F_T(h)$ and $^0F_A(h)$

by ${}^{0}\phi_T(h)$ and ${}^{0}F_A(h)$, respectively, the expression for experimentally measurable quantities in wavelength dependent diffraction will assume the form:

$$| {}^{\lambda}F_T(h) |^2 = |{}^{0}F_T(h)|^2 + a(\lambda)\, |{}^{0}F_A(h)|^2 + b(\lambda)\, |{}^{0}F_T(h)|\, |{}^{0}F_A(h)| \cos [{}^{0}\phi_T(h) - {}^{0}\phi_A(h)] + c(\lambda)\, |{}^{0}F_T(h)|\, |{}^{0}F_A(h)| \sin [{}^{0}\phi_T(h) - {}^{0}\phi_A(h)] \quad (3)$$

where

$$a(\lambda) = (f^{\Delta}/f^{0})^2,\ b(\lambda) = 2(f'/f^{0}) \quad (4A)$$

and

$$c(\lambda) = 2(f''/f^{0}) \quad (4B)$$

and

$$f\Delta = (f'^2 + f''^2)^{1/2} \quad (4C)$$

the magnitude of anomalous scattering.

It can be seen, from Eq. (3) that all wavelength-dependent information is confined to coefficients of Eq. (3), given in Eq. (4), and these coefficients can be determined without prior knowledge of the structure, since they are exclusive functions of f' and f'', which can be experimentally measured, and f^0, which can be calculated *(15)*. Thus, in case of a structure containing a single kind of anomalous diffractor, there are three independent unknowns to be determined: $|{}^{0}F_T(h)|$, $|{}^{0}F_A(h)|$ and $\Delta\phi = [{}^{0}\phi_T(h) - {}^{0}\phi_A(h)]$. Therefore, if one has a sufficient number of measurements at $\pm h$ and at different wavelengths, one can construct a system of equations analogous to Eq. 3, and solve them for the three unknowns. Then, using $|{}^{0}F_A(h)|$ thus obtained, positions of anomalous diffractors can be determined, which in turn enables determination of ${}^{0}\phi_A(h)$. Since both $\Delta\phi$ and ${}^{0}\phi_A(h)$ are now known, ${}^{0}\phi_T(h)$ may be directly determined, and combined with already known $|{}^{0}F_T(h)|$ to compute an electron density map. The map may then be interpreted in terms of a structural model of the macromolecule under investigation. In principle, this formalism may be extended to structures containing more than one kind of anomalous diffractor, although the number of independent parameters to be determined increases as $2n + 1$, where n is the number of kinds of anomalous diffractor. Although it is theoretically possible to determine the unknown parameters by measurements of $+h$ at different wavelengths, using measurements made at both $+h$ and $-h$ at different wavelengths takes advantage of the orthogonality of information implied by the trigonometric terms in Eq. 3.

3.2. Experimental Protocol

Structure determination by crystallographic techniques presupposes the existence of suitable crystals, and since methods used to obtain such crystals are detailed in several publications *(16)*, they will not be addressed here. Although the utility of multiple-wavelength radiation from laboratory sources for MAD phasing has been demonstrated *(12,18)*, and instruments to implement MAD phasing in a laboratory setting are being developed *(16,17)*, most successful MAD phasing experiments have thus far been carried out using synchrotron sources, and this seems to be the most practical route at the present time. Synchrotrons generate X-rays in a fundamentally different way from that of laboratory X-ray generators *(21)*. The generated radiation at synchrotrons has a high intensity from ≈0.3 Å on the low-wavelength side to well into the UV region on the high-wavelength side. Thus, one may choose appropriate wavelengths from this broad spectrum by using a suitable monochromator *(21)*. Although MAD experiments primarily exploit the tunability of synchrotron radiation, other kinds of crystallographic experiments exploit other properties of the generated radiation. Excellent reviews that detail the use of its high brilliance *(22)*, wide bandwidth *(23)*, and time structure *(24,25)* are available. Assuming availability of suitable crystals, and a source of multiwavelength X-radiation, a stepwise MAD experimental protocol may be devised as follows:

1. Introduce one or more suitable anomalous diffractor(s) into crystals of the macromolecule if it does not already have one associated with it. This may be done either by diffusing suitable anomalous diffractors into preformed crystals or by crystallizing suitably modified macromolecules.
2. Obtain values of f' and f'' as a function of wavelength to permit choice of appropriate wavelengths for diffraction data acquisition.
3. Measure diffraction data at a sufficient number of wavelengths to overdetermine the least-squares problem.
4. Process the measured data to reduce it to a set of structure factor amplitudes at each wavelength and then obtain the required differences. These include differences between Friedel mates measured at the same wavelength, as well as differences between the same reflection measured at different wavelengths.
5. Determine the three unknown quantities (assuming a single kind of anomalous diffractor) $|^0F_T(h)|$, $|^0F_A(h)|$, and $\Delta\phi = [^0\phi_T(h) - {}^0\phi_A(h)]$ for each reflection.
6. Determine positions of anomalous diffractors using the $|^0F_A(h)|$.

7. Use the positions of the anomalous diffractors to evaluate the ${}^{0}\phi_A(h)$.
8. Use values of the ${}^{0}\phi_A(h)$ and $\Delta\phi$ to determine values of ${}^{0}\phi_T(h)$ from the equation $\Delta\phi = {}^{0}\phi_T(h) - {}^{0}\phi_A(h)$.
9. Compute an electron density map from the now-known values of $|{}^{0}F_T(h)|$ and ${}^{0}\phi_T(h)$.
10. Interpret the map in terms of a molecular model, and refine resulting atomic positions.

4. Notes

4.1. Introduction of Suitable Anomalous Diffractors

Although the bandwidth of the X-radiation emitted by synchrotrons is large, owing to current technical limitations, the useful wavelength range for MAD experiments is from ≈0.5 Å to ≈3.0 Å. Thus, atoms with absorption edges in this wavelength range would serve as useful anomalous diffractors. This limitation on the range of usable wavelengths confines choice of anomalous diffractors to a subset of atoms in the periodic table: to those with atomic numbers in the range 20 (Ca) to 47 (Ag), and 50 (Sn) to 92 (U). Some macromolecules, for example metalloproteins, are naturally associated with metal ions and crystallize with these specifically associated metal ions. Such crystals may be directly used in MAD experiments. Table 1 lists a number of metal ions likely to be liganded with proteins, along with their absorption edges and associated f' and f'' values. The table also lists the molecular weight of the macromolecule, which, when associated with a single such metal ion, would generate a dispersive signal of at least 2.5% and a Bijvoet signal of at least 2.5%. These limiting values of signals are chosen because they are the smallest signals that have so far been measured and used in successful MAD phasing of a structure *(26)*; they may or may not represent a lower limit for measurability and exploitability.

Although a glance at Table 1 might make Ca seem to be an attractive metal ion to use, its absorption edge is at a somewhat long wavelength (3.07 Å), where severe problems owing to nonspecific absorption are expected to be encountered. Most of the other ions listed in Table 1, except Mo and Cd, are expected, in principle, to be useful for macromolecules in the mol-wt range of 8–14 kDa. Of course, if more than one metal ion is specifically associated with a macromolecule, anomalous diffraction signals go up correspondingly by a factor of $\sqrt{N_A}$ where N_A is the total number of anomalous diffractors per molecule. It is also frequently possible *(27)* to replace native metal ions in macromolecules by others

Table 1
Expected Signal Levels for Some Metal Ions of Biological Significance

Metal ion	K or L[a] Edge	f'^{b}	f''^{b}	Maximal molecular mass[c,d]
Mg	9.51	—	—	—
Ca	3.07	–10.5	4.06	17,000
Mn	1.89	–5.54	3.94	8000
Fe	1.74	–6.11	3.94	10,000
Co	1.60	–4.91	3.90	6000
Ni	1.48	–6.32	3.87	10,000
Cu	1.38	–7.35	3.90	14,000
Zn	1.28	–5.52	3.87	8000
Se	0.98	–7.17	3.72	13,000
Mo	0.61	–3.46	3.59	3000
Cd	0.46	–4.11	3.54	4000
Hg	1.01	–17.3	10.18	77,000
Eu	1.78	–16.9	10.64	73,000

[a]Position of absorption edge in angstroms; represents the K edge for all execept Hg and Eu for which L edge positions are given.

[b]Expressed in electrons.

[c]Expressed in Daltons, assuming substitution of a single metal ion per molecule of protein.

[d]Numbers in this column are derived as follows: initial estimates of the molecular weight were obtained using the squared and transposed forms of Eqs. (5) and (6) (i.e., $N_P = (N_A/2S_{\text{Bij}}^2)\,(2f''/Z_{\text{eff}})^2$ and $N_P = (N_A/2S_{\text{Dis}}^2)\,(f\lambda_1' - f_{\lambda 2}'/Z_{\text{eff}})^2$ respectively, with Z_{eff} set to 6.7 and S_{Bij} and S_{Dis} set to 0.025, *see text*). These numbers were multiplied by 13.4 to convert them to atomic weight units, revised upward by 7% to account for the hydrogen atoms, and the resulting numbers rounded to the nearest thousand.

that generate larger anomalous signals. Table 1 also lists the effects on expected anomalous signals, of using Hg or Eu ions, which have a much larger anomalous effect associated with their several L absorption edges. It can be seen that such a substitution would enable measurement of potentially usable anomalous signals from macromolecules of fairly large molecular weights.

Metal ions may also be incorporated into macromolecules that do not naturally have metal ions specifically associated with them using the procedures used for making heavy atom derivatives *(3)*. Since structure determination by MAD procedures involves utilization of diffraction data from such derivatized crystals alone rather than from a collection of isomorphous crystals, substantial nonisomorphism can be tolerated *(27)*. Structure of the native macromolecule may be derived from that of the derivative by direct substitution of the derivative structure into the native unit cell *(26,27)*, followed by subsequent refinement or by use of molecular replacement techniques *(28)*.

A general method of incorporating anomalous diffractors into proteins that possess methionine residues is the replacement of these methionine residues by their Se analogs. As can be seen from Table 2, Se is an excellent anomalous diffractor and is known to substitute for S in small molecules without significant structural distortion *(12)*. Thus, it has been proposed *(29)* and has been experimentally demonstrated *(30–32)* that seleno-proteins, made by replacement of methionine residues by seleno-methionine residues, form excellent phasing vehicles. Replacement by seleno-methionine generally involves the utilization of the techniques of molecular biology, consisting of cloning the protein of interest into a methionine auxotroph, growing the auxotrophic strain in a medium containing seleno-methionine, purifying the expressed seleno-protein, and crystallizing it. In most cases where such experiments have been performed, the seleno-protein has crystallized isomorphously with the corresponding sulfur protein and is very nearly iso-structural *(26,30–32,32a, 32b)*, with differences localized to the region of Se substitution. Substitution by brominated bases has been successfully used *(33)* as a way of incorporating suitable anomalous diffractors into nucleic acids and might provide a general procedure for this class of macromolecule. In addition to these general methods, specific methods might be available for particular molecules owing to their unique properties, such as use of ligands carrying good anomalous diffractors *(26)*.

4.2. Data Measurement

Since anomalous diffraction signals are frequently rather small fractions of the total macromolecular diffraction signal, exercise of extreme care and ingenuity in measurement and treatment of diffraction data is essential. The objective is to acquire data to as high a precision as possible, in as short a time as possible. Although these requirements conflict with each other to some extent, suitable compromises can usually be made to obtain a usable set of data. It is advantageous to calculate the signal levels to be expected before committing oneself to carrying out diffraction experiments, not only to determine feasibility of applying MAD techniques, but also as an aid in planning subsequent data measurement strategy. Formulae have been derived *(14)* and may be directly used to calculate expected signal values using approximate values of

Table 2[a]

Expected Anomalous Signal

λ[b]	Fe scattering factor strengths					S scattering factor strengths					Anomalous scattering factors			
											Fe		S	
	1.9000	1.8000	1.7419	1.7390	1.5000	1.9000	1.8000	1.7419	1.7390	1.5000	f'	f''	f'	f''
1.9000	1.80[c]	1.60[d]	6.98	5.56	0.63	1.64[c]	0.01[d]	0.02	0.02	0.06	–1.93	0.59	0.37	0.82
1.8000		1.06	5.41	3.99	2.20		1.48	0.01	0.01	0.05	–3.50	0.53	0.36	0.74
1.7419			3.84	1.42	7.61			1.40	0.00	0.04	–8.91	1.92	0.35	0.70
1.7390				8.18	6.19				1.40	0.04	–7.49	4.09	0.35	0.70
1.5000					6.80					1.06	–1.30	3.40	0.31	0.53

Diffraction ratios[e]

λ[b]	Fe					S					Total				
	1.9000	1.8000	1.7419	1.7390	1.5000	1.9000	1.8000	1.7419	1.7390	1.5000	1.9000	1.8000	1.7419	1.7390	1.5000
1.9000	1.8[f]	2.5[g]	10.8	8.6	1.0	4.0[f]	0.0[f]	0.0	0.0	0.0	4.4[h]	2.5[i]	10.8	8.6	1.0
1.8000		1.6	8.4	6.2	3.4		3.2	0.0	0.0	0.0		3.6	8.4	6.2	3.4
1.7419			6.0	2.2	11.8			3.1	0.0	0.0			6.8	2.2	11.8
1.7390				12.7	9.6				3.1	0.0				13.1	9.6
1.5000					10.5					2.3					10.7

[a]Reproduced with permission from ref. *9*.

[b]Wavelength in angstroms.

[c]Diagonal elements, $2f''$.

[d]Off diagonal elements, $|f'_{\lambda i} - f'_{\lambda j}|$.

[e]Expressed as a % of $(|F|)$.

[f]Diagonal elements, $<(|\Delta F(\pm h)|> = (1/\sqrt{2})\ [N_A/N_P]^{1/2} 2f''/Z_{eff}$ where N_A = number of anomalous scatterers, 8Fe, 16S; N_P = number of protein atoms, 371; Z_{eff} = effective atomic number of protein atoms, 6.7.

[g]Off diagonal elements, $<(|\Delta F_{\Delta\lambda}|)> = (1/\sqrt{2}[N_A/N_P]^{1/2}|(f'_{\lambda i} - f'_{\lambda j})|Z_{eff}$.

[h]Diagonal elements, $<(|\Delta F(\pm h)|>_{multi} = [\Sigma<|\Delta F(\pm h)|>_i^2]^{1/2}$.

[i]Off diagonal elements, $<(|\Delta F_{\Delta\lambda}|)>_{multi} = [\Sigma<|\Delta F(\Delta\lambda)|>_i^2]^{1/2}$.

theoretically computed *(34)* f' and f'' for relevant anomalously diffracting species. These formulae express expected signal levels as a fraction of average structure factor amplitudes and are very useful guides. The expected Bijvoet signal is given by:

$$S_{Bij} = \langle|\Delta F_{\pm h}|\rangle / \langle|F|\rangle = (1/2)^{1/2} (N_A/N_P)^{1/2} (2f''/Z_{eff}) \quad (5)$$

and the expected dispersive signal by:

$$S_{Dis} = \langle|\Delta F \Delta\lambda|\rangle / \langle|F|\rangle = (1/2)^{1/2} (N_A/N_P)^{1/2} (|f'_{\lambda 1} - f'_{\lambda 2}|/Z_{eff}) \quad (6)$$

where:

- $\langle|\Delta F_{\pm h}|\rangle$ is the average difference between Friedel mates at a given wavelength λ;
- $|\Delta F_{\Delta\lambda}|\rangle$ is the average difference between the average of ${}^{\lambda}F(+h)$ and ${}^{\lambda}F(-h)$ at two different wavelengths $\lambda 1$ and $\lambda 2$ differing by Δ_{λ};
- $\langle|F|\rangle$ is the average structure factor amplitude;
- N_A is the number of anomalously diffracting atoms per molecule;
- N_P is the total number of nonhydrogen atoms in the molecule; and
- Z_{eff} is the average atomic number of nonhydrogen atoms (≈ 6.7 for proteins).

The calculation carried out for *Clostridium acidi-urici* ferredoxin at a number of relevant wavelengths is given in Table 2 along with values used for various quantities occurring in Eqs. (5) and (6).

To maximize the rate of data acquisition, area detectors are preferred over single reflection counters. Single reflection counting has, however, been used in some cases *(26,31)* and may be used when radiation damage to crystals is not a serious problem, since potentially attainable precision in measurements using diffractometers is greater. X-ray film is by definition an area detector, but has not been used frequently because of its inherent disadvantages *(35)*. Multiwire area detectors *(6,9,26,36)* or imaging plates *(32,33)* have been successfully used for MAD experiments. Other types of area detectors *(37)* may also be suitable, although their successful use for such experiments has so far not been reported.

A desired level of counting precision may be achieved by accumulating a sufficiently large number of counts. The relevant formula is given by $\sigma F/F \approx 0.5\ \sigma I/I \approx \sqrt{N}/N$, where σF and σI are standard deviations in the $|F|$ and I, respectively, and N is the number of counts measured for a particular reflection. Thus, depending on values of expected anomalous signals, an appropriate value for *N* may be chosen. Another point worth keeping in mind is that it is possible to attain a higher level of precision in derived differences than in the measurements themselves because of cancellation of systematic errors on forming differences.

It is possible to minimize systematic errors like absorption by arranging for the diffracted rays for Friedel mates to follow similar optical paths, hence suffering similar attenuation owing to absorption. If the unit cell possesses rotational axes of symmetry, then it is possible to choose a suitable direction of rotation, such that Friedel mates are measured on the same frames of the area detector. Even for triclinic crystals, it is possible to measure frames after rotating the crystal through suitable angles between successive frames. Suitable strategies might also be devised for diffractometric data acquisition. To minimize differences owing to radiation damage, small blocks of reflections, perhaps 15–25, may be measured on the diffractometer immediately followed by their Friedel mates. Differences resulting from radiation damage may also be minimized by measuring the relevent data close together in time. Thus, frames or reflections could be measured at all chosen wavelengths before going on to the next frame or reflection. Residual errors can be reduced by suitable computational scaling procedures *(38–40)*.

4.3. Choice of Wavelengths

Absorption edges for various atomic species have been tabulated *(41)*, and approximate positions and magnitudes of edges may also be theoretically computed *(34)*. It is, however, advisable to experimentally determine these wavelengths because of the fact that both positions and the magnitudes of edges depend on the specific molecular geometry around the anomalously diffracting species *(42)*. Measurements of X-ray absorption spectra can be made employing fluorescence techniques *(43)*, using the very crystals under investigation; one such spectrum is shown in Fig. 2. Such measurement unambiguously determines both position and magnitude of the absorption edge under the experimental conditions that are to be used for acquisition of diffraction data.

Given such a measurement, one may readily choose a set of suitable wavelengths for accumulation of diffraction data. There are several recommendations available *(45–47)* for making such a choice; following the simple dictum of maximizing both dispersive and Bijvoet signals, however, leads to a particularly simple procedure. The Bijvoet signal can be maximized by choosing the wavelength corresponding to the absorption peak (the point marked "a" in Fig. 2). The dispersive signal can be maximized by choosing a pair of wavelengths, one of which corresponds to the minimum of the f' curve (the point marked "b" in Fig. 2), and the other being far removed from it (the points marked "c" or "d" in Fig. 2), to give a large $\Delta f'$. The number of wavelengths at which one chooses to measure diffraction data depends on the number of parameters that need to be determined, which in turn depends on the number of kinds of anomalous diffractors present in the unit cell. The number of independent parameters that need to be determined varies as $2n + 1$, where n is the number of kinds of anomalous diffractors in the unit cell. Since unknown parameters are determined by a least-squares procedure, larger numbers of observations lead to parameters that are better determined in a statistical sense. The minimum number of wavelengths that need to be used if the above strategy for choosing the number of wavelengths is followed is three (assuming a single kind of anomalous diffractor in the unit cell); this seems to work well in practice *(26,31–33)*.

Decisions on the extent of data to be measured depend on size, diffraction quality, and radiation sensitivity of the crystals, as well as on time available for measurement. Experimental strategies, such as cooling crystals to liquid nitrogen temperatures, may be used to reduce the rate of radiation damage dramatically *(48)*. Electron density maps computed with MAD data at as low a resolution as 3 Å have been successfully interpreted in terms of molecular models *(26)*, and subsequently refined against data measured at a single wavelength on a laboratory source of X-radiation. Even with data of substantially lower resolution than 3 Å, it might be possible to use emerging refinement and phase extension techniques *(49,50)* to arrive at a suitable initial structural model that might later be improved by higher-resolution data measured at a single wavelength.

4.4. Data Processing and Handling of Errors

Measured data are processed using techniques appropriate to either the diffractometric or area detector method used for acquiring them. All corrections conventionally applied in single-wavelength diffraction experiments are just as relevant here. Thus, a set of observed |F| values is obtained at each of the wavelengths at which the diffraction data are

Table 3[a]
Statistics for Data Sets

Wavelength Å	R_{BU}[b]	Signals[c]
1.9000	0.066 (0.063)	0.7 (0.4)
1.8000	0.057 (0.053)	1.1 (0.6)
1.7419	0.060 (0.042)	1.2 (0.6)
1.7390	0.160 (0.042)	2.7 (0.5)
1.5000	0.119 (0.017)	2.4 (0.5)

[a]Reproduced with permission from ref. *9*.
[b]$R_{BU} = \Sigma \{|(F^+ - F^-)|1/2(F^+ + F^-)\}$; centric values in parentheses.
[c]Signal = RMS(ΔF) – RMS(σF); RMS(σF) in parentheses.

measured by using very similar strategies to those used in single-wavelength experiments *(3)*. The quality of processed data and the signal levels obtained for *C. acidi-urici* ferredoxin *(9)* are given in Table 3.

Absorption corrections are, as in single-wavelength experiments, a serious concern. In addition, the dependence of absorption on the X-ray wavelength used is a further complication. When one is using wavelengths in the vicinity of an absorption edge, as happens in a multiple-wavelength experiment, absorption effects are larger than when experimental wavelengths are remote from absorption edges. Corrections can, however, be computed and applied as for single-wavelength experiments. Widely used empirical methods are available to evaluate and apply absorption corrections to diffractometer data *(51)*, and processing of area detector data implicitly corrects them for absorption *(52)*. Data measured at different wavelengths are scaled and corrected for absorption separately to take into account differences in absorption as a function of wavelength as well as anisotropic anomalous diffraction effects *(53)*. Theoretical bases of wavelength-dependent scaling have been adequately discussed *(40)*. It is also feasible to measure an absorption curve on a diffractometer for a crystal for which the diffraction data were measured on an area detector and apply the correction thus computed after making necessary geometrical transformations to relate the two systems *(9)*. Thus, although differences owing to absorption may be minimized by a suitable choice of crystal orientation, residual systematic differences owing to absorption that will inevitably remain may be reduced by applying appropriate corrections determined as already mentioned.

Table 4[a]
Observed Anomalous Signal

λ Å	1.9000	1.8000	1.7419	1.7390	1.5000
1.9000	26.1[b] (28.6)	23.9[c]	38.7	30.1	23.1
1.8000		41.5 (41.8)	30.5	21.9	23.5
1.7419			26.9 (10.7)	22.4	45.8
1.7390				57.2 (10.4)	35.5
1.5000					52.6 (18.5)

[a]Reproduced with permission from ref. *9*.
[b]Diagonal elements, $\{(|^{\lambda}F(h)| - |^{\lambda}F(-h)|)^2\}^{1/2}$.
[c]Off diagonal elements, $\{(\langle|^{\lambda i}F|\rangle - \langle|^{\lambda i}F|\rangle)^2\}^{1/2}$; $\langle|^{\lambda}F|\rangle = 1/2(|^{\lambda}F(h)| - |^{\lambda}F(-h)|)$.

Systematic errors owing to decay of the crystal in the X-ray beam are also minimized by measuring all data that enter into the determination of a single phase close together in time. This still leaves unaddressed the problem of relating later measurements made on a crystal partially damaged by X-radiation to those made at earlier times in the experiment on a crystal that was less damaged or not damaged. There are methods available that allow one to make such corrections, either empirically *(54)* or based on a kinetic model of the decay process *(55,56)*, both for diffractometer and area detector data.

Residual errors are usually minimized by computational procedures, such as local scaling *(38)*. Choice of local region might be based on variables tied to experimental geometry *(54)* or on a general anisotropic parameterization *(39)*. Signal levels obtained for *C. acidi-urici* ferredoxin derived from processed diffraction data *(9)* are given in Table 4 and can be seen to approximate closely the expected signals, which are shown in Table 2.

4.5. Determination of the Unknowns

The three independent unknown parameters, $|^0F_T(h)|$, $|^0F_A(h)|$, and $\Delta\phi = {}^0\phi_T(h) - {}^0\phi_A(h)$, for a single anomalous diffractor case are determined as mentioned earlier by a least-squares procedure. Currently, this determination is made using the program MADLSQ *(9,14)*, which implements the solution of the nonlinear MAD equation (Eq. [3]) in two stages. First, the linear least-squares problem:

$$f = \lambda_{\pm h} (y_o - y_c)^2 \quad (7)$$

where f is the function to be minimized, and y_o and y_c are observed and calculated values of structure factor amplitudes for a reflection, respectively. The observed value of the structure amplitude, y_0, is derived from diffraction measurements made at various chosen wavelengths [$y_0(h,\lambda) = |^{\lambda}F(\pm h|^{2]}$, and the calculated value, y_c, is given by:

$$y_c(h,\lambda) = p_1(h) + a(\lambda)\, p_2(h) + b(\lambda)\, p_3(h) + c(\lambda) p_4(h) \quad (8)$$

where the p's bear obviously recognizable relationships with corresponding symbols in Eq. (3), and a, b, and c are coefficients that would have been determined before performing the diffraction experiment from measurements of relevant anomalous scattering factors. Solving Eq. (7) gives initial values for the determinable parameters for a nonlinear constrained fit, using the equation $\mathbf{G} = f + \gamma g$, where γ is a Lagrange multiplier, used to apply appropriate constraints *(14)*. Currently applied constraints are nonnegativity of intensity and the trigonometric identity involving phase angles; $\cos^2\Delta\phi + \sin^2\Delta\phi = 1$. The unknown quantities may then be determined by the following straightforward relations: $|^0F_T(h)| = \sqrt{p1}$, $|^0F_A(h)| = \sqrt{p2}$, and $\Delta\phi = {}^0\phi_T(h) - {}^0\phi_A(h) = \tan^{-1}(p4/p3)$. The precision of the determined parameters can be judged from standard deviations that may be derived from elements of the inverted least-squares matrix.

4.6. Determination of the Position(s) of the Anomalous Diffractor(s)

The $|^0F_A(h)|$ values determined in the previous step can be used, in a direct way, to determine position(s) of anomalous diffractor(s). Before using $|^0F_A(h)|$ values computed, it is important to filter out the obviously incorrect ones (for instance, those reflections for which $|^0F_A[h]| > [F_{A000}]$), as well as those values that are statistically less reliable (perhaps those reflections for which $|^0F_A(h)| < 2\sigma\, |^0F_A[h]|$), to reduce noise in subsequent computations.

Position(s) of the anomalous diffractor(s) may be determined either by using $|^0F_A(h)|^2$ as coefficients in a Patterson synthesis and subsequent deconvolution of the vector map *(9,26,36)*, or by using the $|^0F_A(h)|$ as input to a direct methods program, such as MULTAN *(57)*, and interpreting the resulting E-map *(9,33)*.

4.7. Derivation of Native Phases

Native phases are determined by executing steps 6 and 7 in Section 2.2. Using the positions of anomalous diffractors, determined in Section

2.2., step 5, the structure factor for each reflection representing the normal diffraction by the anomalous diffractors can be readily computed. This involves evaluation of the right hand side of Eq. (1) for each reflection h from known values of positions of anomalous diffractors, x_j, their normal scattering factors, f_j, and temperature factors, B_j. Calculation of these structure factors makes available both amplitudes, $|^0F_A(h)|_{calc}$, and phases, $^0\phi_A(h)_{calc}$, for normal scattering by the anomalous diffractor(s). Determination of the $^0\phi_A(h)$ makes known two out of three quantities in the relationship, $\Delta\phi = {}^0\phi_T(h) - {}^0\phi_A(h)$ ($\Delta\phi$ having been determined in Section 2.2., step 4), enabling the determination of $^0\phi_T(h)$, the native phase for each reflection. It is also possible to derive *(58)* and use *(26,59)* a phase probability analysis, similar to that used in MIR applications *(3)*, to provide figures-of-merit for native phases, and a set of Hendrickson-Lattman coefficients *(59,60)*. The figures-of-merit can be used in computing weighted electron density maps and the coefficients for phase combination.

Since all the information that is required—amplitudes $|^0F_T(h)|$, and phases $^0\phi_T(h)$—is now available, one can compute an electron density map, although an ambiguity pertaining to choice of the correct enantiomorph remains. This might be resolved in the usual manner *(3)* by assuming an arbitrary hand for the anomalous diffractor distribution, calculating phases, and a subsequent map based on these phases and assaying the map for appearance of expected macromolecular features. Choice of correct hand for anomalous diffractor distribution will yield a map that is readily interpretable in terms of known chemical connectivity of the macromolecule under investigation. Additional criteria, such as appearance of expected noncrystallographic symmetry, may also be used instead of, or in addition to, the map interpretability criterion *(9)*. An atomic model of the macromolecule may then be built into electron density and the resulting set of coordinates refined by suitable methods. Excellent reviews documenting the details of these processes are available *(61–64)*.

5. Prospects

Although a number of structures have already been determined using MAD phasing and more are being worked on, one of the limitations to the wider applications of the MAD method is paucity of experimental time at synchrotron installations. Efforts are under way to build more synchrotron sources *(65,66)*, as well as more MAD stations at existing synchrotron sources *(67)*. Efforts are also being made to develop laboratory sources of multiple-wavelength X-radiation *(19,20)* for use in MAD phasing. It has been mentioned earlier that absorption edges of atoms with fewer electrons than Ca are not suitable for MAD experiments for technical reasons. One of these reasons is the large absorption cross-section of atoms that are normally found in biological macromolecules for X-radiation in the wavelength range corresponding to their absorption edges *(68)*. If the absorption edge of sulfur (5.07 Å) can be used, then a substantial generalization of the MAD method becomes possible for naturally occurring macromolecules. Crystals of proteins that contain methionine residues could then be used directly for MAD experiments without Se substitution and, moreover, the S atoms in cysteine residues would also become potentially useful. The elegant work on crambin *(39)* has demonstrated the considerable power of anomalous diffraction from native sulfur as a phasing vehicle, even though diffraction measurements in that case were made far from a wavelength optimal for maximization of anomalous diffraction signals. Hence, it seems plausible to expect that macromolecules with a far smaller relative abundance of sulfur than crambin would become amenable to MAD analysis, if diffraction measurements could be made closer to the sulfur absorption edge. Similarly, if the absorption edge of phosphorus (5.8 Å) can be used, nucleic acids and nucleic acid-containing macromolecular systems with a small number of P atoms become potentially accessible to MAD phasing experiments. Very promising results in this direction have already been obtained *(69)*.

Acknowledgments

I thank Wayne Hendrickson for an informal review of the manuscript. Thanks are also owed to Charles Grubmeyer, K. Balendran, and K. Usha for helpful comments. The work on *Clostridium acidi-urici* ferredoxin referred to was supported by grants GM34102 (to Hendrickson) and GM28358 (to Orme-Johnson), and was carried out at the Stanford Synchrotron Radiation Laboratory, whose facilities are supported by the US Department of Energy.

References

1. Durbin, S. D. and Feher, G. (1990) Studies of crystal growth mechanisms of proteins by electron microscopy. *J. Mol. Biol.* **212,** 763–774.
2. McPherson, A. (1982) *Preparation and Analysis of Protein Crystals.* Wiley, New York.

3. Blundell, T. L. and Johnson, L. N. (1976) *Protein Crystallography.* Academic, London.
4. Singh, A. K. and Ramaseshan, S. (1968) The use of neutron scattering in crystal structure analysis. I. Non-centrosymmetric structures. *Acta. Cryst.* **B24,** 35–39.
5. Karle, J. (1980) Some developments in anomalous dispersion for the structural investigation of macromolecular systems in biology. *Int. J. Quant. Chem: Quant. Biol. Symp.* **7,** 357–367.
6. Kahn, R., Fourme, R., Bosshard, R., Chiadmi, M., Rister, J. L., Dideberg, O., and Wery, J. P. (1985) Crystal structure study of *Opsanus tan* parvalbumin by multiwavelength anomalous diffraction. *FEBS Lett.* **179,** 133–137.
7. Phizackerly, R. P., Cork, C. W., and Merritt, E. A. (1986) An area detector data acquisition system for protein crystallography using multiple energy anomalous dispersion techniques. *Nucl. Instrum. Methods Phys. Res.* **A246,** 579–595.
8. James, R. W. (1982) *The Optical Principles of the Diffraction of X-rays.* Ox Bow Press, Woodbridge, CT.
9. Krishna Murthy, H. M., Hendrickson, W. A., Orme-Johnson, W. H., Merritt, E. A., and Phizackersly, R. P. (1988) Crystal structure of *Clostridium acidi-urici* ferredoxin at 5-Å resolution based on measurements of anomalous X-ray scattering at multiple wavelengths. *J. Biol. Chem.* **263,** 18,430–18,436.
10. Ramaseshan, S. (1964) The use of anomalous scattering in crystal structure analysis, in *Advanced Methods in Crystallography* (Ramachandran, G. N., ed.), Academic, London, pp. 67–95.
11. Fourme, R. and Hendrickson, W. A. (1990) Analysis of macromolecular structures by the method of multiwavelength anomalous diffraction, in *Synchrotron Radiation and Biophysics* (Hasnain, S. S., ed.), Ellis Harwood, Chichester, pp. 156–175.
12. Hendrickson, W. A. (1991) Determination of macromolecular structures from anomalous diffraction of synchrotron radiation. *Science* **254,** 51–58.
13. Hendrickson, W. A., Smith, J. L., and Sheriff, S. (1985) Direct phase determination based on anomalous scattering, in *Methods in Enzymol.*, vol. 115 (Wyckoff, H. W., Hirs, C. H. W., and Timasheff, S. N., eds.), Academic, New York, pp. 41–55.
14. Hendrickson, W. A. (1985) Analysis of protein structure from diffraction measurements at multiple wavelengths. *Trans. Am. Cryst. Assoc.* **21,** 11–21.
15. Cromer, D. T. and Waber, J. T. (1974) Atomic scattering factors for X-rays, in *International Tables for X-ray Crystallography,* vol. IV (Lonsdale, K. D., ed.), Reidel, Dordrecht, Holland, pp. 71–147.
16. McPherson, A. (1985) Crystallization of macromolecules: General principles, in *Mets. in Enzymol.*, vol. 114 (Wyckoff, H. W., Hirs, C. H. W., and Timasheff, S. N., eds.), Academic, New York, pp. 112–120.
17. Hoppe, W. and Jakubowski, V. (1975) The determination of phases of erythrocruorin using the two wavelength method with iron as anomalous scatterer, in *Anomalous Scattering* (Ramaseshan, S. and Abrahams, S. C., eds.), Academic, New York, pp. 437–461.
18. Hendrickson, W. A., Troup, J. M., Swepston, P. N., and Zdanski, G. (1986) Structure of D-selenolanthionine determined directly from multiwavelength anomalous diffraction of bremstrahlung. *Abstract Amer. Cryst. Assoc.* **Ser. 2, 14,** 48.

19. DeTitta, G. T., Swenson, D. C., Han, F., and Pangborn, W. A. (1990) Preliminary results with a dual target sealed X-ray tube as a tool for anomalous scattering at multiple wavelengths. Abs # C01, Am. Cryst. Assoc. Meeting, April 1990.
20. Xuong, N.-H., Sullivan, D., Nielson, C., Dai, X., and Ashford, V. (1990) A multi-wavelength diffractometer using the L emission lines of a heavy metal. Abstract # C02, Am. Cryst. Assn. Meeting, April 1990.
21. Greenhough, T. J. and Helliwell, J. R. (1983) The uses of synchrotron X-radiation in the crystallography of molecular biology. *Prog. Biophys. Mol. Biol.* **41,** 67–123.
22. Fourme, R. and Kahn, R. (1985) A rotation camera used with a synchrotron radiation source in, *Methods in Enzymol.*, vol. 114, (Wyckoff, H. W., Hirs, C. H. W., and Timasheff, S. N., eds.), Academic, New York, pp. 281–299.
23. Moffat, K. (1989) Time resolved macromolecular crystallography. *Ann. Rev. Biophys. Biophys. Chem.* **18,** 309–332.
24. Mills, D. M. (1984) Time-resolution experiments using X-ray synchrotron radiation. *Phys. Today* **37(4),** 22–30.
25. Gruner, S. M. (1987) Time-resolved X-ray diffraction of biological materials. *Science* **238,** 305–312.
26. Hendrickson, W. A., Pahler, A., Smith, J. L., Satow, Y., Merritt, E. A., and Phizackerly, R. P. (1989) Crystal structure of core streptavidin determined from multiwavelength anomalous diffraction of synchrotron radiation. *Proc. Natl. Acad. Sci.* **86,** 2190–2194.
27. Weis, W. I., Kahn, R., Fourme, K., Drickamer, K., and Hendrickson, W. A. (1991) Structure of the calcium-dependent domain from a rat mannose binding protein determined by MAD phasing. *Science* **254,** 1608–1615.
28. Rossmann, M. G. (1990) The molecular replacement method. *Acta. Cryst.* **A46,** 73–82.
29. Hendrickson, W. A., Horton, J. R., Krishna Murthy, H. M., Pahler, A., and Smith, J. L. (1990) Multiwavelength anomalous diffraction as a direct phasing vehicle in macromolecular crystallography, in *Synchrotron Radiation in Biology* (Sweet, R. W., ed.), Plenum, New York, pp. 317–324.
30. Hendrickson, W. A., Horton, J. R., and LeMaster, D. M. (1990) Selenomethionine proteins produced for analysis by multiwavelength anomalous diffraction (MAD): a vehicle for direct determination of three-dimensional structure *EMBO J.* **9,** 1665–1672.
31. Graves, B. J., Hatada, M. H., Hendrickson, W. A., Miller, J. K., Madison, V. S., and Satow, Y. (1990) Structure of interleukin 1 α at 2.7-Å resolution. *Biochemistry* **29,** 2679–2684.
32. Yang, W., Hendrickson, W. A., Crouch, R. J., and Satow, Y. (1990) Structure of ribonuclease H phased at 2 Å by MAD analysis of the selenomethionine protein. *Science* **249,** 1398–1403.
32a. Korszun, Z. R. (1987) The tertiary structure of azurin from pseudomonas denitrificans as determined by Cu resonant diffraction using synchrotron radiation. *J. Mol. Biol.* **196,** 413–419.
32b. Ramakrishnan, V., Finch, J. T., Graziano, V., Lee, P. L., and Sweet, R. M. (1993) Crystal structure of globular domain of histone H5 and its implications for nucleo-

Crystal structure of globular domain of histone H5 and its implications for nucleosome binding. *Nature* **362,** 219–223.

33. Ogata, C., Hendrickson, W. A., Gao, X., Satow, Y., and Amemia, Y., Structure of 23 chromomycin:DNA complex based on bromine MAD data detected with imaging plates, in preparation.
34. Cromer, D. T. and Lieberman, D. (1970) Relativistic calculation of anomalous scattering factors for X-rays. *J. Chem. Phys.* **53,** 1891–1898.
35. Elder, M. (1985) Photographic science and microdensitometry in X-ray diffraction data collection, in *Mets. in Enzymol.*, vol. 114 (Wyckoff, H. W., Hirs, C. H. W., and Timasheff, S. N., eds.), Academic, New York, pp. 199–210.
36. Guss, J. M., Merritt, E. A., Phizackerly, R. P., Hedman, B., Murata, M., Hodgson, K. O., and Freeman, H. C. (1988) Phase determination by multi-wavelength X-ray diffraction: Crystal structure of a basic "blue" copper protein from cucumbers. *Science* **241,** 806–811.
37. Arndt, U. W. (1985) Television area X-ray detectors, in *Methods in Enzymology*, vol. 114 (Wyckoff, H. W., Hirs, C. H. W., and Timasheff, S. N., eds.), Academic, New York, pp. 472–485.
38. Weissman, L. (1982) Strategies for extraction of isomorphous and anomalous signals, in *Computational Crystallography*, Sayre, D., ed., Clarendon, Oxford, pp. 56–64.
39. Hendrickson, W. A. and Teeter, M. M. (1981) Structure of the hydrophobic protein crambin determined directly from the anomalous scattering of sulphur. *Nature* **290,** 107–113.
40. Karle, J. (1984) The relative scaling of multiple-wavelength anomalous dispersion data. *Acta. Cryst.* **A40,** 1–4.
41. Lonsdale, K. D. (ed.) (1985) *International Tables for X-ray Crystallography, vol. III.* Reidel Publishing Company, Dordrecht, Holland.
42. Brown, G. S. (1980) Extended X-ray absorption fine structure in condensed materials, in *Synchrotron Radiation Research* (Winnick, H. and Doniach, S., eds.), Plenum, New York, pp. 387–400.
43. Strohr, J. (1980) EXAFS and surface EXAFS: Principles, analysis and applications. SSRL report 80/07, SSRL, Stanford University, CA.
44. Cusatis, C. and Hart, M. (1975) Dispersion correction measurements by X-ray interferometry, in *Anomalous Scattering* (Ramaseshan, S. and Abrahams, S. C., eds.), Academic, New York, pp. 57–68.
45. Phillips, J. C. and Hodgson, K. O. (1980) The use of anomalous scattering effects to phase diffraction patterns from macromolecules. *Acta. Cryst.* **A36,** 856–864.
46. Narayan, R. and Ramaseshan, S. (1981) Optimum choice of wavelengths in the anomalous scattering technique with synchrotron radiation. *Acta. Cryst.* **A37,** 636–641.
47. Arndt, U. W., Greenhough, T. J., Helliwell, J. R., Howard, J. A. K., Rule, S. A., and Thompson, A. W. (1982) Optimized anomalous dispersion in crystallography: a synchrotron X-ray polychromatic simultaneous method. *Nature* **298,** 835–838.
48. Hope, H. (1988) Cryocrystallography of Biological Macromolecules: a Generally Applicable Method, *Acta. Cryst.* **B44,** 22–26.

(Wyckoff, H. W., Hirs, C. H. W., and Timasheff, S. N., eds.), Academic, New York, pp. 112–117.

50. Wang, B.-C. (1985) Resolution of phase ambiguity in macromolecular crystallography, in *Methods in Enzymology*, vol. 115, (Wyckoff, H. W., Hirs, C. H. W., and Timasheff, S. N., eds.), Academic, New York, pp. 90–112.
51. North, A. C. T., Phillips, D. C., and Mathews, F. S. (1968) A Semi-empirical method of absorption correction, *Acta. Cryst.* **A24,** 351–359.
52. Howard, A. J., Nielson, C., and Xuong, N. H. (1985) Software for a diffractometer with a multiwire area detector, in *Methods in Enzymology*, vol. 114 (Wyckoff, H. W., Hirs, C. H. W., and Timasheff, S. N., eds.), Academic, New York, pp. 452–472.
53. Templeton, L. K. and Templeton, D. H. (1988) Biaxial tensors for anomalous scattering of X-rays in Selenolanthionine, *Acta. Cryst.* **A44,** 1045–1051.
54. Sowadski, J. M., Foster, B. A., and Wyckoff, H. W. (1981) Structure of alkaline phosphatase with zinc/magnesium cobalt or cadmium in the functional metal sites. *J. Mol. Biol.* **150,** 245–272.
55. Hendrickson,W. A. (1976) Radiation Damage in Crystallography. *J. Mol. Biol.* **106,** 889–893.
56. Fletterick, R. J. and Sygusch, J. (1985) Measuring X-ray diffraction data from large proteins with X-ray diffractometry, in *Methods in Enzymology*, vol. 114, (Wyckoff, H. W., Hirs, C. H. W., and Timasheff, S. N., eds.), Academic, New York, pp. 386–397.
57. Germain, G., Main, P., and Woolfson, M. M. (1971) The application of phase relationships to complex structures III. The optimum use of phase relationships. *Acta. Cryst.* **A27,** 368–376.
58. Pahler, A., Smith, J. L., and Hendrickson, W. A. (1990) A probability representation for phase information from multiple wavelength anomalous dispersion. *Acta. Cryst.* **A46,** 537–540.
59. Chiadmi, M., Kahn, R., De La Fortelle, E., and Fourme, R. (1993) Derivation by stastistical methods of phase information from multiple-wavelength anomalous diffraction data. Basic questions, "best" electron-density map, implementation and tests. *Acta. Cryst.* **D49,** 522–529.
60. Hendrickson, W. A. and Lattman, E. E. (1970) Representation of phase probability distributions for simplified combination of independent phase information. *Acta. Cryst.* **B26,** 136–143.
61. Jones, T. A. (1985) Interactive computer graphics: FRODO, in *Methods in Enzymology*, vol. 115 (Wyckoff, H. W., Hirs, C. H. W., and Timasheff, S. N., eds.), Academic, New York, pp. 157–171.
62. Richardson, J. S. and Richardson, D. C. (1985) Interpretation of electron density maps, in *Methods in Enzymology*, vol. 115 (Wyckoff, H. W., Hirs, C. H. W., and Timasheff, S. N., eds.), Academic, New York, pp. 179–206.
63. Hendrickson, W. A. (1985) Stereochemically restrained refinement of macromolecular structures, in *Methods in Enzymology*, vol. 115 (Wyckoff, H. W., Hirs, C. H. W., and Timasheff, S. N., eds.), Academic, New York, pp. 252–271.
64. Deisenhofer, J., Remington, S. J., and Steigemann, W. (1985) Experience With Various Techniques for The Refinement of Protein Structures, in *Methods in Enzy-*

mology, vol. 115 (Wyckoff, H. W., Hirs, C. H. W., and Timasheff, S. N., eds.), Academic, New York, pp. 303–324.
65. Hamilton, D. P. (1990) Advanced photons. *Science* **249,** 21.
66. Sietman, R. (1990) Doris gets a face-lift. *Science* **249,** 26.
67. Staudenmann, J.-L., Hendrickson, W. A., and Abramowitz, R. (1989) The synchrotron resource of the Howard Hughes Medical Institute. *Rev. Sci. Instrum.* **60,** 1939–1942.
68. Boulin, C., Buldt, G., Dauvergne, F., Gabriel, A., Goerigk, G., Munk, B., and Stuhrmann, H. B. (1990) Anomalous scattering in membrane studies, in *Synchrotron Radiation in Biology* (Sweet, R. W., ed.), Academic, New York, pp. 83–92.
69. Lehman, M. S., Muller, H. H., and Stuhrmann, H. B. (1993) Protein single-crystal diffraction with 5 Å synchrotron X-rays at the sulphur K-absorption edge. *Acta. Cryst.* **D49,** 308–310.

CHAPTER 6

Structure Determination Using Isomorphous Replacement

Sherin S. Abdel-Meguid

1. Introduction

1.1. General

The determination of the three-dimensional structure of molecules using single crystal X-ray diffraction techniques requires the measurement of amplitudes and the calculation of phases for each diffraction point (maximum). Although amplitudes can be directly measured from diffracting crystals, phases are indirectly determined, because there are no lenses that can bend and focus X-rays. Thus, methods were developed to calculate phases from the intensities of the diffracted waves. Isomorphous replacement is the most widely used method for *ab initio* phase determination of macromolecules. Its first successful application to large biological molecules was undertaken in 1954 by Perutz and coworkers *(1)* while studying hemoglobin. Since then, this method has played a central role in the determination of almost all unique protein and nucleic acid structures, and it is likely to retain such a role in the foreseeable future.

The technique of isomorphous replacement requires the introduction of atoms of high atomic number (heavy atoms), such as mercury, platinum, uranium, and so forth, into the macromolecule under study without disrupting its structure or packing in the crystal. Thus, a perfect isomorphous derivative is one in which the only change between it and the native molecule is the incorporation of one or more heavy atoms. This is commonly done by soaking crystals of native molecules in a solution containing the

From: *Methods in Molecular Biology, Vol. 56: Crystallographic Methods and Protocols*
Edited by: C. Jones, B. Mulloy, and M. Sanderson Humana Press Inc., Totowa, NJ

desired heavy atom. The binding of these atoms to functional groups of macromolecules is facilitated by the presence of large liquid channels in protein and nucleic acid crystals into which these functional groups protrude.

The addition of one or more heavy atoms to a macromolecule introduces differences in the diffraction pattern of the derivative relative to that of the native. If this addition is truly isomorphous, these differences will represent the contribution from the heavy atoms only; thus, the problem of determining atomic positions is initially reduced to locating the position of a few atoms. Once the positions of these atoms are accurately determined, they are used to calculate a set of phases for data measured from the native crystals. Although theoretically one needs only two isomorphous derivatives to determine the three-dimensional structure of biological macromolecules, in practice one needs more than two, owing to errors in data measurement and scaling and in heavy-atom positions, as well as lack of isomorphism.

The steps required for the determination of the three-dimensional structure of macromolecules using the isomorphous replacement technique are as follows:

1. The growing of crystals (*see* Chapter 2).
2. Characterization of crystals (space group and lattice constants) (*see* Chapter 3).
3. Preparation of heavy-atom derivatives.
4. Measurement of diffraction data from native and derivative crystals (*see* Chapter 4).
5. Reduction, correction, and scaling of data.
6. Determination of heavy-atom positions.
7. Refinement of heavy-atom positions and phase determination.
8. Calculation of electron density.
9. Interpretation of the structure and chain tracing.
10. Model building.
11. Structure refinement (Chapter 9).

This chapter, an overview of the isomorphous replacement technique, will focus on steps 3, 6, and 7.

1.2. Theoretical Aspects

This section is a brief overview of the theoretical aspects of isomorphous replacement. In theory, isomorphous replacement phasing of biological macromolecules requires the measurement of three X-ray diffraction data sets: a native and two derivatives. Each diffraction maximum is reduced to a structure factor amplitude with unknown phases. Native and derivative structure factors *(F)* are related as shown in the following equation:

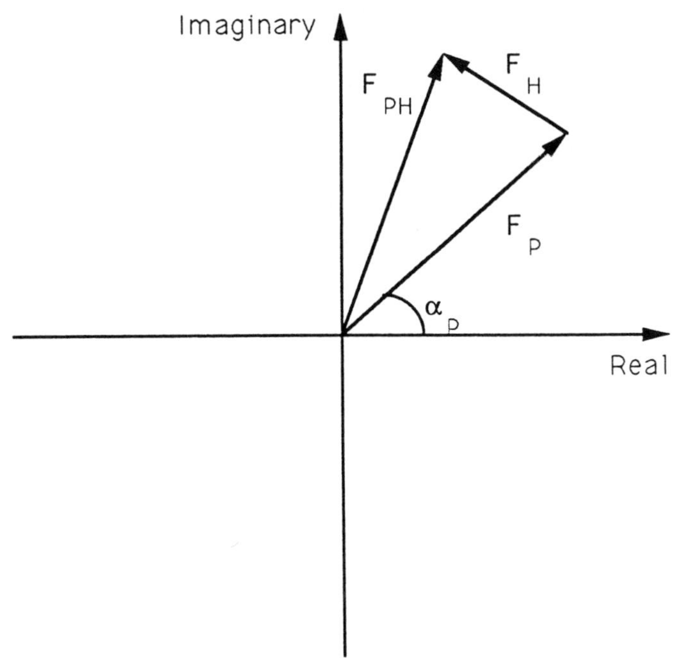

Fig. 1. Vector diagram (Argand diagram) showing the relationships between heavy-atom derivative (F_{PH}), native protein (F_P), and heavy-atom (F_H); α is the phase angle for the native protein. The vectors are plotted in the complex plane.

$$F_{PH} = F_P + F_H \quad (1)$$

where F_{PH}, F_P, and F_H are the structure factors of the derivative, the native protein, and the heavy atom, respectively. Each of these quantities is a vector (Fig. 1) and can be described in terms of an amplitude F and a phase α. Once the heavy-atom position has been determined, its structure factor amplitude F_H and phase α_H can be calculated. Since the structure factor amplitudes for the native (F_P) and derivative (F_{PH}) are experimentally measured quantities, it is thus possible to calculate the protein phase angle α_P from the following equation:

$$F_{PH}^2 = F_P^2 + F_H^2 + 2F_PF_H \cos(\alpha_P - \alpha_H) \quad (2)$$

or

$$\alpha_P = \alpha_H + \cos^{-1}[(F_{PH}^2 - F_P^2 - F_H^2)/2F_PF_H] = \alpha_H \pm \alpha' \quad (3)$$

From Eq. (3) and Fig. 2A it is clear that with only one heavy-atom derivative (single isomorphous replacement; SIR), the resultant phase

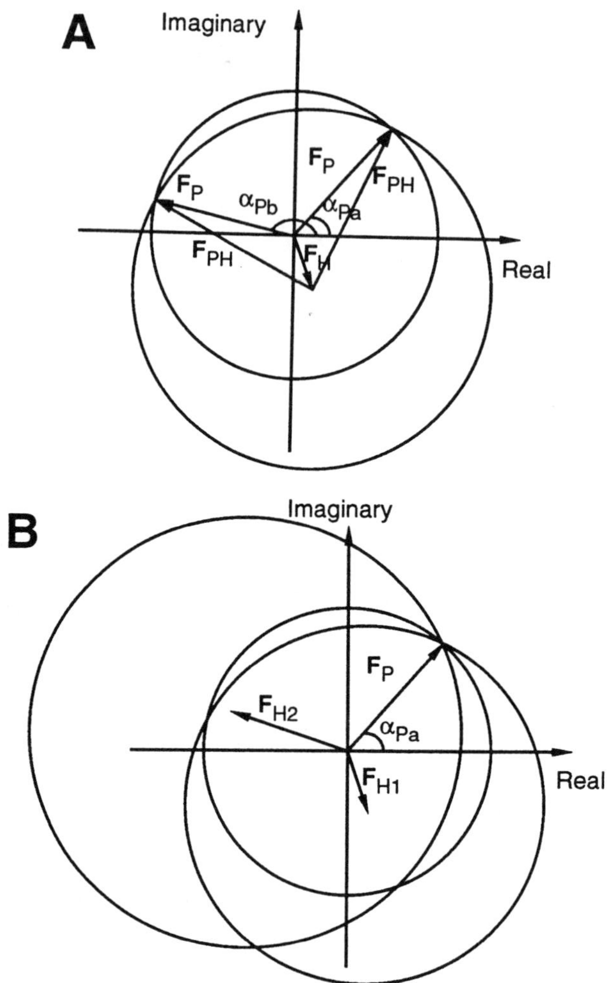

Fig. 2. Isomorphous replacement phase determination (Harker construction). **(A)** Single isomorphous replacement. The circle with radius F_{PH} represents the heavy-atom derivative, whereas that with radius F_P represents the native protein. Note that the circles intersect at two points causing an ambiguity in the phase angle; α_{Pa} and α_{Pb} represent the two possible values. **(B)** Double isomorphous replacement. The same construction as that in single isomorphous replacement except that an additional circle with radius F_{PH2} (vector not shown for simplicity) has been added to represent a second heavy-atom derivative. Note that all three circles (in the absence of errors) intersect at one point thus eliminating the ambiguity in the protein phase angle α_P. F_{H1} and F_{H2} represent the heavy atom vectors for their respective derivatives.

will have two values (α_{Pa} and α_{Pb}); one of these phases will represent that of one structure and the other of its mirror image. But since proteins contain only L-amino acids, this phase ambiguity must be eliminated using a second derivative, as shown diagrammatically in Fig. 2B.

Although theoretically one needs only two derivatives, in practice three or more derivatives are required, owing to errors introduced in the data and heavy-atom positions. Hence, this method is commonly referred to as multiple isomorphous replacement (MIR). However, one or more of these derivatives may be replaced by the anomalous component of the heavy atom or by solvent leveling (*2*; Chapter 4). It is thus possible to determine a macromolecule structure using SIR phasing, given a heavy atom with an anomalous scattering component that gives large differences, accurate data measurement, and solvent leveling treatment.

Once the phase angle α_P has been determined for every diffraction maximum *(hkl)*, a Fourier synthesis is used to compute the electron density (ρ) at each position *(xyz)* in the unit cell (the repeating unit forming the crystal lattice) using equation:

$$\rho(xyz) = 1/V \, \Sigma_h \Sigma_k \Sigma_l \, F_P(hkl) \, e^{-2\pi i(hx+ky+lz)} \tag{4}$$

where V is the volume of the unit cell, i is the imaginary component $\sqrt{-1}$, and

$$F_P = F_P e^{i\alpha_P} = F_P \cos\alpha_P + iF_P \sin\alpha_P \tag{5}$$

for every *(hkl)*. This electron density map (ρ *[xyz]*) can then be interpreted in terms of a three-dimensional atomic model.

2. Materials

2.1. Protein Crystallization

A comprehensive review and discussion of the materials required for protein crystallization can be found in Chapter 2 and in ref. *3*.

2.2. Heavy-Atom Derivatives

An assortment of reagents containing heavy atoms is needed. Table 1 lists many of the heavy atoms used in the successful determination of the structures of macromolecules. Reagents containing each of these elements are commercially available.

2.3. Measurement of Diffraction Data

For investigators who do not have available a laboratory well equipped for X-ray data acquisition, it is best to collaborate with a protein crystal-

Table 1
Elements Commonly Used in Heavy-Atom Derivative Preparation[a]

Elements	Atomic number	Elements	Atomic number
Palladium (Pd)[b]	46	Platinum (Pt)	78
Silver (Ag)[b]	47	Gold (Au)	79
Iodine (I)[c]	53	Mercury (Hg)	80
Lanthanides (La–Lu)[d]	57–71	Lead (Pb)	82
Rhenium (Re)	75	Thorium (Th)	90
Osmium (Os)	76	Uranium (U)	92
Iridium (Ir)	77		

[a]Many of the reagents that contain the elements below are listed in ref. *5*.

[b]Because of their low atomic numbers, these are only recommended for small proteins.

[c]Has been used to iodinate tyrosines *(22–24)*. However, often causes crystal disorder and can be difficult to locate in a difference Patterson owing to its difuse electron cloud.

[d]Many can be used to replace bound magnesium or calcium.

lographer or obtain time from one of the national synchrotron facilities. There are a number of such facilities around the world equipped for protein crystallographic studies. For a review of the equipment required for data acquisition, *see* the section on data collection in Chapter 5 and Wyckoff et al. *(4)*. The minimum requirements are a diffractometer or an area detector, a precession camera, and an X-ray generator.

3. Methods

3.1. Isomorphous Heavy-Atom Derivatives

The reader may wish to consult two additional references regarding the preparation of derivatives and selection of reagents; namely, the chapter on preparation of heavy-atom derivatives in Blundell and Johnson *(5)*, those in Petsko *(6)* for proteins, and by Kim et al. *(7)* and Holbrook and Kim *(8)* for nucleic acids.

3.1.1. Preparation of Heavy-Atom Derivatives

The mechanics of derivative preparation are simple; they involve the transfer of one or more native crystals to a solution containing the desired heavy atom. The solution is usually the same as that used to store native crystals (mother liquor), differing only in the presence of a compound containing a heavy atom. Soaking times are usually on the order of 1–4 d, but can be as short as a few hours. If no binding is detected in such a time, longer soaking times may be necessary; soaking of several weeks have been reported. Soaking times are dependent on temperature and heavy-atom compound concentration; at lower temperatures and heavy-atom concentrations, it may be necessary to soak for longer periods of time. The concentration of the heavy-atom reagent used for derivative preparation will depend on its solubility in the mother liquor; however, 1 m*M* is a good starting value. Concentrations as low as 0.05 m*M* and as high as 100 m*M* have been reported. The ideal derivative is arrived at by varying soaking time and heavy-atom compound concentration. The latter variable is more useful, as mass action can force the formation of a derivative, even in the case of weak binding functional groups. Much of the initial scanning can be done visually, using small crystals, by observing deterioration (cracking) of the crystals. Concentrations of the heavy-atom reagent and soaking times should be adjusted to ensure that the crystals do not show serious cracks (minor surface cracks may not be detrimental to some crystals). Soaking times as short as 1 h combined with concentrations of 0.3 m*M* were reported *(9)* to produce good mercury derivatives of iron superoxide dismutase. For successful derivative preparation, two additional points should be considered: composition of mother liquor and pH. Many of the buffers, additives, and precipitants used in mother liquors, such as tris, phosphate, citrate, β-mercaptoethanol, dithiothreitol, and NH_3 derived from ammonium sulfate at high pH may compete with the protein for heavy-atom binding. It may be necessary at times to transfer crystals into a more appropriate mother liquor before derivative preparation. For example, crystals grown out of ammonium sulfate may be transferred to lithium sulfate to avoid the formation of metal ammonia complexes, and salts may possibly be replaced by polyethylene glycol (PEG). Such changes in mother liquor are best done incrementally and slowly, to avoid shocking the crystals. Also, one should recognize that the solubility of heavy-atom reagents in the mother liquor and their binding to functional groups on the protein are pH dependent. Ideal pH range is 6–8; lower pH may result in protonation of glutamic and aspartic acids of proteins, whereas at higher pH many heavy-atom reagents are labile and form insoluble hydroxides.

Although the soaking method for heavy-atom derivative preparation is by far the simplest and most common method, it is not the only one used. One can first derivatize the macromolecule, then crystallize. This procedure is less frequently used because of possible drawbacks, such as the inability to produce isomorphous crystals owing to the disruption of intermolecular contacts by the heavy atoms. Other frequent problems are the introduction of additional heavy-atom sites (a potential complicating factor in phasing) by exposing sites hidden by crystal contacts, and change in the solubility of the derivatized macromolecule, thus having to search again for conditions suitable for isomorphous crystallization. However, this method is preferable if one wishes to derivatize proteins using a ligand, substrate, or inhibitor containing a heavy atom, assuming that it is large and cannot intercalate in the crystal solvent channels.

The aforementioned two methods for derivative preparation have been successfully used in the phasing of both nucleic acids and proteins; an additional method, however, has been used for phasing of nucleic acids, in which the heavy atom is synthetically incorporated into the molecule. For example, Drew et al. *(10)* determined the structure of $d(CGCG)_2$ by incorporating 5-bromocytosine into the synthesis of their nucleic acid and then using the bromine atoms for phasing.

3.1.2. Selection of Heavy-Atom Reagents

Both the size and chemical composition of the molecule under investigation are important criteria to consider when selecting heavy-atom reagents for derivatization. One's choice must ensure that the differences in diffraction amplitudes owing to heavy-atom contributions are larger than the errors in data measurement. The size of the heavy atom (atomic number) and the number of sites required for successful phasing are proportional to the size (mol wt) of the macromolecule. Thus, larger molecules may require not only atoms of high atomic number, but also more than one heavy atom per molecule. In the case of proteins, especially small ones, inspection of the amino acid composition can give valuable insights into which reagents should be tried first. For example, if the protein contains no free cysteines or histidines it may be best to start soaking with compounds other than mercurials, or to genetically engineer heavy-atom binding sites, as was done with the catalytic domain of λδ resolvase *(11,12)*. However, assuming normal distribution of amino acids, one should begin with platinum compounds such as K_2PtCl_4 (the most widely successful heavy atom reagent), which binds mainly to methionine, histidine, and cysteine residues. Petsko et al. *(13)* have described the chemistry of this reagent in a variety of crystal mother liquors. They also concluded that most other platinum compounds react with proteins in a similar fashion, except for those containing $Pt(CN)_2^{2-}$ which bind to positively charged residues. Mercurial compounds are the second most successful group of reagents in derivative preparation; most mercurials either bind to cysteine sulfurs or histidine nitrogens. In addition to platinum and mercurial reagents, many compounds containing palladium, silver, rhenium, osmium, iridium, gold, lead, lanthanides, or uranium have been used successfully in isomorphous replacement phasing (Table 1). The latter two, in addition to platinum and mercury, have also been used successfully in nucleic acid phasing.

3.1.3. Assessment of Derivative Formation

As might be expected, not every crystal soaked in a solution containing a heavy-atom reagent will be a derivative. In order to determine whether the macromolecules in the crystals at hand have been derivatized, a series of diffraction experiments must be undertaken. In such experiments, a crystal of the putative derivative is inserted approx 1 cm from one end of a thin glass capillary (about 1.0 mm in diameter); at the other end a small drop of mother liquor is deposited to ensure that the crystal does not dry, and the capillary is sealed (*see* Chapter 3). The capillary is then positioned on a precession camera, a diffractometer or an area detector in such a way that the crystal always remains in the path of the X-ray beam. If a precession camera is used, a set of alignment photographs are taken to insure that a particular orientation (zone) of the crystal is presented to the photographic film, then a precession photograph is taken of that orientation (for a detailed description of precession photography, *see* Chapter 3 and ref. *14*). The diffraction pattern on this photograph is then compared with that of the exact orientation obtained from a native crystal. If side-by-side inspection of the two photographs reveals differences in the intensity of some diffraction spots, then the derivatization of the macromolecule has been successful. Similar assessment of derivative formation can be accomplished (using a diffractometer or an area detector instead of a precession camera) by measuring a set of low-resolution data from a crystal of the putative derivative and comparing it to a set obtained from a native crystal. Significant differences between the two data sets indicate successful derivative formation.

3.2. Determination of Heavy-Atom Positions

By far, the most common procedure for the determination of heavy-atom positions is the difference Patterson method; it is often used in combination with the difference Fourier technique to locate sites in second and third derivatives.

3.2.1. Difference Patterson Method

The Patterson function *(15)* is a phaseless Fourier summation similar to that in Eq. (4) but employing F^2 as coefficients, thus it can be calculated directly from the experimentally measured amplitudes (F_P) without the need to determine the phase angle. In the case of macromolecules, $(F_{PH}-F_P)^2$ are used as coefficients in Eq. (4) to produce a Patterson map (hence the name difference Patterson). Such a map contains peaks of vectors between atoms (interatomic vectors). Thus, in the case of a difference Patterson of macromolecules, it is a heavy-atom vector map. For example, if a structure has an atom at position (0.25,0.11,0.32) and another atom at position (0.10,0.35,0.15), there will be a peak in the Paterson map at position (0.25–0.10,0.11–0.35,0.32–0.15), namely a peak at (0.15,–0.24,0.17).

The interpretation of Patterson maps requires knowledge of crystallographic symmetry and space groups; Chapter 3 offers a concise review of this topic. The ease of interpretation of these maps depends on the quality of the data, the degree of isomorphism, the number of heavy-atom sites per macromolecule, and the degree of substitution for each heavy atom. An ideal case is one in which:

1. The native and derivative data are of very good quality;
2. The derivative shows a high degree of isomorphism;
3. Only one highly substituted heavy atom is present per macromolecule; and
4. The heavy atom is of sufficiently high atomic number to give significant differences.

The following is a simple example of how to interpret a Patterson map. Assume that the macromolecule under investigation crystallizes in the orthorhombic space group P222, the derivative contains only one heavy-atom site per molecule and the heavy atom is located at position a in the unit cell with atomic coordinates (0.1,0.2,0.3). Since in the P222 space group there are four molecules per unit cell related by three mutually perpendicular twofold axes, then the other three atoms, *b*, *c*, and *d*, will be at positions (–0.1,–0.2,0.3), (0.1,–0.2,–0.3), and (–0.1,0.2,–0.3), respectively. Thus, in a Patterson map, one should find three prominent unique peaks at (0.2,0.4,0), (0,0.4,0.6), and (0.2,0,0.6), representing the vectors *(a-b)*, *(a-c)*, and *(a-d)*. Note that each of these vectors is found on sections perpendicular to unit cell axes at 0 (Harker sections). Also, note that other vectors such as *(b-c)*, *(b-d)*, and *(c-d)* are related to the above three by symmetry and that self vectors will pile up at the origin (0,0,0). Thus, if inspection of the crystallographically unique portion of the Patterson map reveals only three prominent peaks at positions (0.2,0.4,0), (0,0.4,0.6), and (0.2,0,0.6), then one can conclude that this derivative contains only one heavy atom at position (0.1,0.2,0.3).

Now, assume that there are two heavy-atom sites per molecule. Then every Harker section will contain two peaks, each representing a vector between a site and its symmetry-related mate. In addition, a set of peaks representing the vectors between the two heavy atoms will be present. These, however, will not necessarily be on Harker sections, but most likely will be in general positions. It should be clear (from the preceding examples) that, as the number of heavy-atom sites increases, the Patterson map becomes more and more complex, and many of the interatomic vectors become indistinguishable from noise. Although these maps may not be interpretable, identification of the heavy-atom positions from such derivatives may still be possible using difference Fourier techniques (*see* Section 3.2.2.), assuming the existence of other derivatives with interpretable Patterson maps. At times, however, it is advisable to try to produce crystals of this derivative that contain less sites per molecule. This can be achieved by soaking the crystals for a shorter time in a less concentrated solution of the heavy-atom reagent.

Figure 3 shows a Harker section from the difference Patterson function for the K_2OsCl_6 single-site derivative of porcine growth hormone *(16)*. This protein crystallizes in the space group $P3_221$ and contains six molecules in the unit cell *(17)*. If one assumes that the single site of this derivative is at position *(x,y,z)*, then the positions of the interatomic vectors representing symmetry-related heavy atoms in the unit cell will be those listed in Table 2. Inspection of the Harker section at 1/3 (Fig. 3) reveals six significant peaks, all of which are related by symmetry; thus this section contains only one unique peak. Table 3 lists the position of this peak (number 2) as well as that of the other strong peaks found in this difference Patterson. Using vector position number 1 (Table 2) and values for peak number 2 (Table 3), one can determine that the values of x and y are 0.16 and 0.59, respectively. If one had chosen another peak from Fig. 3, the values for x and y would be different, but related, to values given above by crystallographic symmetry. The above procedure can also be carried out using the Harker section at 2/3, which should produce the same or symmetry-related values for x and y. Once x and y

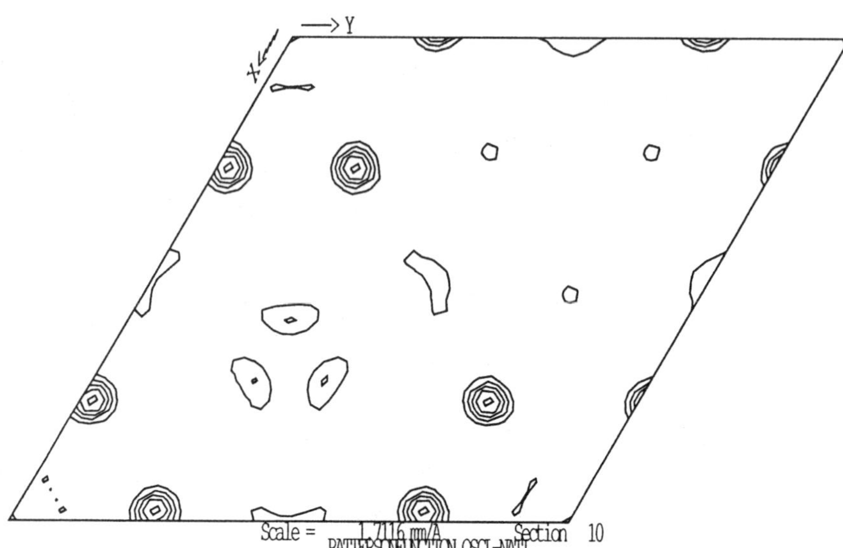

Fig. 3. The Harker section at 1/3 from the difference Patterson function for the K_2OsCl_6 single-site derivative of porcine growth hormone *(16)*. Peaks are contoured at equal intervals with the first contour at one standard deviation of the entire map. *See* Table 3 for peak heights and positions.

Table 2
Position of Interatomic Vectors
Representing Symmetry-Related Heavy Atoms
in a Unit Cell of Space Group $P3_221$

Number	Symmetry operations used	Vector positions
1	(x,y,z) – $(-y,x-y,z+2/3)$	$(x+y,-x+2y,1/3)$
2[a]	(x,y,z) – $(-x+y,-x,z+1/3)$	$(2x-y,x+y,2/3)$
3	(x,y,z) – $(y,x,-z)$	$(x-y,-x+y,2z)$
4	(x,y,z) – $(x-y,-y,-z+1/3)$	$(y,2y,2z+2/3)$
5	(x,y,z) – $(-x,-x+y,-z+2/3)$	$(2x,x,2z+1/3)$

[a]Vector position number 2 is related to that of number 1 by symmetry, thus it is not unique.

are known, the value of z can be obtained by identifying strong peaks at general positions in the difference Patterson, which would independently satisfy vector position numbers 3, 4, and 5 of Table 2. Using peak number; 3, 4, and 5 (Table 3), the value of $z = 0.09$ can be obtained, thus the heavy-atom position is at (0.16,0.59,0.09). Note that peak 1 of Table 3 is

Table 3
Ten Strongest Unique Peaks in the Difference Patterson of the Single-Site K_2OsCl_6 Derivative of Porcine Growth Hormone *(16)*

Peak number	Peak height[b]	Peak position[a] *u*	*v*	*w*
1[c]	15,719	0.00	0.00	0.00
2[d]	2,269	0.75	1.02	0.33
3[d]	2,094	–.43	0.43	0.18
4	1,902	0.32	0.16	0.51
5[d]	1,777	0.59	1.18	0.85
6	1,006	0.00	0.39	0.00
7	975	0.00	0.00	0.29
8	868	0.58	0.29	0.33
9	814	0.14	0.27	0.18
10	768	0.04	0.08	0.23

[a]Peak positions are in fractions of one unit cell (fractional coordinates).
[b]Peak heights are in arbitrary relative units; map standard deviation is 418.
[c]Origin peak representing the pile up of all interatomic vectors in the difference Patterson.
[d]The positions listed for peaks number 2, 3, and 5 are translated by plus, minus, and plus one unit cell, the actual values in the difference Patterson were (0.75,0.02,0.33), (0.57,0.43,0.18), and (0.59,0.18,0.85), respectively. These translations are necessary to obtain the correct heavy-atom position that must be consistent with all the *(u,v,w)* values of peaks number 2, 3, 4, and 5.

the origin peak and represents the pile up of all interatomic vectors in this difference Patterson and that peaks beyond peak number 5 are noise representing <2.5 SD of this map.

3.2.2. Difference Fourier Technique

As can be seen from Eqs. (4) and (5), a Fourier synthesis requires phase angles as input, thus it cannot be used to locate heavy-atom positions in a derivative if no phase information exists. However, it can be used to determine such positions in a derivative if phases are already available from one or more other derivatives. As in the case of a difference Patterson, the Fourier synthesis here also employs difference coefficients. They are of the form:

$$m\,(F_{PH} - F_P)\,e^{i\alpha_P}$$

where F_{PH} is the structure factor amplitude of the derivative in which the heavy atoms are to be located, α_P is the protein phase angle calculated from other derivatives and *m* (figure of merit; whose value is between 0 and 1) is a weighting factor related to the reliability of the phase angle. The success of this technique is highly dependent on the correctness of α_P, since it has been demonstrated clearly that Fourier summations with correct phases but wrong amplitudes can result in correct structure, whereas having incorrect phases even with correct amplitudes results in incorrect structure.

Difference Fourier techniques are most useful in locating sites in a multisite derivative when a Patterson map is too complicated to be interpretable. The phases for such a Fourier must be calculated from the heavy-atom model of other derivatives in which a difference Patterson map was successfully interpreted, and should not be obtained from the derivative being tested, in order not to bias the phases. Also, difference Fourier techniques can be used to test the correctness of an already identified heavy-atom site, by removing that site from the phasing model and seeing whether it will appear in a difference Fourier map. Again, the success of this feedback technique depends on the correctness of the phasing model.

3.3. Refinement of Heavy-Atom Positions

Once the positions of the heavy atoms have been determined, their accuracy can be significantly improved through least squares refinement. This is achieved by allowing the heavy-atom positions, thermal parameters, and occupancy to vary, while minimizing the difference between the structure factor amplitudes calculated from the heavy-atom model and experimentally measured amplitudes. It is important to note that thermal parameters and occupancy are often correlated and should not be refined in the same cycle. A good strategy in heavy-atom refinement is to obtain the best possible heavy-atom model for each derivative alone, then add one derivative at a time while refining, to ensure that the best derivative does not dominate phasing to the exclusion of others. Also, it is better to omit minor sites from the initial refinement if there is doubt about their existence. These sites can be included in later steps of refinement when it becomes more certain that they are real.

Although heavy-atom refinement programs employ different refinement strategies (*see* Blundell and Johnson *[5]* for an explanation of some of these strategies), most of them produce a large number of similar statistics. In practice, the monitoring of these statistics is what directs one in the decision of which variable to vary or which derivative to add in a given set of refinement cycles. The definition of many of these statistics can be found in the Chapter by Watenphaugh on isomorphous replacement in *Methods in Enzymology (18)*. Below are definitions of some of the most commonly used statistics and their magnitudes:

1. Figure of merit *(m)*. It is a quantity that represents the unimodality and sharpness of the phase probability distribution. It can be defined as the mean value of the cosine of the error in phase angle for a particular reflection. In theory, a value of $m = 1$ corresponds to a 0° error in phase angle, whereas a value of $m = 0$ corresponds to a 90° error.
2. Cullis R-factor (R_c). This is a reliability index (R-factor) calculated from centric data only *(19)*, and is defined as follows:

$$R_C = \Sigma \,||\, F_{PH} \pm F_P \,| - F_H \,|/\Sigma \,|\, F_{PH} - F_P|$$

where F_{PH} and F_P are the observed structure factor amplitudes for the derivative and parent crystals, respectively, and F_H is the calculated heavy-atom structure factor amplitude; it is commonly reported as $\%R_c$. In theory, an index of 0 indicates perfect agreement between observed and calculated structure factors; however, in practice, its value for successful structure determinations ranges from 30 to 65%, depending on the quality of data, resolution, and degree of substitution.
3. Lack of closure error (E). This value represents the error in the closure of the phase triangle. The mean square isomorphous lack of closure error for a data set is defined as:

$$E^2_{(iso)} = \Sigma[F_{PH(obs)} - F_{PH(calc)}]^2/N$$

where N is the number of reflections used. The mean square anomalous lack of closure error is defined as:

$$E^2_{(ano)} = \Sigma[(F^+_{PH(obs)} - F_{PH(obs)}) - (F^+_{PH(calc)} - F^-_{PH(calc)})]^2/N$$

4. Ratio of root mean squared calculated heavy-atom structure factor amplitude to root mean squared lack of closure error [rms(F_H)/rms(E)]. This value is a good indicator of the phasing power of a heavy-atom derivative; it should be >1 and should become larger as the phasing model improves.

It is important to inspect many of these statistics while refining heavy-atom parameters, as there is no single parameter that is clearly indicative of the quality of phases. Table 4 shows typical values for some of these statistics as a function of resolution for the single-site K_2OsCl_6 derivative of porcine growth hormone *(16)*. Note that in SIR phasing, where anomalous scattering is used to resolve phase ambiguity, both possible

Table 4
Porcine Growth Hormone K_2OsCl_6 Heavy-Atom Derivative Statistics[a]

	Resolution limit, Å								
	10.7	7.8	5.9	4.9	4.1	3.6	3.1	2.8	Total
$rms(F_H)/rms(E)$[b]	2.10	2.05	2.64	3.97	2.78	2.49	1.97	1.54	2.44
R_c[c]	0.49	0.43	0.44	0.38	0.55	0.60	0.68	0.66	0.53
Number of reflections	89	212	367	563	801	1049	1270	1446	5797
Mean figure of merit[d]	0.67	0.70	0.61	0.76	0.61	0.52	0.39	0.28	0.48

[a]Reproduced by permission from ref. *16*. Abbreviations: $rms(F_H)$, the root mean squared calculated heavy-atom structure factor amplitude; rms(E), the root mean square lack of closure determined from centric reflections only; R_c, Cullis R-factor.

[b]The ratio (F_H-rms/E-rms) should get larger as the phasing model improves.

[c]A reliability index (R) of zero indicates perfect agreement between observed and calculated structure factors.

[d]A figure of merit (*m*) of one indicates no error in phase angle.

hands of the single heavy-atom site give the same statistics, one set of which is wrong. The selection of the correct hand can be done only by inspection of electron density maps.

3.4. Conclusion

In principle, the determination of the structure of macromolecules using the isomorphous replacement method is simple. In practice, it is full of pitfalls. Many of the difficulties associated with this technique are a result of:

1. The inability to form heavy-atom derivatives;
2. Errors in data acquisition;
3. Lack of isomorphism;
4. Low heavy-atom substitution;
5. Variable substitution from crystal to crystal of the same derivative;
6. Crystal deterioration in the X-ray beam; and
7. The presence of too many heavy-atom sites for a successful interpretation of the Patterson.

However, given an ample supply of good crystals, a state-of-the-art data acquisition facility, and enough time, insight, and patience, these difficulties can be overcome.

4. Notes

1. Before attempting to prepare derivatives, it is important to recognize that heavy-atom reagents are very toxic and must be handled with the utmost care.
2. Some crystals are very sensitive to a variety of heavy-atom reagents. They tend to shatter and lose their ability to diffract even if the reagents are dilute and soak times are short. This can be overcome by crosslinking the crystals with glutaraldehyde before soaking.
3. At times, derivative crystals show large differences in lattice constants compared to native crystals. For example, crystals of a ribonuclease-resistant fragment of *Escherichia coli* 5 S ribosomal RNA *(20)* showed as much as 10% change in their *c* axis, on derivative formation. Such crystals cannot be used for isomorphous replacement phasing.
4. In addition to the difference Patterson method, direct methods have been used for the determination of heavy-atom positions; this method was first used by Steitz *(21)* in 1968 while studying carboxypeptidase. Although it is the most widely used technique for determining the structure of small molecules, it has never gained wide popularity for use with macromolecules. One reason is the successfulness of the differencc Patterson method.
5. There are numerous computer programs available to assist in data reduction and processing, Fourier summation calculation, and heavy-atom

refinement; many of these can be obtained from their authors or from other protein crystallographers. Some of these programs have been incorporated into multiprogram packages, Such as PROTEIN (available from W. Steigemann, Max-Planck Institute for Biochemistry, Martinsried, Germany), ROCKS (available from G. N. Reeke, Jr., The Rockefeller University, New York, NY), and CCP4 (available from Daresbury Laboratories, Warrington, Cheshire, UK) packages. For advice on which computer programs to select, how to use them, and how to interpret the results, it is best to consult with a protein crystallographer.

Acknowledgments

I am grateful to Krishna Murthy and Susan Dendinger for valuable comments after reading a draft of this chapter.

References

1. Green, D. W., Ingram, V. M., and Perutz, M. F. (1954) The structure determination of heamoglobin: IV. Sign determination by the isomorphous replacement method. *Proc. R. Soc. London* A225, 287–307.
2. Wang, B. C. (1985) Resolution of phase ambiguity in macromolecular crystallography, in *Methods in Enzymology*, vol. 115 (Wyckoff, H. W., Hirs, C. H. W., and Timasheff, S. N., eds.), Academic, New York, pp. 90–112.
3. McPherson, A. (1982) *Preparation and Analysis of Protein Crystals.* John Wiley & Sons, New York.
4. Wyckoff, H. W., Hirs, C. H. W., and Timasheff, S. N., eds. (1985) *Methods in Enzymology*, vol. 114: *Diffraction Methods for Biological Macromolecules.* Academic, New York.
5. Blundell, T. L. and Johnson, L. N. (1976) *Protein Crystallography.* Academic, London.
6. Petsko, G. A. (1985) Preparation of isomorphous heavy-atom derivatives, in *Methods in Enzymology*, vol. 114 (Wyckoff, H. W., Hirs, C. H. W., and Timasheff, S. N., eds.), Academic, New York, pp. 147–156.
7. Kim, S.-H., Shin, W. C., and Warrant, R. W. (1985) Heavy metal ion-nucleic acid interaction, in *Methods in Enzymology*, vol. 114 (Wyckoff, H. W., Hirs, C. H. W., and Timasheff, S. N., eds.), Academic, New York, pp. 156–167.
8. Holbrook, S. R. and Kim, S.-H. (1985) Crystallization and heavy-atom derivatives of polynucleotides, in *Methods in Enzymology*, vol. 114 (Wyckoff, H. W., Hirs, C. H. W., and Timasheff, S. N., eds.), Academic, New York, pp. 167–176.
9. Ringe, D., Petsko, G. A., Yanahura, F., Suzuki, K., and Ohmori, D. (1983) Structure of iron superoxide dismutase from *Pseudomonas ovalis* at 2.9 Å resolution. *Proc. Natl. Acad. Sci. USA* **80,** 3879–3883.
10. Drew, H., Takano, T., Takana, S., Itakura, K., and Dickerson, R. E. (1980) High-salt d(CpGpCpG), a left-handed Z' DNA double helix. *Nature (London)* **286,** 567–573.
11. Abdel-Meguid, S. S., Grindley, N. D. F., Templeton, N. S., and Steitz, T. A. (1984) Cleavage of the site-specific recombination protein γδ resolvase: the smaller of the two fragments binds DNA specifically. *Proc. Natl. Acad. Sci. USA* **81,** 2001–2005.

12. Hatfull, G.F., Sanderson, M. R., Freemont, P. S., Raccuia, P. R., Grindley, N. D. F., and Steitz, T. A. (1989) Preparation of heavy-atom derivatives using site-directed mutagenesis: introduction of cysteine residues into γδ resolvase. *J. Mol. Biol.* **208,** 661–667.
13. Petsko, G. A., Phillips, D. C., Williams, R. J. P., and Wilson, I. A. (1978) On the protein crystal chemistry of chloroplatinite ions: general principles and interactions with triose phosphate isomerase. *J. Mol. Biol.* **120,** 345–359.
14. Buerger, M. J. (1964) *The Precession Method.* Wiley, New York.
15. Patterson, A. L. (1934) A Fourier series method for the determination of the components of interatomic distances in crystals. *Phys. Rev.* **46,** 372–376.
16. Abdel-Meguid, S. S., Shieh, H.-S., Smith, W. W., Dayringer, H. E., Violand, B. N., and Bentle, L. A. (1987) Three-dimensional structure of a genetically engineered variant of porcine growth hormone. *Proc. Natl. Acad. Sci. USA* **84,** 6434–6437.
17. Abdel-Meguid, S.S., Smith, W.W., Violand, B.N., and Bentle, L.A. (1986) Crystallization of methionyl porcine somatotropin, a genetically engineered variant of porcine growth hormone. *J. Mol. Biol.* **192,** 159,160.
18. Watenphaugh, K. D. (1985) Overview of phasing by isomorphous replacement, in *Methods in Enzymology*, vol. 115 (Wyckoff, H. W., Hirs, C. H. W., and Timasheff, S. N., eds.), Academic, New York, pp. 3–15.
19. Cullis, A.F., Muirhead, H., Perutz, M.F., Rossmann, M.G., and North, A.C.T. (1962) The structure of haemoglobin VIII. A three-dimensional Fourier synthesis at 5.5 Å resolution: determination of phase angles. *Proc. R. Soc. London* **A265,** 15–38.
20. Abdel-Meguid, S. S., Moore, P. B., and Steitz, T. A. (1983) Crystallization of a ribonuclease-resistant fragment of *Escherichia coli* 5 S ribosomal RNA and its complex with protein L25. *J. Mol. Biol.* **171,** 207–215.
21. Steitz, T. A. (1968) A new method of locating heavy atoms bound to protein crystals. *Acta Cryst.* **B24,** 504–507.
22. Kretsinger, R. H. (1968) A crystallographic study of iodinated sperm whale metmyoglobin. *J. Mol. Biol.* **31,** 315–318.
23. Wright, C. S., Alden, R. A., and Kraut, J. (1969) Structure of subtilisin BPN' at 2.5 Å resolution. *Nature* **221,** 235–242.
24. McPherson, A., Jurnak, F. A., Wang, A. H. J., Molineux, I., and Rich, A. (1979) Structure at 2.3 Å resolution of the gene 5 product of bacteriophage fd: a DNA unwiding protein. *J. Mol. Biol.* **134,** 379–400.

CHAPTER 7

Molecular Replacement Using Known Structural Information

Ian J. Tickle and Huub P. C. Driessen

1. Introduction

When identical or similar structures exist in different crystallographic environments, similarities between their diffraction patterns, which are directly related to their Fourier transforms, would be expected. The technique of Molecular Replacement in protein X-ray crystallography *(1–7)* exploits this similarity to determine phases. The dominant application is the case of identical or similar proteins crystallizing nonisomorphously in different space groups, and where one of the structures is already known. The proteins may have been cocrystallized with a cofactor, inhibitor, or other protein. There may be more than one subunit in the crystallographic asymmetric unit. The similarity may be only partial so that a fragment is used. Examples include related structures, site-directed mutants, structures with ligands, and Fab-antigen complexes.

When comparing two identical objects, x_2 (the model) and x_1 (the target), in different environments, they can be related by six rigid body parameters, three angles of rotation defining the rotation matrix, R, and three translation components defining the translation vector, t.

$$x_1 = \mathrm{R} \cdot x_2 + t \qquad (1)$$

In principle, a six-dimensional search would therefore solve this relationship. Unfortunately, even with the fastest computers, this type of search is only practicable in a limited number of cases *(8)*. Hoppe *(9,10)*

From: *Methods in Molecular Biology, Vol. 56: Crystallographic Methods and Protocols*
Edited by: C. Jones, B. Mulloy, and M. Sanderson Humana Press Inc., Totowa, NJ

considered the Patterson function as the sum of two functions. One is represented by the intramolecular vector or self-vector set, which is centered at the origin of the unit cell, and has a structure and orientation that are determined by the structure and orientation of the molecule. These vectors lie within the largest intramolecular distance, r_{max}, from the origin. The second function is represented by the intermolecular vector or cross-vector sets between molecules related by the crystal symmetry that are centered at the intermolecular origin vectors of the pairs of molecules from which they originate. They are dependent on the molecular structure, orientation, and crystal packing, and on average are longer than the intramolecular ones, with many larger than r_{max}. The Patterson function, which is readily calculated from the observed structure factors, is therefore a useful vehicle for simplifying the six-dimensional problem to two three-dimensional problems. The relative orientation of molecules must be determined first by correlating the intramolecular vector set of the model structure with the native Patterson of the target using a rotation function. No knowledge of the translation is needed. After applying the rotation found to the model, the cross-vector set can be correlated with the native Patterson using a translation function.

2. Rotation Functions

Two functionally equivalent types of "search" functions are in use, one in real space where the rotated Patterson of the model, as a map of an array of vectors, is compared with the Patterson of the target, and the other in reciprocal space where the Fourier transform of the model is compared with the observed diffraction data.

2.1. Rotation Function Computer Programs

2.1.1. Rossmann-Blow Rotation Function

In a classic paper of protein crystallography, Rossmann and Blow *(11)* expressed the rotation function, C, of the rotation, Ω, by the integral:

$$C(\Omega) = \int_U P_1(r)\ \Re P_2(r)\ dr \qquad (2)$$

where P_1 is the target Patterson function, and $\Re P_2$ the rotated model Patterson function. Note that $\Re P(r) = P(R^T \cdot r)$, where R is the rotation matrix previously defined. The integration is usually performed over all points, r, in a spherical volume, U, centered on the origin that will contain all self-vectors. This product function, which is insensitive to the relative scaling of the Pattersons, will have a maximum value when the two self-vector sets are equivalently oriented. It can be expressed *(11)* as a double summation in terms of reciprocal lattice vectors h and p.

$$C(\Omega) = \Sigma_h \{ \Sigma_p | F_h |^2 | F_p |^2 G_{h,h'} \} \quad (3)$$

where the nonintegral lattice vector is $h' = R \cdot p$, and $G_{h,h'}$ is an interference function that is 0 unless h is close to h'. The computer program coding this rotation function is very slow *(12)*. Therefore, usually a coarse search is performed first, followed by a fine search around peak positions.

2.1.2. Lattman Rotation Function

Lattman and Love *(5,13)* proposed a faster version for the Rossmann-Blow rotation function. The Patterson function P_2 of an isolated molecule, M, will fall to 0 for a radius, r, equal to the maximum dimension of the molecule. The rotation function can then be written as:

$$C(\Omega) = \Sigma_h |F_o(h)|^2 |F_M(R^T \cdot h)|^2 \quad (4)$$

where F_o is the observed structure factor, and $F_M(R^T \cdot h)$ the Fourier transform of the model, which is finely sampled in order to bring the rotated vector $R^T \cdot h$ close to an integral index of the target. The advantages of Lattman's rotation function are that it is faster to compute than Rossmann and Blow's, that no radius of integration is required, and that sampling can be as precise as possible. Disadvantages are that all intermolecular vectors in the observed data shorter than the molecular diameter are present, and that the cell for the model must be large, at least two times the maximum diameter. The Lattman rotation function is available in the computer program RATFINC *(5)* and the option LATSUM of the MERLOT package *(14)*. Since this rotation function is still slow, it is often used for a fine grid optimization of the angular parameters of peaks found with the fast rotation function (MERLOT).

2.1.3. Huber Real Space Rotation Function

Following on the work of Hoppe, Huber *(15,16)* implemented the rotation function in real space. To limit calculations, only those peaks of the model Patterson that are generally built up by overlapping vectors are used. A point-by-point product correlation *(17,18)* of two Pattersons with interpolation for nonintegral values is a slow process. The procedure usually followed is to do first a search of the asymmetric unit of the rotation function with 5° steps, followed by a fine search of 1–2° around peaks. The real space rotation function is generally available as the option

SEARCH of the PROTEIN package *(19)*, which is used regularly. It has also been implemented in the program X-PLOR *(20)*.

2.1.4. Crowther Fast Rotation Function

In 1972, Crowther showed, in another classic paper of protein crystallography *(21)*, that instead of using Cartesian Fourier coefficients $|F_h|^2$, it is possible to use fast Fourier transform techniques by expanding each Patterson function within a spherical volume as a product sum of spherical harmonics and spherical Bessel functions *(22,23)*.

$$C(\beta,\alpha,\gamma) = \Sigma_{mm'} [\Sigma_l (\Sigma_n a_{lmn}{}^* b_{lm'n}) d^l_{m'm}(\beta)] \exp^{-im\alpha} \exp^{-im'\gamma} \quad (5)$$

Here the rotation Ω is specified by the Eulerian angles (α,β,γ). The rotation function has been separated into three steps. The coefficients a_{lmn} and $b_{lm'n}$ of the stationary and rotating Patterson functions, respectively, only depend on a particular pair of Patterson densities and the radius of integration, and not on the rotation. The coefficients $d^l_{m'm}(\beta)$ refer to rotations of spherical harmonics and are independent of the particular Patterson densities. They need to be calculated only once for each β. In the third step, a two-dimensional fast Fourier transform can be used for the m and m' summations to calculate the rotation function for β-sections. This factorization gives the fast rotation function its speed and makes it the most frequently used algorithm.

The fast rotation function is generally available as the program ALMN in the CCP4 suite *(24)* and as the options HARMCO/CROSUM of the MERLOT package *(14)*. A spherical polar coordinates representation *(25)* is available in the program POLARRFN (Kabsch implementation) in the CCP4 suite *(24)* and options HARMCO/CROSUM of the MERLOT package *(14)*. These programs run more slowly than the Eulerian angle versions.

The usual procedure with ALMN is to calculate first a coarse rotation function (5°) over the asymmetric unit, followed by a fine search (2.5°). With MERLOT, a coarse rotation function by HARMCO/CROSUM is followed by a fine grid Lattman rotation function (LATSUM).

2.1.5. Other Rotation Functions

Navaza *(26–28)* has written a modified fast rotation function (ROTING) that uses a numerical integration rule to calculate the coefficients a_{lmn} and a more robust way of determining the $d^l_{m'm}(\beta)$. Yeates *(29)* has proposed a full-symmetry rotation function based on the fast rotation function, which may remove bias for models with approximate symmetry coinciding with a symmetry operation of the target.

2.2. Practical Notes on Rotation Functions

2.2.1. Rotation Angles

The rotation applied in the Rossmann-Blow function, in PROTEIN and in X-PLOR, is specified by Eulerian angles (θ_1, θ_2, θ_3) in the zxz-convention *(11)*. In order to specify the rotation uniquely, it is necessary to define about which axes the individual rotation steps are to take place (e.g., zxz or zyz), the order in which the angles are to be applied (e.g., θ_1, θ_2, θ_3 or θ_3, θ_2, θ_1), the positive sense of an angle (clockwise or counter-clockwise when viewed from a specified end of the axis), whether world axes or molecular axes (equivalent to coordinates) are to be rotated, and whether the individual rotations are performed around rotated or fixed axes (e.g., z,y',z" or z,y,z). Certain combinations of these alternatives produce the same rotation, however. Unfortunately, there appears to be considerable confusion about the representation of the Eulerian rotations in the literature. The commonly used convention of rotating world axes with respect to fixed molecular axes corresponds to considering the coordinates of the "rotating" molecule as fixed with respect to the rotating "stationary" molecule! It is therefore important to take great care when comparing the results from different programs. The Crowther fast rotation function in ALMN and CROSUM uses Eulerian angles (α, β, γ) in the zyz-convention *(21,30)*. In our opinion, the rotation of coordinates is most easily viewed with respect to the fixed world axes (Fig. lA), but it is usually explained by reference to the rotated molecular axes (Fig. 1B). The rotated coordinates are given by:

$$\begin{bmatrix} \cos\alpha & -\sin\alpha & 0 \\ \sin\alpha & \cos\alpha & 0 \\ 0 & 0 & 1 \end{bmatrix} \begin{bmatrix} \cos\beta & 0 & \sin\beta \\ 0 & 1 & 0 \\ -\sin\beta & 0 & \cos\beta \end{bmatrix} \begin{bmatrix} \cos\gamma & -\sin\gamma & 0 \\ \sin\gamma & \cos\gamma & 0 \\ 0 & 0 & 1 \end{bmatrix} \begin{bmatrix} x \\ y \\ z \end{bmatrix}$$

It is wise always to check the rotation function solution by computing a new function derived from the rotated coordinates verifying that the solution is at the origin in rotation space.

The great advantage of these conventions is that the rotation function exhibits some or all of the point group symmetry, so that the search volume is reduced *(11,31)*. A disadvantage is that the use of fixed grid steps in the three Eulerian angles produces uneven and inefficient sampling

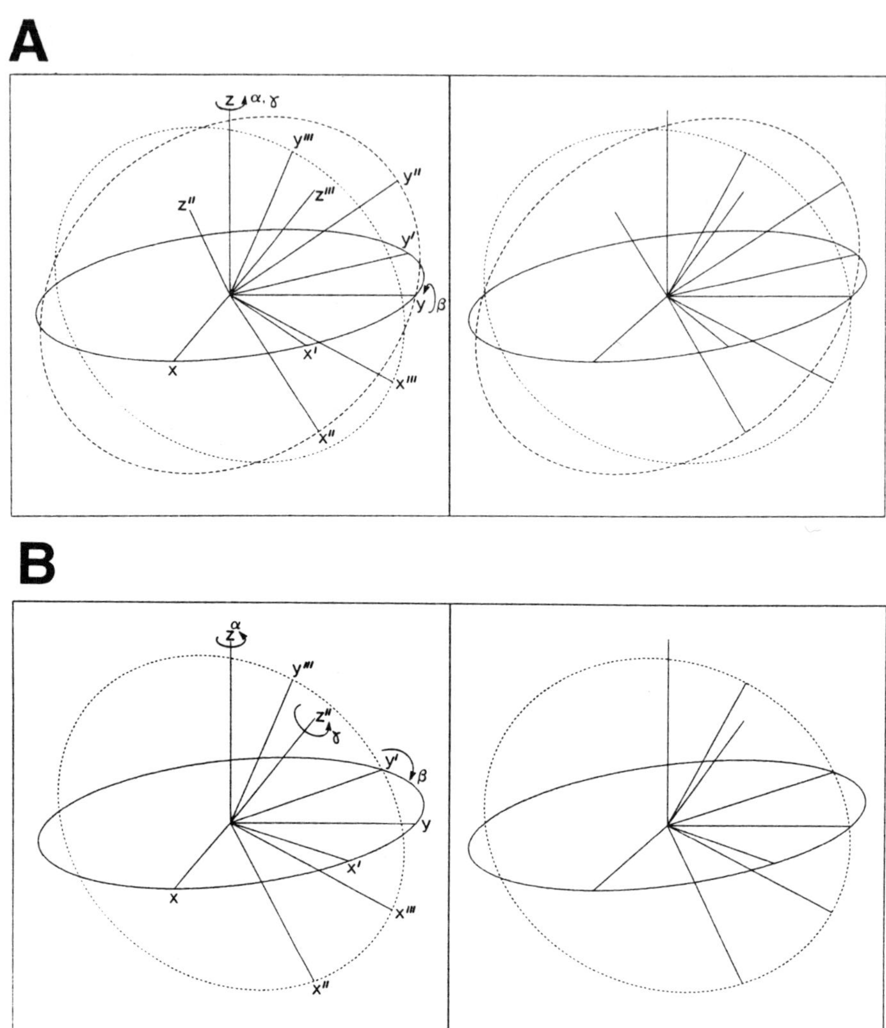

Fig. 1. Stereodiagrams of the Eulerian angle convention used in Crowther's FRF. **(A)** The diagram shows the local axes of the search molecule being rotated with respect to fixed world axes. The coordinates of the search molecule are rotated in the order γ about the fixed axis z, β about fixed y, and α about fixed z, where a positive rotation is counterclockwise when looking from the positive direction of the corresponding axis toward the origin. Rotation steps are indicated by ', '', and '''. **(B)** The same operation can be viewed equivalently as rotations of the coordinates around the rotated molecular axes in the same direction, first of α about fixed z, then β about rotated y (y'), and then γ about rotated z (z''). Note that although the final positions are the same, the intermediate positions (z', x'', y', y'', z'') are not.

(32). When θ_2 or β is close to 0 or 180, the other two angles are rotations of nearly parallel axes and are coupled, producing distortion in a θ_1/θ_3 grid plot. The pseudo-orthogonal Eulerian angles ($\theta_+ = \theta_1 + \theta_3$, $\theta_- = \theta_1 - \theta_3$, θ_2) give uniform sampling *(32)*. This convention is used in RATFINC and X-PLOR. The Eulerian angles (ψ, θ, φ) *(16)* in PROTEIN cause the equivalence of first and last rotations to occur at $\theta = \pm 90$, which may be useful when the peak is close to the origin *(33)*.

Another way of describing the rotations is in terms of the spherical polar coordinates (φ, ψ, κ) where φ and ψ define the rotation axis with an angle of rotation, κ, around it *(11)*. These angles are useful when searching in a self-rotation function for specific noncrystallographic symmetry axes. This convention is used in POLARRFN, MERLOT, PROTEIN, and X-PLOR.

2.2.2. Symmetry

The symmetry data for the rotation space groups have been tabulated *(34,35)*. Pseudo-rotation axes parallel to z produce the most helpful rotation group symmetry, because they generate pseudotranslational symmetry. Note that the orthogonalization convention must be taken into account *(23)*. This will enable the related peaks to be viewed on the same β-section with the same Eulerian distortion *(35–37)*. Another point is that peaks near mirror planes may give inaccurate positions because of coalescence, especially at lower resolution *(36)*. It is unclear how far the new refinement methods correct for this, but a simple remedy is to reorient the model coordinates.

2.2.3. Model

Most often complete model structures without modifications are used. Residues likely to be different may be removed *(38–40)*. Occasionally models are stripped to Cα-atoms only *(38,39,41,42)* or backbone with Cβ-atoms *(38,41–43)*. However, this may not improve the accuracy of the peak position. NMR-models have been used *(33)* as well as model-built energy-minimized structures *(44)*. A refined structure sometimes succeeds where an unrefined one fails *(40,45)*. The atomic temperature factors of the model have been used *(36)* or have been set to twice the value of the original ones *(46)*, whereas for other models, only an overall temperature factor has been applied *(41)*. When rigid body movements are significant, as in multidomain or multisubunit proteins, models may have to be split into fragments to achieve success, as in the case of Fabs where the C and V domains as well as their subdomains have been used separately *(37,42)*. To ascertain the significance of a peak, a series of homologous models may be used *(37,40,42)*.

2.2.4. Model Cell

The standard cell for the search model is triclinic with $\alpha = \beta = \gamma = 90°$ and with the cell dimensions twice the molecular diameter, in order to exclude any intermolecular vectors *(12,47)*. This approach is most common with a tendency for factors to be >2. However, to reduce computations, it may be desirable to decrease the size of this largely empty cell. Using as longest search radius half the diameter of the molecule *(48)* allows a cell dimension of 1.5 times the diameter *(47)*. This works well for spherical molecules, but less well for highly aspherical ones. Then the molecular axes may be roughly aligned along the cell edges of the triclinic cell, so that cell dimensions may be three times the radius in each of the three directions. Such a reduced cell is also used frequently. Structure factors can be calculated directly from the coordinates using SFALL (CCP4), STRFAC (MERLOT), or FC (PROTEIN).

For the Lattman rotation function LATSUM, the radius of integration is implicitly the maximum diameter, and a finely sampled transform needs to be calculated from the structure factors with WRITTF (MERLOT). In this case, the cell dimensions should be two to three times the maximum molecular diameter. However, factors as low as 1.5 appear to work *(30)*.

2.2.5. Radius of Integration

Although in theory, a search radius equal to the diameter of the molecule would give all intramolecular vectors, the practical choice of the search radius should maximize the number of intramolecular and minimize the number of intermolecular vectors. The target cell does not usually allow a clean separation, since molecules are closely packed, although the model cell does. As a good starting value for the radius of integration, half the diameter of the molecule is often used *(47,48)*. Alternatively, for a spherical molecule 75–80% of the diameter could be used *(49)*. For ellipsoidal molecules, the geometrical mean of the ellipsoid semiaxes can be used as the radius *(47)*. The radius of integration is usually varied to examine the significance of a potential solution. Note that when using the fast rotation function, the ratio radius/resolution is subject to program constraints.

2.2.6. Origin Removal

For searches with radii greater than any of the cell dimensions of the target, the origin peak of the next cell will be included, giving a large but useless contribution to the rotation function *(13)*. This will normally only happen for elongated molecules and cells with a relatively short dimension. In such cases, the origin must be removed. This can be done by a user-defined radial cutoff in the real space rotation function and in ALMN. Alternatively, the squared amplitude averaged in shells can be subtracted from the squared amplitude of each reflection *(13)*. MERLOT *(30)* uses one shell for all data, so that the origin is not removed effectively. Note that weak reflections should not be removed prior to this operation. Dodson *(23)* suggests that the cutoff should be at least equal to the resolution limit. In practice, such a rule does not appear to be followed.

2.2.7. Resolution

The rotation function is proportional to the fourth power of the amplitude so that it will be dominated by the large terms. Since the search model does not have a representation for the solvent structure, low-resolution terms must be removed *(11)*. Blow *(49)* suggests 7–8 Å. Navaza *(26–28)* proposes the removal of the low-order coefficients (1 = 2, 4, 6) in the fast rotation function. Too high a resolution might emphasize the molecular differences between target and model, although this appears to be less of a problem in practice. Blow *(49)* suggests at least two to three times the expected mean coordinate difference for a high-resolution cutoff.

To speed up calculations, the smaller intensities are generally omitted in the calculations *(12)*, which on average means a downweighting of higher resolution data, although the noise might be reduced as well. There are different ways of ensuring an even distribution of Fs against resolution. Although some studies use resolution shells as narrow as 1 Å *(36)*, most use much wider shells. In practice, the number of terms used varies enormously (500–10,000), also depending on the cell sizes involved, on the maximum resolution of the data, and on computer program limitations. Some studies use several shells to determine the significance of peaks *(36,41,42,50)*. However, none of these methods are totally satisfactory, since they still do not lead to an even representation over the whole resolution range. Dodson *(23)* advocates the use of sharpening the intensities with an artifical temperature factor. Tickle *(51)* proposes the use of normalized structure factors, E-values, which do have this property. For a maximum of 1500 terms they have been shown to be superior to Fs *(36)*. Their use has recently become more widespread *(50,52)*. They can be calculated from the Fs with the CCP4 program ECALC. Note that a finely sampled transform gives many similar structure factors, so that because of program limitations on storage of structure factors, the effective number of independent Fs may be smaller than the number actually stored.

It is important to note that the fast rotation function has a limitation on the resolution for a given radius *(22)*. For a radius of integration, r (Å), the maximum resolution d_{min} (Å) possible for a given l_{max} is defined by the ratio r/d_{min}. For $l_{max} = 30$ and 60, this ratio is 5.83 and 10.83, respectively. The version as used in MERLOT *(30)* has a maximum of $l = 30$ only, which limits the resolution to about 4 Å for a radius >23 Å.

2.2.8. Completeness of Data

Recent studies *(28,53)* show that a systematic lack in the completeness of the observed data may have serious consequences for accuracy and success. Alzari and Navaza *(28)* showed that the ROTING program was more robust with respect to this problem than the standard fast rotation function.

2.2.9. Sampling

For LATSUM (MERLOT) and SEARCH (PROTEIN), the sampling limit is determined by the available computer memory and time only. Patterson maps for SEARCH are often calculated in 1-Å grid steps. This allows fine grid searches around peaks found in coarse grid searches. The sampling is fixed by space group symmetry in HARMCO/CROSUM (MERLOT), so that in low-symmetry space groups, it can be quite coarse. The program uses 5°/NSYM as a step size, where NSYM is the rotational symmetry about the z-axis, and compresses the intermediate array of coefficients before passing it on to the FFT. ALMN has a choice of 5 or 2.5° in the sampling of α and γ, which is independent of the space group; sampling in β can be chosen freely. The coarse and/or uneven sampling in the MERLOT version means that LATSUM has to be used to optimize the angles after an initial fast rotation function run. The sampling suggested for the fine transform used in LATSUM is 1/3 of the resolution *(5)*.

2.2.10. Significance

Normally peak heights are quoted in units of r.m.s. value of the rotation function map. However, the CCP4 and MERLOT rotation functions do not calculate these values correctly since no account is taken of the uneven sampling *(26,28,30,54)*. This is especially important when different models with differing orientations are used. The latest version of ALMN has been corrected. The signal/noise ratio should be quoted as the r.m.s. value for the correct peak relative to that of the highest noise peak. To check for significant peaks, the usual approach is still to repeat the calculations many times, varying parameters that are considered important, and to look for consistency *(28,42)*.

3. Rotation Function Refinement

The residual errors in the rotation function solution will determine success or failure in the translation function. In recent years, the models have become increasingly dissimilar from the target. A common situation is exemplified by the Fab structures, where the large range in elbow angles necessitates fragmentation into constant and variable domains and subdomains as separate models *(37,42)*. The rotation parameters for an incomplete model may be biased by the absence of the missing part, and signal-to-noise ratios may be low. Even with solutions optimized by fine grid searches, it may still be difficult if not impossible to obtain reliable translation parameters. The highest peak in the rotation function may not be the correct one *(36,42)*. The translation function result may be outside the radius of convergence of the final six-parameter rigid body refinement. Recent developments therefore have introduced a new step between the rotation and translation function to allow "refinement" of potential rotation function solutions.

3.1. BRUTE

The six rigid body parameter search procedure BRUTE of Fujinaga and Read *(55)* allows translation searches with small grid step variation of the parameters of a rotation function solution *(56)*. Alternatively, rotational parameters only can be "refined" by specifying P1 symmetry and adjusting them in small steps (0.5–1.0°) for maximum correlation *(57,58)*. The use of P1 "refinement" is problematic, as it is comparable with doing a rotation function in the presence of all intermolecular vectors. Furthermore, BRUTE cannot refine a fragment without knowing the orientation/translation of the others.

3.2. X-PLOR

Brünger proposed *(20)* that selected potential rotation solutions should be subjected to refinement before the translation function is attempted, in order to avoid the need for a translation function for each new orientation. The target function for the refinement consists of an effective Patterson energy term that is the negated standard linear correlation coefficient between the observed and calculated squared normalized amplitudes, the latter for one molecule of the model, plus optionally a suitably scaled empirical energy function that describes the geometry and nonbonded energy of the model. The model is in a triclinic cell identical to that of the target. Rigid body PC refinement is more efficient than a fine grid search as used in BRUTE, since it only explores the shifts that improve the agreement. However, the problem of the intermolecular vectors as noted for BRUTE is still present.

3.3. INTREF

Yeates and Rini have proposed *(59)* an intensity-based refinement method (INTREF) for rotation function solutions that does take into account the intermolecular vector problem.

4. Translation Functions

Translation functions are used in the second stage of the Molecular Replacement method to position the correctly oriented molecules of a search model relative to the crystallographic space group symmetry elements. In cases of noncrystallographic symmetry, where there are several subunits within the crystallographic asymmetric unit, the symmetry elements ambiguously define the origin of the crystal unit cell, and so translation functions are also needed to position the subunits relative to each other. In some cases, the structural homology between the target and model structures may extend to the partial or full quaternary structure, so that the search model may consist of two or more subunits that are already correctly positioned relative to each other. Therefore, the positioning problem is then considerably reduced.

There seems to have been a popular belief in the past *(60)* that translation functions are inherently less reliable than rotation functions. However, in retrospect, this seems to have stemmed from the fact that it was not always appreciated that for the translation function to be successful, accurate orientation parameters are essential *(55)*, and consequently, the effort required to obtain such accuracy was not always expended. There also may be other factors, such as degree of structural homology, completeness of the model, and symmetry (crystallographic and noncrystallographic), which will impact on the reliability of both the rotation and translation functions. Because of this general lack of confidence in translation functions, it is only in the last few years that the method has gained widespread acceptance as a technique to be used routinely for structure solution.

Generally, the available programs work by effectively stepping the search model over a three-dimensional grid, with a step size determined by the resolution of the data, generating the symmetry-equivalent molecules, and finally computing some measure of fit at each trial position between the observed and calculated structure factor amplitudes, or their equivalent in Patterson or real space. The translation functions in the literature can be classified according to the type of function used to measure the fit between the observed and calculated structures. Conceptually, the functions are variously defined in real or reciprocal space, but in practice, they are invariably transformed to and computed in reciprocal space, using structure factors. Many of the various translation functions are actually much more similar than may be apparent *(6)* and often differ only in minor detail.

4.1. Translation Function Computer Programs

4.1.1. MERLOT

A widely used program package for Molecular Replacement is undoubtedly MERLOT *(14)*, and this contains two translation function programs, RVAMAP and TRNSUM. RVAMAP computes a map of the standard crystallographic R-factor *(61–63)*. A problem with the R-factor is that it is sensitive to the F_{obs}/F_{calc} scale factor. In contrast, the product and correlation functions described below are not. TRNSUM uses the T_1 product function *(64)* (implemented by Lattman), and employs the technique of calculating a set of two-dimensional T_1 functions, one for each symmetry element, with optional subtraction of the intramolecular vectors. TRNSUM also has an option to use the phase-modulated Langs *(65)* formulation for the translation function, in place of that of Crowther and Blow. The intermolecular origin vectors have to be extracted manually from the set of maps.

The structure factor calculation used in MERLOT (STRFAC) is a conventional one, but there is a program (WRITTF) to generate a finely sampled transform from the calculated P1 structure factors, from which the structure factors of the rotated molecules can be interpolated *(13)*. To calculate structure factors for the symmetry-equivalent molecules, the P1 structure factors or their interpolated values are phase-shifted *(63)*, thus saving time in computing additional sets of structure factors. Other versions of the T_1 translation function program are available *(5,33)*.

4.1.2. CCP4

Another popular package is CCP4 *(24)*. This has recently been updated so that it now contains two translation function programs, TFFC and TSEARCH. TSEARCH is functionally identical to RVAMAP. TFFC *(51,54)* computes the crystallographic and noncrystallographic components of the three-dimensional full-symmetry T_2 product function, with subtraction not only of the intramolecular vectors *(64)*, but also of all known intermolecular vectors. The full-symmetry T_2 function effectively performs a three-dimensional sum function of the individual symmetry-element T_1 functions, and gives the desired translation vectors directly without manual analysis of the peak coordinates. Thus, the redundancy inherent in the T_1 functions is used to improve the signal/noise ratio in the T_2 function. Beurskens et al. *(6)* reported that "in several cases the correct t was present in the combined map [T_2] but absent in one of the individual maps [T_1]. We never found that the correct peak was present in one or more individual maps and not present in the combined map."

In the T_2 function programs, normalized structure amplitudes, Es, are used in place of Fs. This avoids the fourth power fall-off owing to the scattering and temperature factors, and considerably improves the signal/noise ratio *(51,60)*. The squared normalized amplitudes are calculated as the ratio of the squared amplitudes to the mean square amplitude in narrow equivolume shells in reciprocal space. The CCP4 translation-function programs use the same phase-shifting technique as MERLOT, but TFFC uses a Fourier transform algorithm for the summations *(60,66)*, which is executed by an FFT program *(67)*. TSEARCH is about 100 times slower to compute than the Fourier transform-based functions. The approach taken to structure factor calculation is also different from MERLOT; an FFT algorithm (SFALL *[68]*) is used in which an intermediate electron density map is generated from the atomic coordinates and scattering factors, so that recalculation from the rotated coordinates is comparable in compute time with interpolation.

4.1.3. BRUTE

Fujinaga and Read *(55)* proposed a six-dimensional search procedure (BRUTE) based on calculation of the standard linear correlation coefficient between the observed and calculated squared amplitudes. A three-dimensional translation search over the asymmetric unit is performed, optionally with at each trial position a few orientations near the rotation function peak. In the six-dimensional search, structure factors have to be recalculated at each orientation, so the whole calculation is computationally very expensive.

4.1.4. X-PLOR

The program package X-PLOR *(20)* employs a somewhat different, although related, strategy. Using PC-refined rotation function solutions, a separate three-dimensional translation search is performed, again using the standard linear correlation coefficient between the observed and calculated squared normalized amplitudes, this time for the full symmetry-generated model, using the same phase-shifting technique employed by the other programs.

4.1.5. Other Translation Functions

All the techniques described so far work by comparing functions of amplitudes. Colman and Fehlhammer *(69,70)* used a reciprocal-space translation function requiring phases, for example, from a poor isomorphous replacement derivative, that did not give an interpretable map by itself, but that was good enough to allow searching for a model structure. Reynolds et al. *(71)* used the real-space equivalent of this (the electron density map) for the same purpose, extending the search to six dimensions. Read and Schierbeek *(72)* proposed a full-symmetry reciprocal-space version of the "phased translation function." Subbiah and Harrison *(73)* performed the six-dimensional real-space search, but reduced the computational requirements by using the "simulated annealing" method of combinatorial optimization. Cygler and Desrochers *(74)* eliminated the requirement for phases by proposing a full-symmetry translation function (RTRANS) based on a technique used in small-molecule direct methods *(75)*, where an electron density map in space group P1 was computed by phasing on the model molecule in an arbitrary position. Again this map can be searched for the symmetry-related occurrences of the model structure.

4.1.6. Cautionary Note

All of the aforementioned programs are space-group general, to the best of the authors' belief. However, the user should be aware that "bugs" relating to symmetry, particularly in less common space groups, such as $P3_121/P3_221$, and $P3_112/P3_212$, were reported in earlier versions of some of the programs, which have generally been fixed in the current versions. The advice is therefore always to obtain the current working version of the programs and to check the results, for example, by generating a test structure in the same space group as the target.

4.2. Theory and Practice of Translation Functions

This section provides a description of the workings of the Fourier transform-based translation functions in greater depth, because many users of these programs seem to find the algorithms much more difficult to comprehend than those employed in R-factor and correlation coefficient-based functions.

4.2.1. The T and T_1 Functions

Crowther and Blow's T function *(64)* is a product function between the Patterson functions of the target (P_o) and model (P_{kl}) structures

$$T(v_{kl}) = \int_V P_o(u)\, P_{kl}(u,t)\, du \tag{7}$$

The integral is over all points u, in the primitive unit cell volume, V. The variable vector, v_{kl}, is the intermolecular origin vector between the local origins of symmetry-generated molecules k and l, which is determined by the unknown translation vector, t, of the search molecule:

$$v_{kl} = (A_k - A_l) \cdot t + (d_k - d_l) \tag{8}$$

where A_k and d_k are the rotation and translation components, respectively, of the k-th space group symmetry operator. The structure factors do not need to be calculated for each symmetry-generated molecule. Rather, they are calculated once for the search molecule in a P1 cell with the same dimensions as the target structure, whereas the structure factors for the other molecules are obtained by applying appropriate index transformations and phase shifts.

Usually a modified T function is used, the T_1 function, where the calculated intramolecular Patterson for all molecules is subtracted. This reduces noise owing to accidental overlap with intramolecular vectors, but does make the assumption that the intramolecular vector set is the same in both structures:

$$T_1(v_{kl}) = \int_V [P_o(u) - \Sigma_k P_{kk}(u)]\, P_{kl}(u,t)\, du \tag{9}$$

The functions T and T_1 are maximal when the two Patterson functions exhibit maximum correspondence. For computational purposes, this equation is transformed into reciprocal space:

$$T_1(v_{kl}) = \Sigma_h\, [\, |F_o(h)|^2 - \Sigma_k |F_k(h)|^2]\, F_k^*(h)\, F_l(h) \exp(-i2\pi h \cdot v_{kl}) \tag{10}$$

where $F_k(h)$ is the structure factor for the k-th molecule. The function is evaluated by a Fourier transform.

The Patterson of the model is based only on the interatomic vectors between two different molecules in crystallographically equivalent positions k and l, which are related by a rotation or screw axis. The component of the translation vector parallel to this axis will affect both molecules equally, and so the model Patterson will only change when the components of the translation vector perpendicular to this axis are varied. Therefore, only one section for each symmetry element relating each pair of different k and l needs to be calculated, but pairs obtained by interchanging k and l need not be considered.

For *m* molecules in the primitive unit cell, there are $m(m-1)/2$ pairs, each of which corresponds to a Harker vector for one molecule per asymmetric unit. For example, there will be 1 pair in monoclinic, 6 pairs in orthorhombic, 28 in $P4_12_12$, 66 in $P6_122$, and 276 in I432. For space groups other than triclinic and monoclinic, it is necessary to correlate the coordinates of the peaks found in the sections to obtain a consistent solution for the translation vectors. In theory, it is only necessary to use a subset of two of the pairs (except in polar space groups where only one is needed) to determine the (x,y,z) components of the translation vector for each molecule in the asymmetric unit. In practice, in high-symmetry space groups, the T_1 functions are very noisy, and the set of pairs chosen may not yield a consistent solution, so it is wise to check other pairs for possible consistent solutions. Common practice *(30,76)* appears to be that $(m-1)$ pairs are checked based on the number of Harker sections. In addition, for the case of *n* subunits per asymmetric unit $(n-1)$ intersubunit pairs are required to determine the relative positions of the subunits, but again it would be wise to check the other intersubunit pairs for consistency.

4.2.2. The T_2 Function

The T_2 differs from the T_1 function in that all intermolecular vector sets are compared simultaneously with the target Patterson. It turns out that the T_2 function is just a three-dimensional sum function of all the $m(m-1)/2$ T_1 functions *(6)*. Crowther and Blow's *(64)* original T_2 function was expressed in terms of all the intermolecular origin vectors and, consequently, could not be cast in terms of a Fourier transform. However, Harada et al. *(60)* showed how by recasting the formula in terms of the single translation vector *t*, the T_2 function could be expressed as a Fourier transform. In Patterson space:

$$T_2(t) = \int_v [P_o(u) - \Sigma_k P_{kk}(u)] \Sigma_k \Sigma_{l<k} P_{kl}(u,t)\, du \quad (11)$$

Expressing this in reciprocal space:

$$T_2(t) = \Sigma_h [|E_o(h)|^2 - \Sigma_k |E_k(h)|^2] \Sigma_k \Sigma_{l<k} E_k{}^*(h) E_l(h) \exp [-i2\pi h \cdot (d_k - d_l)] \exp[-i2\pi h \cdot (A_k - A_l) \cdot t] \quad (12)$$

This is a Fourier series in the index $h \cdot (A_k - A_l)$. Here, instead of Fs, normalized structure factors, Es are used *(51)*.

Driessen at al. *(54)* have extended the T_2 function to the case of multisubunit structures, each subunit with a separate translation vector. A simultaneous search for all the translation vectors would be the optimum strategy *(66)*, but this would be impractical for more than two subunits. The feasible strategy is to determine the positions of the subunits in a stepwise fashion, using all the information already obtained, and deciding on the optimum sequence by trial. At each step, the intermolecular vectors between already determined subunits, as well as all intramolecular vectors, are subtracted.

4.2.3. Symmetry

As an example of symmetry considerations, in space groups belonging to point group 222, such as $P2_12_12_1$, the calculated Patterson function does not change if any combination of 1/2 unit cell translations along any of the principal axes is applied to the asymmetric unit. The effect of this is to reduce the volume to be searched for the molecule or first subunit to a fraction of the unit cell, in this case 1/8. In the case of two or more subunits, to find the remaining subunits, the whole of the primitive cell, that is 1/2 the unit cell in I222, for example, must be searched. In polar space groups, such as $P2_1$ or $P6_5$, where there is a unique rotation or screw axis, the calculated Patterson function for the first or only subunit is independent of any translation parallel to that axis, so the translation function is reduced to the plane perpendicular to the axis. For two or more subunits, the calculated Patterson function varies with all the relative translations of the subunits, and so the whole of the primitive cell must again be searched.

It is necessary to repeat the calculation for ambiguous or nonenantiomorphic pairs of space groups, for example, $I222/I2_12_12_1$ or $P6_122/P6_522$, and determine the correct solution. For efficiency, the phase shifts owing to the crystallographic translational symmetry should be applied in the translation function program itself.

4.2.4. Resolution and Amplitude Limits

The choice of resolution limits varies according to whether Fs or Es are used. Tickle *(51)* concluded that the best discrimination was obtained in a T_2 function (TFFC) when no amplitude cutoffs were applied, when intramolecular vectors were subtracted, and when Es were used in a wide

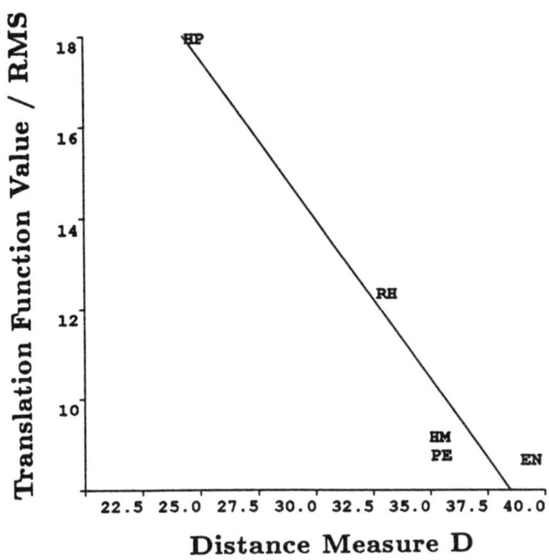

Fig. 2. Translation function maxima vs Structural Distance Measure for a homologous series of aspartic proteinases (Newman, unpublished).

resolution range (20–3 Å). The discrimination was drastically reduced with Fs, although this improved again when a narrow resolution range (6–3 Å) was used. Cygler and Anderson *(56)* concluded that using Fs in a correlation coefficient search (BRUTE) and in the T_1 function (TRNSUM), the best discrimination was obtained with data of intermediate resolution (8–4 Å and 8–4.8 Å, respectively). Weak reflections were very important for discrimination, and the best results were obtained when all reflections in the shell were included. They also noted that subtraction of intramolecular vectors is deleterious when the model orientation is in error by more than a few degrees; this is because the intramolecular vector sets of the target and model will then no longer overlap.

4.2.5. Homology and Completeness

Newman (unpublished) analyzed a series of T_2 translation functions (TFFC) of aspartic proteinases of varying structural homology as quantified by the Structural Distance Measure *(77)* and found a good anticorrelation with the signal/noise ratio (Fig. 2). Homology is undoubtedly therefore the most significant single variable determining success or failure. Dealwis (unpublished) found that for the inhibitor-complexed mouse submaxillary renin structure, which has 4 molecules/asymmetric

Table 1
Signal/Noise Ratios for T_2 Functions for Mouse Renin (Dealwis, unpublished)[a]

Molecule	1	2	3	4
Original model	1.0	2.0	0.7	1.9
After refinement	10.6	26.2	29.8	44.9

[a]The values given are for the correct peak relative to the highest noise peak.

unit in $P2_1$, the signal/noise ratio for each molecule increased considerably when the translation functions were repeated with the refined models (Table 1). Most of the improvement is the result merely of the optimization of the rigid body orientation. Since for each molecule the information from all the previous molecules is used, one would expect the signal/noise ratio to increase with the completeness as more molecules are determined. This clearly occurs in the case of the refined model, but not for the original model. A possible way around the problem of accuracy of the orientation has been suggested *(50,55,56)*: multiple translation searches are run using orientations systematically varied around the maxima of the rotation function, with a small step size.

5. Packing Functions

Packing functions are rarely used as the sole method for structure solution, but are used in conjunction with translation functions to eliminate regions of the translation vector space from further consideration. This is particularly important in the case of the computationally less efficient R-factor and correlation coefficient-based functions. Using the FFT-based functions, there is, however nothing to be gained, except perhaps to verify the correctness of the solution. Verification can also conveniently be performed on a computer graphics system, and several programs are available for this purpose: TOM/FRODO CRYS option *(78)*, PROPACK *(79)*, and MOLPACK *(80)*.

ULTIMA *(8)* uses a six-dimensional search, eliminating a large proportion of the trial structures by virtue of overlaps between ellipsoids fitted to the molecular envelopes. Then a standard R-factor based on a very low resolution subset of the data is calculated for all trial structures passing the overlap tests. The structure factor calculation is based on spherically averaged group scattering factors for maximum efficiency. The program is used often for solving small DNA structures.

MERLOT has a packing function program (PAKFUN), based on an algorithm of Bott and Sarma *(81)*, that translates the model and its symmetry equivalents over a three-dimensional grid, calculates intermolecular contacts, and eliminates a trial solution as soon as a specified number of bad contacts have been logged. The search time is reduced by considering atoms in increasing order of depth from the surface. Louie et al. *(82)* suggested the similar strategy of omitting interior atoms in the search for bad contacts. Mande and Suguna *(83)* showed how to increase the efficiency of the search procedure, but their algorithm appears to be practical only in low-symmetry space groups.

X-PLOR *(20)* uses a packing algorithm *(84)* that is based on logical operations between molecular volumes represented on a bit grid; this is considerably faster than specific calculation of interatomic distances. However, Brünger reported that for the Fab 26-10 structure *(46)*, because the solvent content is about 60%, packing considerations were of no help in the structure solution.

It is worth pointing out that, provided the orientation is precisely determined, the subtraction of intramolecular vectors in the T_1 and T_2 functions does provide discrimination against false solutions that imply molecular overlap, a fact that is not commonly appreciated. This can be understood as follows: the intramolecular-vector subtracted target Patterson will not have significant peaks in the neighborhood of the origin. The calculated Patterson consists of sets of peaks that move relative to each other as the translation vectors are varied, but do not change height unless different sets happen to overlap. If there is molecular overlap, then this Patterson will have high density values near the origin and correspondingly reduced density elsewhere. Hence, the product function value will be reduced when there is molecular overlap. When used to position multiple subunits, the T_2 function in the program TFFC subtracts known intermolecular, as well as intramolecular, vectors. The effect of this is to discriminate against false solutions in which known and unknown parts share intermolecular vectors.

6. Rigid Body Refinement

Traditionally, the six rigid body parameters of a Molecular Replacement solution are refined with a least-squares refinement program as CORELS *(85)*, TRAREF *(86)*, X-PLOR *(87)*, RESTRAIN *(88)*, or ROTLSQ *(89)*. Derewenda *(90)* has described a useful variation. Alter-

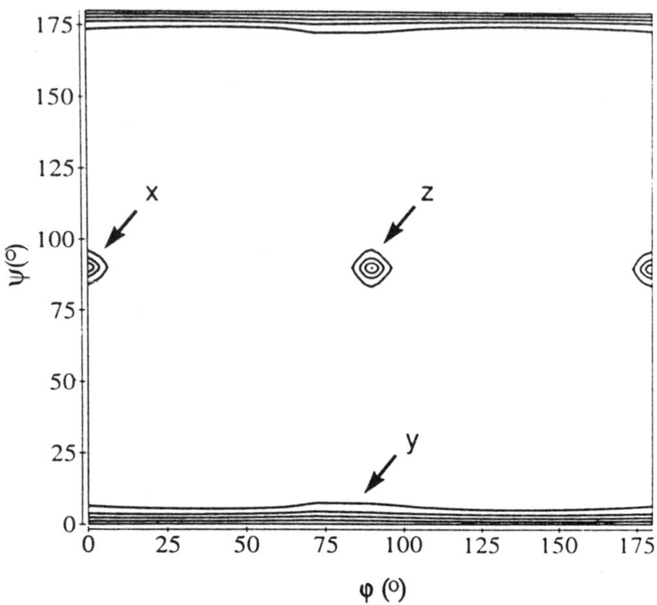

Fig. 3. Fab fragment of mutated antidigoxin antibody. Section κ = 180° of a self-rotation function (X-PLOR). The step size in the spherical polar angles was 2°. The crystallographic twofold y is at ψ = 0, whereas the noncrystallographic twofold is at ψ = 90, φ = 0, and symmetry-related φ = 90, ψ = 0. Reprinted from ref. *46* with permission.

natively, a stepwise variation and R-factor calculation program, like BRUTE *(55)* and RMINIM (MERLOT) *(91)*, is used.

7. Examples

7.1. Mutated Murine Antidigoxin Fab (26–10)

Brünger used X-PLOR *(20)* to solve the structure of an Fab fragment of a mutated monoclonal antibody *(46)*. The space group is $P2_1$ with two molecules in the asymmetric unit, and detwinned data to 2.5 Å were 70% complete for reflections with $|F_o| > 1.5\sigma$. The native Patterson map was calculated on a 1.1-Å grid, resolution 15–4 Å, and the 4000 strongest vectors between 5 and 24 Å were selected. A real space self-rotation function (Fig. 3) showed that the crystal exhibits pseudo-orthorhombic symmetry. The model atomic temperature factors were doubled, and the model was put in a triclinic cell with twice the dimensions of the model in each direction. The model Patterson map was calculated on a 1.1 Å

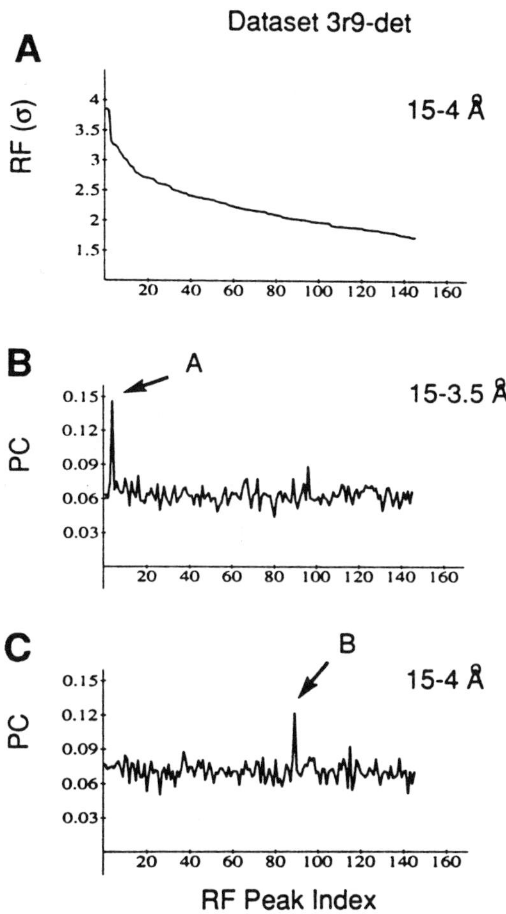

Fig. 4. Fab fragment of mutated antidigoxin antibody. Crossrotation search results (X-PLOR) are given as sorted peak values in units of r.m.s. value above the mean for the crossrotation search **(A)**. PC refinement correlation coefficients are shown for 15–3.5Å **(B)** and 15–4Å resolution **(C)**. Reprinted from ref. *46* with permission.

grid, resolution 15–4 Å, and the 3200 vectors between 4 and 24 Å 3.0σ above the mean of the map were selected. A selected subset of the peaks from the crossrotation function (Fig. 4A) was subjected to PC refinement. Initially, the rigid body parameters of the whole fragment, and then of the four Fab domains V_H, V_L, C_H1, and C_L were refined (Fig. 4B,C). Molecule A rotated by as much as 25°, whereas its elbow angle changed by 10°. The presence of the two solutions at different resolutions may be caused by the limited radius of convergence of the conjugate gradient method used. The positions for molecules A and B were obtained by six correlation coefficient searches, four of which were one-dimensional. Brünger suggests that using the highest peaks of the crossrotation function would not have produced an accurate solution. At the same time, PC refinement of the individual domains was essential for the translation function.

7.2. Bovine Eye Lens βB2-Crystallin

Driessen et al. *(54)* used the CCP4 suite *(24)* to solve the structure of the tetrameric lens protein βB2. The space group is C222 with two dimers in the asymmetric unit, and data to 3.3 Å were 90% complete for all reflections. Amplitudes of target and model were normalized with ECALC. A βB2-dimer was used as a model with its atomic temperature factors. The crossrotation function with ALMN gave two unambiguous solutions. For translation functions, data between 20.0 and 3.3 Å were used both for model and target. A three-dimensional crystallographic translation function with TFFC gave a clear solution for dimer B. Knowing B, the relative position of dimer A was determined by the sum function (Fig. 5) of the three-dimensional crystallographic and noncrystallographic translation functions. This clearly discerned among the four potential origins for A.

8. Computer Programs

BRUTE Translation function using a linear correlation coefficient. Contact R. Read, MRC Group in Protein Structure and Function, Department of Biochemistry, University of Alberta, Edmonton, Alberta, Canada T6G 2H7.

CCP4: ALMN, CAD, ECALC, SFALL, TFFC, TSEARCH. The S.E.R.C. Collaborative Computing Project No. 4. A Suite of programs for protein crystallography. Contact Daresbury Laboratory, Warrington WA4 4AD, UK.

INTREF Intensity-based domain refinement of oriented but unpositioned molecular replacement models. Contact J.M. Rini, Research Institute of Scripps Clinic, 10666 North Torrey Pines Road, La Jolla, CA 92037.

MERLOT "An Integrated package of computer programs for the determination of crystal structures by molecular replacement." Contact P. M. D. Fitzgerald, Merck, Sharp and Dohme Research Laboratories, P.O. Box 2000, RY80M203, Rahway, NJ 07065.

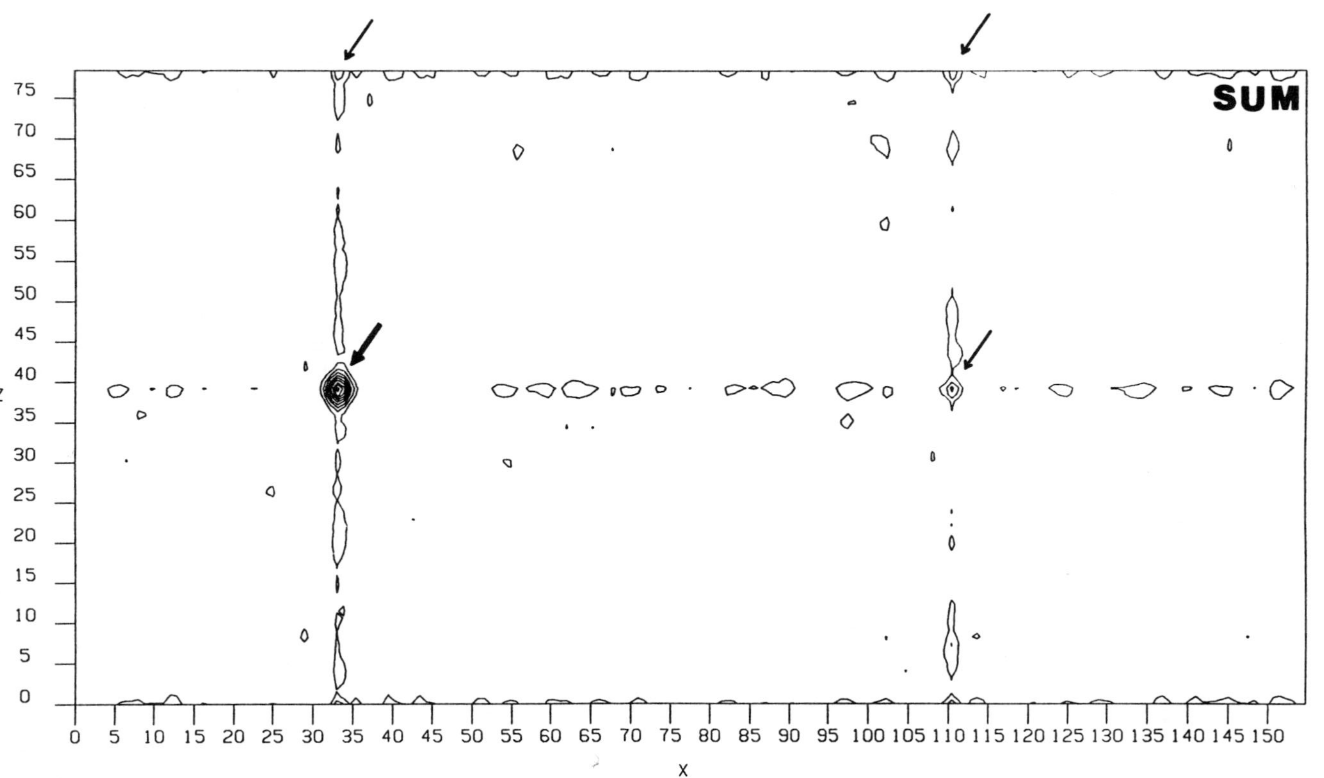

Fig. 5. Bovine eye lens βB2-Crystallin. Determination of the relative position of dimer A with respect to B by the sum function of the three-dimensional crystallographic and noncrystallographic translation functions (TFFC). The noncrystallographic asymmetric unit containing four crystallographic origins is shown.

PROTEIN "A program system for the crystal structure analysis of proteins." Contact W. Steigemann, Max-Planck-Institut fuer Biochemie, 8033 Martinsried bei München, FRG.

RATFINC A package for molecular replacement. Contact E. Lattman, Department of Biophysics, School of Medicine, The Johns Hopkins University, Baltimore, MD 21205.

ROTING A fast rotation function using radial quadrature instead of Bessel expansion and stable recurrence relationship for the rotation matrices. Contact J. Navaza, Unité d'Immunology Structurale, Institut Pasteur, 25 Rue du Dr. Roux, 75724 Paris, France.

X-PLOR "An integrated package for three-dimensional structure determination and refinement using crystallographic diffraction data or nuclear magnetic resonance data." Contact A. Brünger, The Howard Hughes Medical Institute and Department of Molecular Biophysics and Biochemistry, Yale University, 260 Whitney Avenue, P.O. Box 6666, New Haven, CT 06511.

Note Added in Proof

The fast rotation function program ROTING has been incorporated into the AMORE package, which also includes a translation function and rigid-body refinement *(92)*.

Acknowledgment

We thank Bob Sarra for reading and criticizing the manuscript.

References

1. Rossmann, M. G., ed. (1972) *The Molecular Replacement Method,* Gordon and Breach, New York.
2. Blow, D. M. (1976) Non-crystallographic symmetry, in *Crystallographic Computing Techniques* (Ahmed, F. R., ed.), Munksgaard, Copenhagen, pp. 229–238.
3. Argos, P. and Rossmann, M. G. (1980) Molecular replacement method, in *Theory and Practice of Direct Methods in Crystallography* (Ladd, M. F. C. and Palmer, R. A., eds.), Plenum, New York, pp. 361–417.
4. Machin, P. A., ed. (1985) Molecular replacement, Proceedings of the Daresbury Study Weekend, February 15-16, 1985, Science and Engineering Research Council, Daresbury Laboratory, Daresbury, UK.
5. Lattman, E. (1985) Use of the rotation and translation functions, in *Methods in Enzymology*, vol. 115 (Wyckoff, H. W., Hirs, C. H. W., and Timasheff, S. N., eds.), Academic, New York, pp. 55–77.
6. Beurskens, P. T., Gould, R. O., Bruins Slot, H. J., and Bosman, W. P. (1987) Translation functions for the positioning of a well oriented molecular fragment. *Zeitschrift f. Kristallographie* **179,** 127–159.

7. Rossmann, M. G. (1990) The molecular replacement method. *Acta Cryst.* **A46,** 73–82.
8. Rabinovich, D. and Shakked, Z. (1984) A new approach to structure determination of large molecules by multidimensional search methods. *Acta Cryst.* **A40,** 195–200.
9. Hoppe, W. (1957) Strukturanalyse über drei-und zweidimensionale Schnitte aus doppelten Pattersonfunktionen. *Acta Cryst.* **10,** 751.
10. Hoppe, W. (1957) Die Faltmolekuelmethode und ihre Anwendung in der röntgenographischen Konstitutionsanalyse von Biflorin ($C_{20}H_{20}O_4$) *Zeits. f. Elektrochem.* **61,** 1076–1083.
11. Rossmann, M. G. and Blow, D. M. (1962) The detection of subunits within the crystallographic asymmetric unit. *Acta Cryst.* **15,** 24–31.
12. Tollin, P. and Rossmann, M. G. (1966) A description of various rotation function programs. *Acta Cryst.* **21,** 872–876.
13. Lattman, E. E. and Love, W. E. (1970) A rotational search procedure for detecting a known molecule in a crystal. *Acta Cryst.* **B26,** 1854–1857.
14. Fitzgerald, P. M. D. (1988) MERLOT, an integrated package of computer programs for the determination of crystal structures by molecular replacement. *J. Appl. Cryst.* **21,** 273–278.
15. Huber, R. (1965) Die automatisierte Faltmolekülmethode. *Acta Cryst.* **19,** 353–356.
16. Huber, R. (1972) Programmed "Faltmolekuel" method, in *The Molecular Replacement Method*, Gordon and Breach, New York, pp. 165–171.
17. Fehlhammer, H. and Bode, W. (1975) The refined crystal structure of bovine β-Trypsin at 1.8 Å resolution. *J. Mol. Biol.* **98,** 683–692.
18. Huber, R. (1985) Experience with the application of Patterson search techniques, in Proceedings of the Daresbury Study Weekend, February 15–16, Science and Engineering Research Council, Daresbury Laboratory, Daresbury, UK, pp. 58–61.
19. Steigemann, W. (1974) Die Entwicklung und Anwendung von Rechenverfahren und Rechenprogrammen zur Strukturanalyse von Proteinen am Beispiel des Trypsin-Trypsininhibitor Komplexes, des freien Inhibitors und der L-Asparaginase. Ph. D. Thesis, Technical University, München, FRG.
20. Brünger, A. T. (1990) Extension of molecular replacement: a new search strategy based on Patterson correlation refinement. *Acta Cryst.* **A46,** 46–57.
21. Crowther, R. A. (1972) The fast rotation function, in *The Molecular Replacement Method*, Gordon and Breach, New York, pp. 173–178.
22. Crowther, R. A. (1973) The fast rotation function–documentation.
23. Dodson, E. J. (1985) Molecular replacement: the method and its problems, in Preceedings of the Darebury Weekend Study, February 15–16, Science and Engineering Research Council, Daresbury Laboratory, Daresbury, UK, pp. 33–45.
24. Collaborative Computing Project No. 4. (1994) The CCP4 Suite: Programs for protein crystallography. *Acta Cryst.* **D50,** 760–763.
25. Tanaka, N. (1977) Representation of the fast-rotation function in a polar coordinate system. *Acta Cryst.* **A33,** 191–193.
26. Navaza, J. (1987) On the fast rotation function. *Acta Cryst.* **A43,** 645–653.
27. Navaza, J. (1990) Accurate computation of the rotation matrices. *Acta Cryst.* **A46,** 619,620.

28. Alzari, P. M. and Navaza, J. (1991) On the use of the fast rotation function, in *Crystallographic Computing 5* (Moras, D., Podjarny, A. D., and Thierry, J. C., eds.), International Union of Crystallography, Oxford University Press, Oxford, UK, pp. 348–360.
29. Yeates, T. O. (1989) Simultaneous search for symmetry-related molecules in cross-rotation functions. *Acta Cryst.* **A45,** 309–314.
30. Fitzgerald, P. M. D. (1988) MERLOT version 2.3–documentation.
31. Tollin, P., Main, P., and Rossmann, M. G. (1966) The symmetry of the rotation function. *Acta Cryst.* **20,** 404–407.
32. Lattman, E. E. (1972) Optimal sampling of the rotation function. *Acta Cryst.* **B28,** 1065–1068.
33. Braun, W., Epp, O., Wüthrich, K., and Huber, R. (1989) Solution of the phase problem in the X-ray diffraction method for proteins with the nuclear magnetic resonance solution structure as initial model. *J. Mol. Biol.* **206,** 669–676.
34. Rao, S. N., Jih, J.-H., and Hartsuck, J. A. (1980) Rotation function space groups. *Acta Cryst.* **A36,** 878–884.
35. Moss, D. S. (1985) The symmetry of the rotation function. *Acta Cryst.* **A41,** 470–475.
36. White, H. E., Driessen, H. P. C., Slingsby, C., Moss, D. S., Turnell, W. G., and Lindley, P. F. (1988) The use of pseudosymmetry in the rotation function of γIVa-crystallin. *Acta Cryst.* **B44,** 172–178.
37. Sheriff, S., Padlan, E. A., Cohen, G. H., and Davies, D. R. (1990) Molecular-replacement structure determination of two different antibody:antigen complexes. *Acta Cryst.* **B46,** 418–425.
38. Gilliland, G. L., Winborne, E. L., Nachman, J., and Wlodawer, A. (1990) The three-dimensional structure of recombinant bovine chymosin at 2.3Å resolution. *Proteins* **8,** 82–101.
39. Bourne, Y., Abergel, C., Cambillau, C., Frey, M., Rougé, P., and Fontecilla-Camps, J.-C. (1990) X-ray crystal structure determination and refinement at l. 9Å resolution of isolectin I from the seeds of *Lathyrus ochrus*. *J. Mol. Biol.* **214,** 571–584.
40. Cooper, J. B., Khan, G., Taylor, G., Tickle, I. J., and Blundell, T. L. (1990) X-ray analyses of aspartic proteinases. II. Three-dimensional structure of the hexagonal crystal form of porcine pepsin at 2.3Å resolution. *J. Mol. Biol.* **214,** 199–222.
41. Schierbeek, A. J., Renetseder, R., Dijkstra, B. W., and Hol, W. G. J. (1985) Investigations into the limitations of a rotation and a translation function, in Proceedings of the Daresbury Study Weekend, February 15–16, Science and Engineering Research Council, Daresbury Laboratory, Daresbury, UK, pp. 16–21.
42. Cygler, M. and Anderson, W. F. (1988) Application of the molecular replacement method to multidomain proteins. 1. Determination of the orientation of an immunoglobulin Fab fragment. *Acta Cryst.* **A44,** 38–45.
43. Mondragón, A., Subbiah, S., Almo, S. C., Drottar, M., and Harrison, S. C. (1989) Structure of the amino-terminal domain of phage 434 repressor at 2.0Å resolution. *J. Mol. Biol.* **205,** 189–200.
44. Teeter, M. M., Ma, X.-Q., Rao, U., and Whitlow, M. (1990) Crystal structure of a protein-toxin α_1-purothionin at 2.5Å and a comparison with predicted models. *Proteins* **8,** 118–132.

45. Gross, P., Betzel, C., Dauter, Z., Wilson, K. S., and Hol, W. G. J. (1989) Molecular dynamics refinement of a thermitase-eglin-c complex at 1.98Å resolution and comparison of two crystal forms that differ in calcium content. *J. Mol. Biol.* **210,** 347–367.
46. Brünger, A. T. (1991) Solution of a Fab(26-10)/Digoxin complex by generalized molecular replacement. *Acta Cryst.* **A47,** 195–204.
47. Lifchitz, A. (1983) On the choice of the model cell and the integration volume in the use of the rotation function. *Acta Cryst.* **A39,** 130–139.
48. Joynson, M. A., North, A. C. T., Sarma, V. R., Dickerson, R. E., and Steinrauf, L. K. (1970) Low-resolution studies on the relationship between the triclinic and tetragonal forms of lysozyme. *J. Mol. Biol.* **50,** 137–142.
49. Blow, D. M. (1985) Introduction to rotation and translation functions, in Proceedings of the Daresbury Study Weekend, February 15–16, Science and Engineering Research Council, Daresbury Laboratory, Daresbury, UK, pp. 2–7.
50. Delarue, M., Samama, J.-P., Mourey, L., and Moras, D. (1990) Crystal structure of bovine antithrombin III. *Acta Cryst.* **B46,** 550–556.
51. Tickle, I. J. (1985) Review of space group general translation functions that make use of known structure information and can be expanded as Fourier series, in Proceedings of the Daresbury Study Weekend, February 15–16, Science and Engineering Research Council, Daresbury Laboratory, Daresbury, UK, pp. 22-26.
52. Varughese, K. I., Ahmed, F. R., Carey, P. R., Hasnain, S., Huber, C. P., and Storer, A. C. (1989) Crystal structure of a papain-E-64 complex. *Biochemistry* **28,** 1330–1332.
53. Kallen, J. and Pauptit, R. (1989) Misleading results of the self-rotation function arising from (systematically) incomplete data. *Joint CCP4 and ESF-EACBM Newsletter on Protein Crystallography* **24,** 63–66.
54. Driessen, H. P. C, Bax, B., Slingsby, C., Lindley, P. F., Mahadevan, D., Moss, D. S., and Tickle, I. J. (1991) Structure of oligomeric βB2 crystallin: an application of the T_2 translation function to an asymmetric unit containing two dimers. *Acta Cryst.* **B47,** 987–997.
55. Fujinaga, M. and Read, R. J. (1987) Experiences with a new translation-function program. *J. Appl. Cryst.* **20,** 517–521.
56. Cygler, M. and Anderson, W. F. (1988) Application of the molecular replacement method to multidomain proteins. 2. Comparison of various methods for positioning an oriented fragment in the unit cell. *Acta Cryst.* **A44,** 300–308.
57. Moews, P. C., Knox, J. R., Dideberg, O., Charlier, P., and Frère, J.-M. (1990) β-Lactamase of *Bacillus licheniformis* 749/C at 2 Å resolution. *Proteins* **7,** 156–171.
58. Sielecki, A. R., Fedorov, A. A., Boodhoo, A., Andreeva, N. S., and James, M. N. G. (1990) Molecular and crystal structures of monoclinic porcine pepsin refined at 1.8Å resolution. *J. Mol. Biol.* **214,** 143–170.
59. Yeates, T. O. and Rini, J. M. (1990) Intensity-based domain refinement of oriented but unpositioned molecular replacement models. *Acta Cryst.* **A46,** 352–359.
60. Harada, Y., Lifchitz, A., Berthou, J., and Jolles, P. (1981) A translation function combining packing and diffraction information: an application to lysozyme (high-temperature form). *Acta Cryst.* **A37,** 398–406.

61. Bhuiya, A. K. and Stanley, E. (1964) Molecular location from minimum residual calculation. *Acta Cryst.* **17,** 746–748.
62. Cutfield, J. F., Cutfield, S. M., Dodson, E. J., Dodson, G. G., and Sabesan, M. N. (1974) Low resolution crystal structure of hagfish insulin. *J. Mol. Biol.* **87,** 23–30.
63. Nixon, P. E. and North, A. C. T. (1976) Crystallographic relationship between human and hen-egg lysozymes. I. Methods for the establishment of molecular orientational and positional parameters. *Acta Cryst.* **A32,** 320–325.
64. Crowther, R. A. and Blow, D. M. (1967) A method of positioning a known molecule in an unknown crystal structure. *Acta Cryst.* **23,** 544–548.
65. Langs, D. A. (1975) Translation vector functions based on a deconvolution of the Patterson function provided by transform methods. *Acta Cryst.* **A31,** 543–550.
66. Vagin, A. A. (1989) New translation and packing functions. *Joint CCP4 and ESF-EACBM Newsletter on Protein Crystallography* **24,** 117–121.
67. Ten Eyck, L. F. (1973) Crystallographic fast Fourier transforms. *Acta Cryst.* **A29,** 183–191.
68. Agarwal, R. C. (1978) A new least-squares refinement technique based on the fast Fourier transform algorithm. *Acta Cryst.* **A34,** 791–809.
69. Colman, P. M., Fehlhammer, H., and Bartels, K. (1976) Patterson search methods in protein structure determination: β-trypsin and immunoglobulin fragments, in *Crystallographic Computing Techniques* (Ahmed, F. R., ed.), Munksgaard, Copenhagen, pp. 248–258.
70. Colman, P. M. and Fehlhammer, H. (1976) The use of rotation and translation functions in the interpretation of low resolution electron density maps. *J. Mol. Biol.* **100,** 278–282.
71. Reynolds, R. A., Remington, S. J., Weaver, L. H., Fisher, R. G., Anderson, W. F., Ammon, H. L., and Matthews, B. W. (1985) Structure of a serine protease from rat mast cells determined from twinned crystals by isomorphous and molecular replacement. *Acta Cryst.* **B41,** 139–147.
72. Read, R. J. and Schierbeek, A. J. (1988) A phased translation function. *J. Appl. Cryst.* **21,** 490–495.
73. Subbiah, S. and Harrison, S. C. (1989) A simulated annealing approach to the search problem of protein crystallography. *Acta Cryst.* **A45,** 337–342.
74. Cygler, M. and Desrochers, M. (1989) A full-symmetry translation function based on electron density. *Acta Cryst.* **A45,** 563–572.
75. Karle, I. L. and Karle, J. (1971) Structure of the chromophore from the fluorescent peptide produced by iron-deficient *Azotobacter vinelandii. Acta Cryst.* **B27,** 1891–1898.
76. Glockshuber, R., Steipe, B., Huber, R., and Plückthun, A. (1990) Crystallization and preliminary X-ray studies of the V_L domain of the antibody McPC603 produced in *Escherichia coli. J. Mol. Biol.* **213,** 613–615.
77. Johnson, M. S., Sutcliffe, M. J., and Blundell, T. L. (1990) Molecular anatomy: phyletic relationships derived from three-dimensional structures of proteins. *J. Mol. Evol.* **30,** 43–59.
78. Roussel, A., Fontecilla-Camps, J. C., and Cambillau, C. (1990) CRYSTALLIZE: A crystallographic symmetry display and handling sub-package in TOM/FRODO. *J. Mol. Graph.* **8,** 17–19.

79. Erickson, J., Neidhart, D. J., VanDrie, J., Kempf, D. J., Wang, X. C., Norbeck, D. W., Plattner, J. J., Rittenhouse, J. W., Turon, M., Wideburg, N., Kohlbrenner, W. E., Simmer, R., Helfrich, R., Paul, D. A., and Knigge, M. (1990) Design, activity, and 2.8Å crystal structure of a C_2 symmetric inhibitor complexed to HIV-1 protease. *Science* **249,** 527–533.
80. Wang, D., Driessen, H. P. C., and Tickle, I. J. (1991) MOLPACK: Molecular graphics for studying the packing of protein molecules in the crystallographic unit cell. *J. Mol. Graph.* **9,** 38, 50–52.
81. Bott, R. and Sarma, R. (1976) Crystal structure of turkey egg-white lysozyme: Results of the molecular replacement method at 5Å resolution. *J. Mol. Biol.* **106,** 1037–1046.
82. Louie, G. V., Hutcheon, W. L. B., and Brayer, G. D. (1988) Yeast iso-1-cytochrome c. A 2.8Å resolution three-dimensional structure determination. *J. Mol. Biol.* **199,** 295–314.
83. Mande, S. C. and Suguna, K. (1989) A fast algorithm for macromolecular packing calculation. *J. Appl. Cryst.* **22,** 627–629.
84. Hendrickson, W. A. and Ward, K. B. (1976) A packing function for delimiting the allowable locations of crystallized macromolecules. *Acta Cryst.* **A32,** 778–780.
85. Sussman, J. L., Holbrook, S. R., Church, G. M., and Kim, S.-H. (1977) A structure-factor least-squares refinement procedure for macromolecular structures using constrained and restrained parameters. *Acta Cryst.* **A33,** 800–804.
86. Huber, R. and Schneider, M. (1985) A group refinement procedure in protein crystallography using Fourier transforms. *J. Appl. Cryst.* **18,** 165–169.
87. Brünger, A. T., Kuriyan, J., and Karplus, M. (1987) Crystallographic R factor refinement by molecular dynamics. *Science* **235,** 458–460.
88. Driessen, H., Haneef, M. I. J., Harris, G. W., Howlin, B., Khan, G., and Moss, D. S. (1989) RESTRAIN: restrained structure-factor least-squares refinement program for macromolecular structures. *J. Appl. Cryst.* **22,** 510–516.
89. Hubbard, S. R., Hendrickson, W. A., Lambright, D. G., and Boxer, S. G. (1990) X-ray crystal structure of a recombinant human myoglobin mutant at 2.8Å resolution. *J. Mol. Biol.* **213,** 215–218.
90. Derewenda, Z. S. (1989) The minimization of errors in the molecular replacement structure solution; the effect of the errors on the least-squares refinement progress. *Acta Cryst.* **A45,** 227–234.
91. Ward, K. B., Wishner, B. C., Lattman, E. E., and Love, W. E. (1975) Structure of deoxyhemoglobin A crystals grown from polyethylene glycol solutions. *J. Mol. Biol.* **98,** 161–177.
92. Navaza, J. (1994) AMORE: an automated package for molecular replacement. *Acta Cryst.* **A50,** 157–163.

CHAPTER 8

Density Modification in X-Ray Crystallography

Alberto D. Podjarny, Bernard Rees, and Alexandre G. Urzhumtsev

1. Introduction

Crystallographic models are built by interpretation of an experimental image, the electron density map. This map is generally calculated from amplitudes measured experimentally and phases obtained with the multiple isomorphous replacement method. This method has poor precision, generating errors in the phases and therefore in the map. If the quality of the map is not sufficient to trace clearly a molecular model, it is necessary to improve the phases in order to obtain an interpretable map. Density modification methods achieve this by the application of physically meaningful constraints in real space, such as positivity, boundedness, electron density histograms, atomicity at high resolution, uniformity of solvent regions, continuity of the bio-polymer chain, and known noncrystallographic symmetry of the density distribution. To impose the physical constraints on an experimental map, an iterative algorithm has been proposed *(1,2)*. It alternates real and reciprocal space operations, and merges gradually the physical constraints with the initial amplitudes and phases.

The procedure is outlined in the flowchart shown in Fig. 1, with the following steps:

1. The electron density map ρ is calculated by Fourier transform from the experimental data F_{obs}, ϕ_{obs}.

From: *Methods in Molecular Biology, Vol. 56: Crystallographic Methods and Protocols*
Edited by: C. Jones, B. Mulloy, and M. Sanderson Humana Press Inc., Totowa, NJ

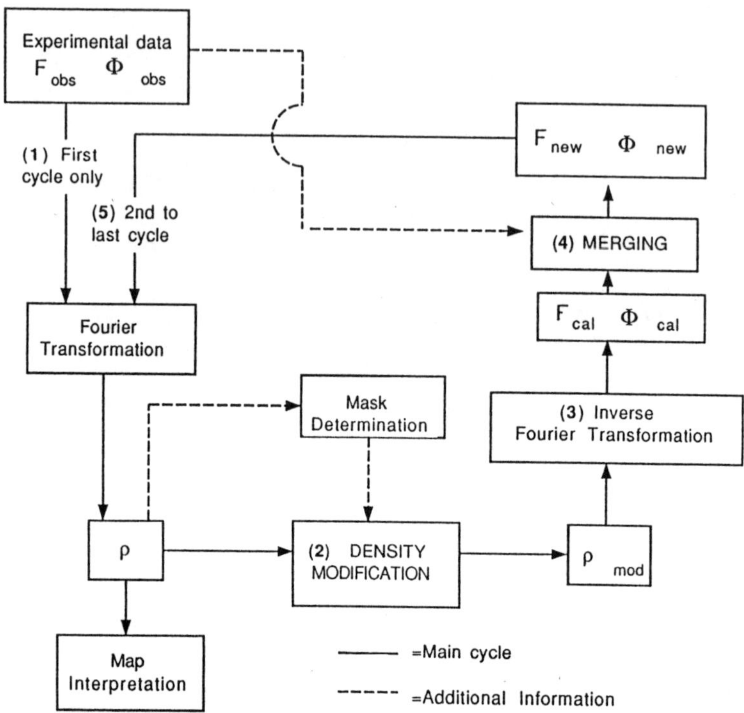

Fig. 1. Flowchart of the iterative scheme of density modification.

2. A modified electron density ρ_{mod} is obtained from the current map by the application of a known physical constraint, as discussed below. In most cases, a molecular mask has to be defined and applied.
3. The map ρ_{mod} is inverted by Fourier transformation to calculate F_{cal}, ϕ_{cal}, which carry information imposed by the physical constraint.
4. The information from the calculated structure factors F_{cal}, ϕ_{cal} is merged with the experimental structure factors to produce a new set of structure factors F_{new}, ϕ_{new}, by one of the merging methods discussed below.
5. A density map is calculated from these merged structure factors. The map is either interpreted or used as input to step 2.

2. Methods

The classical method includes two major steps, density modification to impose a physical constraint in real space and structure factor merging to combine the resulting information with the experimental data in reciprocal space. The two Fourier transformation steps (direct and inverse) of each cycle are necessary to alternate between real and reciprocal space.

The algorithm normally stops when convergence is reached according to some predefined criteria (Section 3.4.).

The result of these iterations can be used to redetermine input parameters, such as heavy atom positions, molecular envelopes, and noncrystallography symmetry operators. These parameters will then lead to a new set of initial phases, which in turn are improved by density modification, defining a larger cycle of iterations. An example of this procedure is discussed in Section 2.4.3.

2.1. Real Space Constraints

The density map is modified point by point by applying one or several of the following physical constraints. In general, the power of a constraint to refine phases increases with the number of density points it affects and the magnitude of the change it imposes.

The constraints applied so far are the following (roughly in chronological order):

1. Atomicity *(3)*.
2. Positivity and boundedness *(2)*.
3. Noncrystallographic symmetry *(4)*.
4. Solvent flatness *(4)*.
5. Map continuity and use of a partial model *(5)*.
6. Maximum entropy *(6)*.
7. Histogram matching *(7)*.

A detailed list of applications of constraints 1–5 can be found in ref. *8* and references therein, and histogram matching is reviewed in ref. *9*. In what follows, we will describe widely used constraints and recent developments in the iterative procedure. Reports on entropy maximization have recently been published by Bricogne *(10)* and Prince *(11)*.

2.1.1. Solvent Flattening

Since most solvent molecules inside a macromolecular crystal are disordered, the corresponding regions should be featureless.* To impose solvent flatness, the density values outside a molecular boundary are set to their mean value. The existence of a uniform solvent region implies

*It should be noted that a correct synthesis at finite resolution does not strictly have a constant level in the solvent region, owing to series termination errors and to partial ordering of the solvent molecules *(12)*. This is clearly shown by electron density histogram analysis *(7,13)*, and better modelization can be obtained, for example, by distance-to-boundary modulation of this density *(14,15)*. However, this effect is much smaller than the noise introduced by phase error, and therefore, the assumption of a flat solvent region is a good first approximation.

strong constraints on the structure factor phases *(4)*, as shown by the following relationship between structure factors *(16)*:

$$F_h = (1/V) \cdot \Sigma_k F_k \cdot \int_U exp[2\pi i(h-k) \cdot r]d^3r \tag{1}$$

where h,k = reciprocal space vectors, r = real space vector, U = molecular volume, and V = unit cell volume.

The smaller and more detailed the molecular volume, the larger the number of structure factors related by this equation. Test calculations *(17)* have clearly shown the importance of the molecular boundary definition. This definition of the mask has been the subject of extensive research, and the following procedures have been used (roughly in chronological order):

1. Hand digitalization of a minimap with the aid of a graphic tablet *(18,19)*.
2. Definition of the molecular region as a set of highest peaks of density, the number of which is related to the number of atoms of the molecule *(20)*. This approach has not yet been applied to the case of macromolecules.
3. Definition of the molecular volume as regions of linked high density *(5,21)*. This method allowed for the first time the fast and automatic calculation of a molecular envelope.
4. Definition of the molecular region as concentration of high density points *(22)*.
5. Definition of the molecular volume by large diffracting elements, such as Gaussian spheres positioned by low-resolution translation searches *(23,24)*.
6. Definition of the molecular envelope by a linked chain of small spheres centered on fake atoms. These fake atoms are obtained from a medium- to high-resolution electron density map *(25–27)*.

The implementation of some of these techniques is discussed in more detail in Section 3.2.

The molecular boundary is crucial for phase improvement. In general, solvent flattening can be applied reliably only if the molecular boundary can be seen clearly in the initial map. Automatic methods, such as those described in steps 3 and 4 should be used only to avoid the labor involved in tracing a visible boundary by hand. If the boundary is not visible because of poor initial data, automatic methods are unreliable and might produce a wrong mask *(17)*. Application of a mask that cuts the molecular region will distort the electron density map instead of improving it, as illustrated in Fig. 2 *(28)*.

2.1.2. Noncrystallographic Symmetry

When noncrystallographic symmetry is present, it is a very powerful constraint relating different density points within a map. Phasing rela-

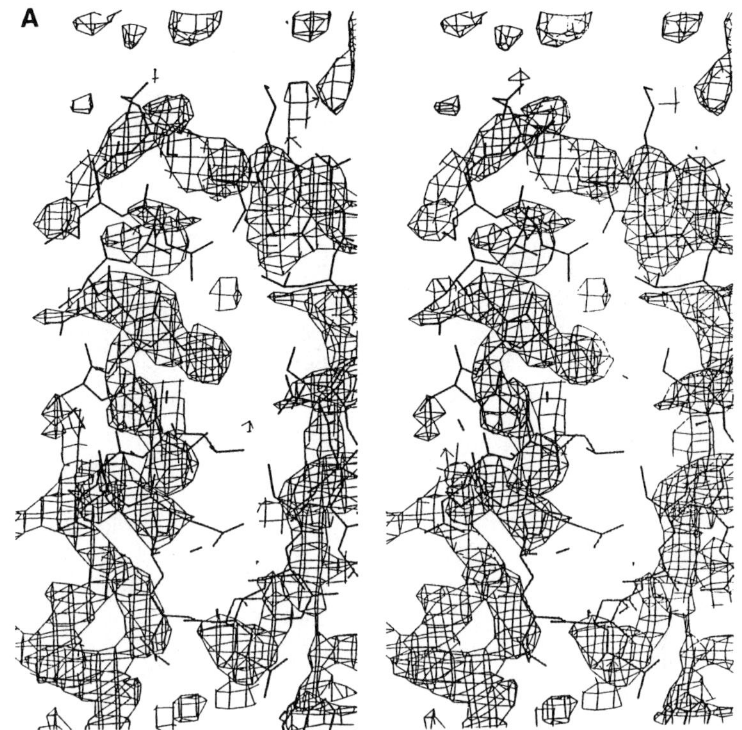

Fig. 2. These stereo maps illustrate the case of aldose reductase *(28)*. The enzyme crystallizes in space group P1, with four copies in the asymmetric unit. The self-rotation function indicated a simple local symmetry (222, or three intersecting twofold axes). The initial phasing was performed using one single mercury derivative, giving an initial SIR map that showed the molecular envelope but was not interpretable **(A)**. This envelope, was enhanced by correlation between monomers, and an averaging was performed using a mask where the protein occupied 55% of the unit cell volume. The resulting map was worse than the original one **(B)**, and in fact it increased the phase error, as calculated once the final refined model was available, from 70 to 90°. The reason for this failure was that the real symmetry was close to, but not exactly 222, the departures being significant. This had the effect that the mask was too small and cut into the real density.

The problem was solved in several steps. First, the mask was extended in real space to a volume of 83% of the unit cell, and the contrast between mask and solvent was improved by the addition of very low resolution terms. This extension solved the main problem, and the averaging now improved the map **(C)**, diminishing the phase error to 55°. This improved map was then used for a partial interpretation, and a rigid body refinement of the partial model revealed the real symmetry. Averaging using this corrected symmetry improved the maps even further **(D)**, diminishing the phase error to 43°. This new phase set was used to rerefine the heavy atom parameters and provide a better initial phase set. Averaging using this phase set produced a very good map **(E)**, with a phase error of 38° and where most of the molecule could be easily traced.

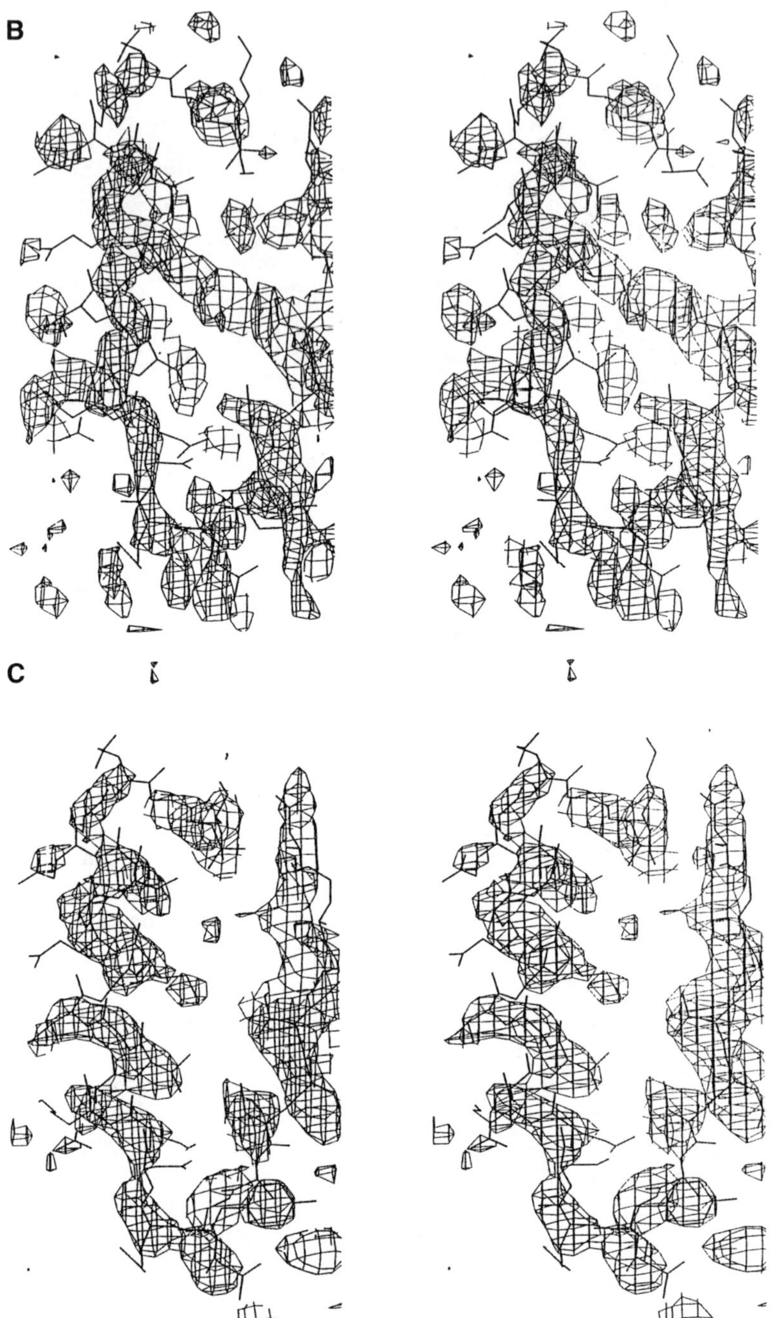

Fig. 2. B,C.

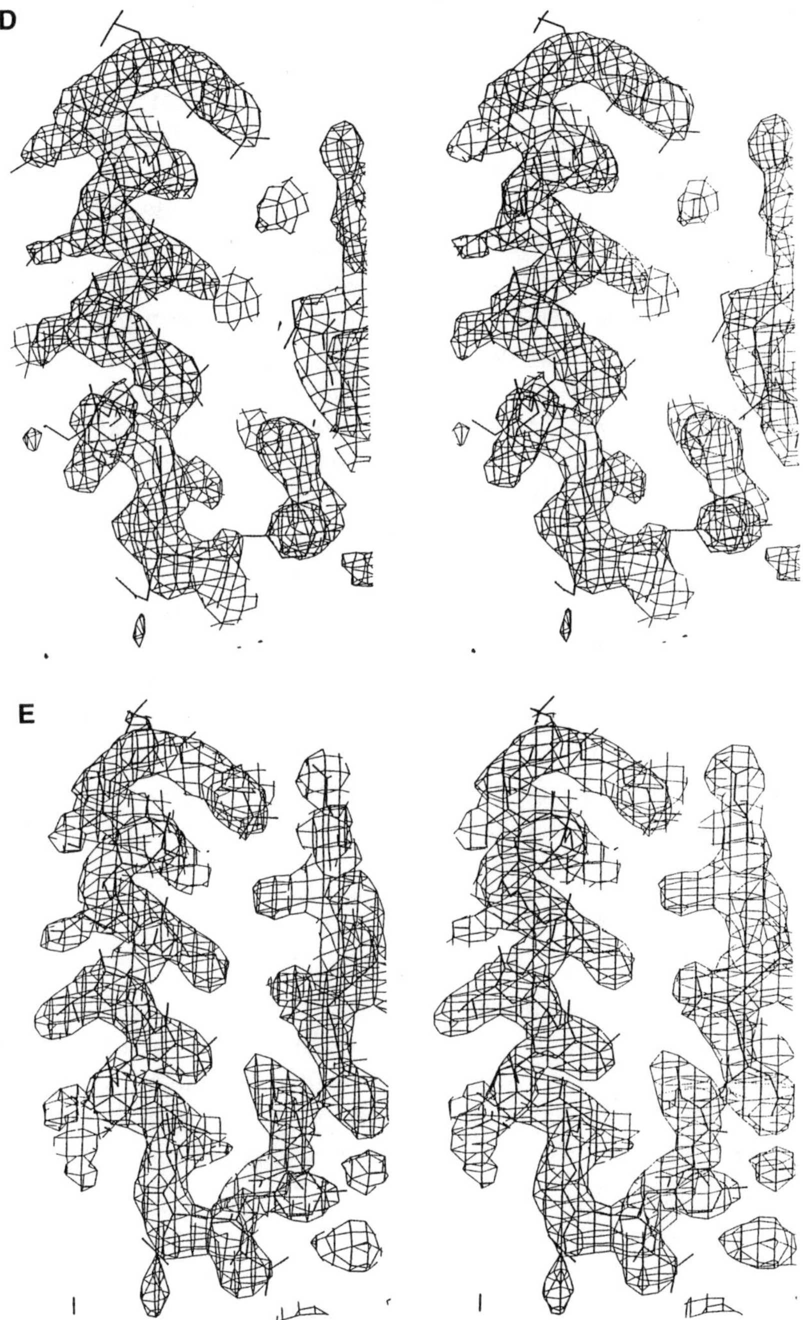

Fig. 2. D,E.

tions arising from this constraint can be formulated either in reciprocal space *(29)* or in real space *(4,30)*, where the method follows the overall scheme of density modification. In some cases (e.g., virus structures), the positioning of the noncrystallographic axes is very accurate and the redundancy of the data is very high; thus, a successful phase refinement and extension is practically assured (*see* examples in ref. *8*). In other cases, the noncrystallographic symmetry operators have to be determined. This may be done in several ways. The orientation and position of a local axis can be deduced from amplitudes alone, through analysis of the rotation and translation functions *(29,31)*. Heavy atom positions are often used to confirm this definition, e.g., in the aldose reductase case *(28)*. A better definition of the noncrystallographic symmetry operators may be obtained from skeletonized maps *(26)* or from partially refined models *(28)*. This will lead to a more interpretable map, as shown in Fig. 2. Difficult cases might also arise from high noncrystallographic symmetry *(32)*. Finally, a refinement may be possible by maximizing the correlation coefficient between the densities at related points within the molecular envelope *(25,27,33)*. The MIR or SIR map used for this refinement must be of sufficient quality.

The definition of a molecular envelope is also required. This is different from the mask used for solvent flattening, because noncrystallographic symmetry is a local constraint and its boundaries must be defined. This point is addressed in Section 3.2.

2.2. Merging of Calculated and Observed Structure Factors

An inverse Fourier transformation generates calculated structure factors and phases (F_{cal}, ϕ_{cal}) from the modified map. In order to obtain new values (F_{new}, ϕ_{new}) and synthesize a new improved map, the calculated structure factors (F_{cal}, ϕ_{cal}) are then combined with the observed amplitudes and phases (F_{obs}, ϕ_{obs}). As discussed in ref. *8*, the following alternatives are possible:

1. $F_{new} = F_{obs}$; $\phi_{new} = \phi_{cal}$
2. $F_{new} = \text{Merge}(F_{obs}, F_{cal})$; $\phi_{new} = \phi_{cal}$
3. $F_{new} = F_{obs}$; $\phi_{new} = \text{Merge}(\phi_{obs}, \phi_{cal})$
4. $F_{new} = \text{Merge}(F_{obs}, F_{cal})$; $\phi_{new} = \text{Merge}(\phi_{obs}, \phi_{cal})$

The MIR method provides not only a phase ϕ_{obs}, but also a probability distribution $\text{Prob}(\phi_{obs})$ *(34)*, which can be multimodal and therefore contains more information than the mere value of ϕ_{obs}. Approaches 1 and 2 neglect this information, and can in fact decrease the quality of the synthesis *(7,35)*. Therefore, the most widely used approach is 3. It keeps the observed amplitudes and merges the phases, through their phase probability distributions Prob(ϕ_{obs}) and Prob(ϕ_{cal}). These two distributions are multiplied, and the merged phase, ϕ_{mer}, is defined as the centroid of the product distribution. The form of the probability distribution for ϕ_{cal} is usually assumed to be the same as the one derived by Sim *(36)* for phases calculated from a partial model. This approach has been used successfully in various phase improvement and extension schemes relying on density modification, as reviewed in ref. *8*. However, Sim's derivation considered a partial, but correct atomic structure. As noted by Read *(37)*, it might be better to use the formula of Srinivasan *(38)*, where errors in the atomic positions are explicitly taken into account. Even in this case, there is a tendency to underestimate phase errors *(35)*.

While testing the possibility of merging both the amplitudes and the phases, Podjarny et al. *(23)* found that the coefficients $F_{new} = 2F_{obs} - F_{cal}$, $\phi_{new} = \phi_{mer}$ led to better results than merging only the phases. Rice *(39)* found a similar result while combining experimental and model structure factors. Stuart and Artymiuk *(40)* have proposed a coefficient of the form $F_{new} = f_{mer} [F_{obs} + Q_{mer} (F_{obs}\text{-}F_{cal})]$, $\phi_{new} = \phi_{mer}$, where Q_{mer} is a function of the figure of merit f_{mer}. Zelwer *(41)* used a merging technique in which the difference in structure factors introduced by the density modification is treated as a heavy atom contribution. In their work at high resolution, Shiono and Woolfson *(42)* proposed the use of E-maps with normalized coefficients instead of the usual F-maps.

A new map is computed using the merged Fourier coefficients, and the entire procedure is iterated until convergence. Optimally, the final map agrees better with the physical constraints because of the density modification step, and therefore, the phase error should be reduced. The merging step is needed to restore, at least partially, the experimental information; this step reduces the bias introduced by the modification.

2.3. Phase Extension

One serious limitation of MIR phasing is the rapid degradation of phase information with increasing resolution. Very often, for the reflections of highest resolution, only the amplitudes are known. Density modification techniques may provide phase information for these reflections *(43)*. The real-space constraints impose relations of the type of Eq. (1) between the structure factors of neighboring reflections. When the resolution limit of phasing is approached, the reflections within the resolution limit, known in both amplitude and phase, can be used to phase reflections just outside the limit. Only a thin shell of additional reflections can be phased in this way, but the procedure may be repeated.

Resolution can thus be progressively extended until the accumulation of errors becomes too large *(44–46)*. However, in a particular case at very high resolution, Shiono and Woolfson *(42)* found that the most successful phase extension could be achieved by including all the available amplitudes from the beginning.

Phase extension can also be used to retrieve unmeasured or unphased reflections at other resolutions. It has been shown that missing reflections, especially those of low resolution, strongly alter the quality of the electron density maps *(7,47,48)*.

2.4. Recent Developments

2.4.1. Histogram Technique

The fraction $\nu(t)dt$ of the unit cell volume with density between t and $t + dt$ can be computed independently of the model coordinates. The function $\nu(t)$, called "electron density histogram," can be used for further synthesis investigation *(49)*. It was found *(7)* that density distributions for all proteins, calculated with exact phases at a given high or middle resolution, have a similar histogram (after proper rescaling). Moreover, this histogram is sensitive to phase errors and missing reflections *(7)*, and thus can be used both as a measure of the accuracy of the synthesis *(50)* and as a tool for nonlinear density modification *(13,51)*. The density-histogram technique is a generalization of standard density modification procedures and can be related to known modification functions coming from the atomicity constraint *(9)*, e.g., $3\rho^2 - 2\rho^3$ for middle resolution *(1,52)* and a step function *(42)* for high resolution.

2.4.2. Continuity

Continuity of the map density, corresponding to chemical connectivity of the biopolymer chain, is a strong constraint that is normally used manually during map interpretation. Attempts to include this constraint in a density modification procedure have been proposed early by Bhat and Blow *(5)*. In the same line of work, Urzhumtsev *(53)* has shown that a "chain envelope" can be obtained by diminishing the averaging radius during envelope calculation, and Zelwer and Ramanoara *(21)* have recently shown that in favorable cases at high resolution, a similar "chain envelope" can be used successfully.

The continuity constraint has been differently implemented in the program PRISM by Wilson and Agard *(54)*. In this program, the density modification step is replaced by the automatic interpretation of the map in terms of a skeleton of dummy atoms to which a connectivity constraint is imposed. Thus, it goes further than the usual density modification in the direction of map interpretation. The power of this method is analyzed by Baker et al. *(55)*, and positive results have been reported by Bystroff et al. *(56)*.

2.4.3. Redetermination of Initial Parameters

The procedure described above generally converges to a map that is strongly dependent on the input parameters, such as MIR heavy atom parameters, molecular envelope, and noncrystallographic symmetry operators. It is therefore important to diminish the errors present in these parameters. The phases after density modification may provide a basis for their refinement. This has been applied to the heavy atom parameters *(57)*. It can resolve ambiguities, such as those present in SIR phasing. To converge to the most interpretable map, this procedure can be applied iteratively by alternating cycles of density modification and parameter refinement, as shown in Fig. 3. Such a technique has been used in the structure determination of the complex of $tRNA^{Gln}$-$tRNA^{Gln}$ synthetase *(58)*, the complex of $tRNA^{Asp}$-$tRNA^{Asp}$ synthetase *(59)*, and aldose reductase *(60)*. This last case is detailed in Fig. 2.

2.4.4. Global Minimization.

As an alternative to standard density modification, a global minimization procedure can be used. To do so, it is necessary to minimize the disagreement between the current density function and an ideal one that fulfills all physical requirements, subject to the constraint of the experimental amplitudes. Sayre and Toupin *(61)* used this approach to solve Sayre's equation (property of atomicity), and later it was generalized to other cases *(62–64)*. In particular, it has been applied in the program SQUASH *(65,66)* in which the simple density modification step is replaced by a Newton-Raphson minimization. In this case, the minimized quantities include not only a residual between the current density map and the modified map, but also one for Sayre's equation and one for the disagreement between observed and calculated structure factors. The resulting phases are then merged with MIR phases, using standard techniques described in Section 2.2. It should be noted that structure factor

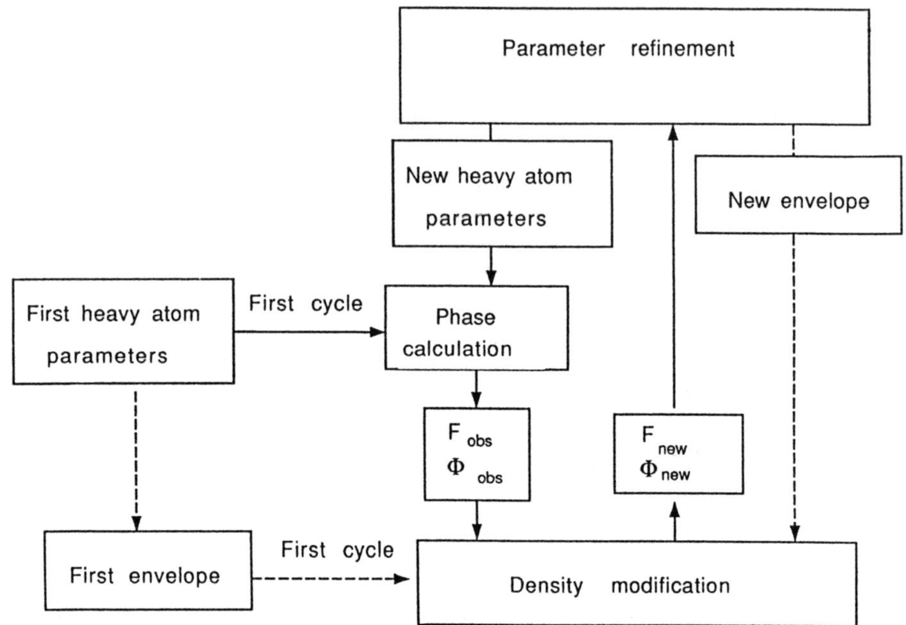

Fig. 3. Flowchart of parameter rerefinement, as applied to the case of the tRNA-synthetase complex *(57)* and of aldose reductase *(28)*. The input to the program is a starting set of heavy atom parameters, used for calculating an initial phase set and envelope. This phase set is refined by density modification, and the new calculated phase set ϕ_{new} is then used to rerefine the heavy atom parameters, which are used to generate an updated ϕ_{obs} phase set. After inspection of the density-modified map, the parameters for envelope calculation can also be varied, and a new envelope generated. The same can be done for the noncrystallographic symmetry operators. A new density modification cycle is then performed, and the procedure is iterated.

information is used twice: once inside the density modification procedure, where the residual between F_{obs} and F_{cal} is minimized, and a second time in the merging of observed and modified structure factors. As noted by the authors, this double use might pose some redundancy problems and is currently being tested.

2.4.5. Very High Resolution Cases

An analysis done by Shiono and Woolfson *(42)* showed that at high resolution (around 1 Å), even a simple density modification with a function which cuts low-density values can drastically improve the phase quality. They also demonstrated that using only the highest density values gives the best results. This technique can be used for *ab initio* phasing.

2.4.6. Ab Initio Phasing

Density modification is an iterative procedure that supposes that an approximate starting point is known, which means that an initial phase estimate is required. However, recent publications on the histogram technique at very low resolution *(67,68)* and low-density elimination at very high resolution *(42)* show that it is possible to find the correct solution by applying density modification techniques to a large number of random-phase sets, meaning that density modification could allow *ab initio* phasing. However, for the time being, this is true only for very particular cases, and MIR remains the main tool for initiating macromolecular phasing.

3. Implementation

Several groups have implemented density modification algorithms. Some of them are available through widely distributed packages. That is the case for the programs of Bricogne *(4)*, Bhat and Blow *(5)*, and Wang *(22)*, available through the CCP4 program package *(69)*. Other implementations by Jones *(25)*, Turk (MAIN) *(26)*, Rossmann et al. *(70)*, Vellieux et al. (DEMON) *(27)*, and Zhang (SQUASH) *(65)* can be obtained directly from their authors, and the programs DENMOD *(19)*, RMOL *(33)*, and DMP/DSF *(71)* can be obtained on request from the authors of the present chapter. Practical aspects concerning all implementations are discussed below.

3.1. Map Sampling and Interpolation Procedures

There are basically two maps in any density modification procedure: the input map from which we fetch the density at predefined points, and the output map used to calculate new phases. Both maps usually extend over an asymmetric unit of the crystal and are sampled over a regularly spaced grid. The optimal step size of this grid is a compromise between accuracy and economy of computing time and storage space. It can be different for the two maps, depending on the type of operations performed on them *(72)*. The output map, used for Fourier transformation, needs to be sampled at intervals only slightly smaller than $d_{min}/2$ (where d_{min} is the resolution limit of the reflections used). The same interval can be used for the input map when no interpolation occurs, as in the simple solvent flattening case, but this is not so when noncrystallographic symmetry averaging is involved. Linear interpolation results in an attenuation of the structure factors computed from the interpolated densities and in random noise. The attenuation, which increases with resolution, can be compensated partially by a Wilson-type scaling on the observed structure factors. A fine grid, typically $d_{min}/5$ or $d_{min}/6$, is needed to keep the noise at an acceptable level. A more elaborate interpolation procedure allows the use of a coarser grid. Model calculations have shown that cubic interpolation in a grid of $d_{min}/3$ is roughly equivalent to linear interpolation in a grid of $d_{min}/5$. In practice, a sampling of about $d_{min}/3$ is thus adequate in most cases provided that noncrystallographic symmetry averaging uses nonlinear interpolation *(70,73)*.

3.2. Molecular Envelope Definition

When the density modification procedures involves only solvent flattening, all that is needed is to mask out the protein regions before replacing the remaining density by a constant. As noted in Section 2.1.1., the definition of the mask can be based on the distinct characteristics of the electron density in the two regions (average density level or amplitude of fluctuations), and an automatic procedure, like that of Wang *(22)*, may be used. In this procedure, a molecular mask is obtained by replacing every point of the map by a weighted average over all positive density neighbors within a sphere of given radius, and then choosing the points above a given level. Such a local map averaging is equivalent to a convolution between the positive portions of the map and a radially weighted sphere. This convolution can be performed by reciprocal space multiplication *(53,74)*. Alternatively, the molecular volume can be defined as regions concentrating low-density points *(75)* or as regions of larger excursions, both positive and negative, from the mean value than in the solvent region *(32,76)*.

In the case of noncrystallographic symmetry averaging, it is not sufficient to distinguish disordered solvent from protein regions, but it is also necessary to separate one protein molecule from another. This is usually done by visual inspection of a low-resolution map. Since the symmetry is local, the averaging of regions not belonging to the oligomeric molecule results in a decrease in the electron density. This may be used to help define the molecular envelope. However, the average of two density points may be large even if the points are very different. Thus, in order to improve the envelope definition, it was found useful to compute a correlated density, defined as $\gamma(r)\ \rho(r)$, where γ is the local correlation coefficient between the electron density at symmetry-related points *(77)*. Once the envelope is defined, it must be checked for overlap with that of the other molecules in the crystal *(70)*. Overlapping regions must then be removed.

If an atomic model is available, it may be used to define (or modify) the envelope by including all grid points inside spheres of suitable radius centered on the atoms. If the model is only partial, Bhat and Blow *(5)* suggested expanding the molecular region by extending it to the closest grid points with high-density values. A similar procedure has been developed for the case where the model does not exist, but the electronic density map is of sufficient quality to allow an automatic interpretation in terms of a continuous skeleton *(25–27)*. This skeleton, which follows the peaks of the density, is meant to represent the biopolymer chain. Discontinuous regions of density are excluded. This representation is particularly useful to refine noncrystallographic symmetry operations *(26)*.

The phasing power of an envelope is related to its precision and to the level of detailed information it carries. This can be seen from (1), since the Fourier transform of the envelope goes to higher resolution if it is more detailed. A better resolution can be achieved by varying parameters of connectivity in the method of Bhat and Blow *(5)* or varying parameters in the envelope definition, as described in refs. *22*, *48*,and *75*. However, if the initial map is not good enough, this might just add noise. Therefore, a compromise has to be found between the resolution and the accuracy of the envelope.

3.3. Combining Different Phase Information

As described above, the combination of phase information from different sources is usually done by multiplying the corresponding probability distributions. This implies statistical independence of the two distributions, which is seldom true. Statistical dependence results in an undue sharpening of the resulting probability and an overestimated figure of merit. Also, the figure of merit of one or both of the initial distributions may already be overestimated. For these reasons, a weighted combination is often used, without much theoretical justification *(72)*:

$$P(\phi_{new}) = P(\phi_{obs})^u\ P(\phi_{cal})^v \tag{2}$$

u and v being empirically adjusted. Unless there are good reasons for a different choice, it is usually desirable to balance the two contributions.

This can be achieved by equating the average figure of merit of the two weighted distributions or the average phase differences $<|\phi_{new} - \phi_{obs}|>$ and $<|\phi_{new} - \phi_{cal}|>$.

3.4. Monitoring the Density Modification Iterations

Two questions must be answered during an iterative density modification computation: did the process converge? Was it successful?

Convergence is easily monitored by the average or rms phase difference between the current cycle n and the preceding cycle $n-1$. Often one has to consider not only the overall average, but also the average over the highest resolution reflections, for which convergence may be slower, especially in a phase-extension process. A more physical criterion of convergence is the correlation coefficient between the density maps ρ_n and $\rho/n-1$. This coefficient is equal to the weighted average of the cosine of the phase difference:

$$[\Sigma(F_n F_{n-1})\cos(\phi_n - \phi_{n-1})]/[\Sigma(F_n^2) \cdot \Sigma(F_{n-1}^2)]^{1/2} \quad (3)$$

where F is the structure factor amplitude weighted by the figure of merit and the summations are carried out over the whole reciprocal sphere.

Whether the density modification was successful is much more difficult to tell. In a successful process, the phases should become more accurately defined, which implies an increase of the figure of merit. The electron density should conform more closely to the external constraints like solvent uniformity or noncrystallographic symmetry. This means a decrease of the R factor between the observed structure factors and those recalculated after density modification, smaller density fluctuations in the solvent regions, an increase of the correlation between noncrystallographically related densities, and so forth. This defines useful numerical criteria, computed after each density modification cycle. However, these conditions are, to some extent, automatically fulfilled. As already pointed out, an artificial increase of the figure of merit results from the nonindependence of the observed and calculated phase information. The absolute value of the R factor is also difficult to assess, since it depends on the degree of constraints imposed. It is normally much smaller than the usual crystallographic R factor. For example, it has been shown that averaging about a completely randomly placed twofold axis results in an expected R value of 0.29, which is half of the R value for a random noncentrosymmetric crystal structure *(78)*.

A good criterion for success would be a better agreement with structural information not used in the density modification procedure, e.g., an increase of the noncrystallographic symmetry correlation coefficient while only solvent flattening is used. The R-free analysis, introduced by Brünger *(79)*, is another way of reserving a part of the information for monitoring purposes. Suggested initially for atomic model refinement, it removes a random subset of experimental data (say 10% of structure factors) from the refinement and uses their comparison with calculated structure factors as a check of the process. For density modification, a similar idea can be applied both in real space *(80)*, where it is closely related to the well-known omit-map technique *(81,82)*, and in reciprocal space *(55,83)*, where it is an excellent indicator of phase error.

The ultimate criterion is the quality of the final electron density map. This means less noise and more continuity (which can be measured automatically *[84]*) and, most importantly, a better interpretability through a chemically reasonable molecular model *(17)*.

4. Conclusions

It is quite clear that density modification has the power to improve an MIR map over a wide range of resolutions. A list of results has been compiled by Tulinsky *(85)*, Wang *(22)*, Podjarny et al. *(8)*, and Podjarny *(86)*, and several particular cases are discussed in detail by Dodson *(87)* and Vellieux et al. *(88)*. Currently used density modification techniques are not geared toward overcoming large errors introduced by almost random initial phases, and their application is generally limited to cases in which initial phases have been determined, most commonly with MIR methods. However, the studies at very high resolution of Shiono and Woolfson *(42)* and, at very low resolution, of Lunin *(9)* show that *ab initio* phasing is possible in special cases, which leads the way toward a generalization of these techniques. Also, it is clear that the standard iteration techniques with fixed input parameters could advantageously be replaced by more general minimization techniques, which at the same time enable a convenient refinement of the input parameters (envelope, heavy atoms, noncrystallographic symmetry operators, and so forth). The work of Zhang *(65)* and Cowtan and Main *(66)* is a step in this direction. Thus, after being widely used in their original formulation, density modification techniques are now being developed toward more powerful implementations.

Acknowledgments

The authors thank F. M. D. Vellieux and V. Yu. Lunin for making available unpublished results, and D. Logan for careful reading of the manuscript.

References

1. Hoppe,W. and Gassmann, J. (1968) Phase correction, a new method to solve partially known structures. *Acta Cryst.* **B24,** 97–107.
2. Barrett, A. N. and Zwick, M. (1971) A method for the extension and refinement of crystallographic protein phases utilising the Fast Fourier transform. *Acta Cryst.* **A27,** 6–11.
3. Hoppe, W. and Gassmann, J. (1964) Phasenbestimmung im Proteinen im Bereich von 2-Å-6bis 1.5-Å-Auflösung. *Ber. Bunsengen. Phys. Chem.* **68,** 808–817.
4. Bricogne, G. (1974) Geometric sources of redundancy in intensity data and their use in phase determination. *Acta Cryst.* **A30,** 395–405.
5. Bhat, T. N. and Blow, D. M. (1982) A density modification method for the improvement of poorly resolved protein electron density maps. *Acta Cryst.* **A38,** 21–29.
6. Collins, D. M. (1982) Electron density images from imperfect data by iterative entropy maximisation. *Nature* **298,** 49–51.
7. Lunin, V. Yu. (1988) Use of the information on electron density distribution in macromolecules. *Acta Cryst.* **A44,** 144–150.
8. Podjarny, A., Bhat, T., and Zwick, M. (1987) Improving crystallographic macromolecular images: The real space approach. *Ann. Rev. Biophys. Biophys. Chem.* **16,** 351–373.
9. Lunin, V. Yu. (1993) Electron-density histograms and the phase problem. *Acta Cryst.* **D49,** 90–99.
10. Bricogne, G. (1993) Direct phase determination by entropy maximisation and likelihood ranking: status report and perspectives. *Acta Cryst.* **D49,** 37–60.
11. Prince, E. (1993) Construction of maximum-entropy maps, and their use in phase determination and extension. *Acta Cryst.* **D49,** 61–65.
12. Badger, J. and Caspar, D. L. D. (1991) Water structure in cubic insulin crystals. *Proc. Natl. Acad. Sci. USA* **88,** 622–626.
13. Zhang, K. Y. J. and Main, P. (1990) Histogram matching as a new density modification technique for phase refinement and extension of protein molecules. *Acta Cryst.* **A46,** 41–46.
14. Cheng, X. and Schoenborn, B. P. (1990) Hydration in protein crystals. A neutron diffraction analysis of carbonmonoxymyoglobin. *Acta Cryst.* **B46,** 195–208.
15. Urzhumtsev, A. G. and Podjarny, A. D. (1995) On the problem of solvent modelling in macromolecular crystals using diffraction data: 1. The low resolution range. *Joint ccpy ESF EACBM Newsletter.* **31,** 12–16.
16. Arnold, E. and Rossmann, M. G. (1986) Effect of errors, redundancy and solvent content in the molecular replacement procedure for the structure determination of biological macromolecules. *Proc. Natl. Acad. Sci. USA* **83,** 5489–5493.
17. Fenderson, F. F., Herriott, J. R., and Adman, E. T. (1990) An evaluation of selected density-modification methods for protein structure determination. *J. Appl. Cryst.* **23,** 115–131.
18. Hendrickson, W. A., Klippenstein, G. L., and Ward, K. B. (1975) Tertiary structure of myohemerythrin at low resolution. *Proc. Natl. Acad. Sci. USA* **72,** 2160–2164.

19. Schevitz, R. W., Podjarny, A. D., Zwick, M., Hughes, J. J., and Sigler, P. B. (1981) Improving and extending the phases of medium and low resolution macromolecular structure factors by density modification. *Acta Cryst.* **A37,** 669–677.
20. Simonov, V. I. (1976) Phase refinement by the method of modification and Fourier transformation of an approximate electron density distribution, in *Crystallographic Computing Techniques* (Ahmed, F. R., Huml, K., and Sedlacek, B., eds.), Munskgaard, Copenhagen, pp. 138–143.
21. Zelwer, Ch. and Ramanoara, E. (1993) The use of a chain envelope combined with the IPD solvent flattening technique to refine the phases of the Met-RS structure. *3rd European Workshop on Crystallography of Biological Macromolecules,* Como (Italy) May 24–28 , 1993, M5.
22. Wang, B. C. (1985) Resolution of phase ambiguity in macromolecular crystallography. *Methods Enzymol.* **115,** 90–112.
23. Podjarny, A. D., Rees, B., Thierry, J. C., Cavarelli, J., Jésior, J. C., Roth, M., Lewitt-Bentley, A., Kahn, R., Lorber, B., Ebel, J. P., Giegé, R., and Moras, D. (1987) Yeast $tRNA^{Asp}$-Aspartyl tRNA synthetase complex: Low resolution crystal structure. *J. Biomol. Struct. and Dynamics* **5,** 187–198.
24. Lunin, V. Yu., Lunina, N. L., Petrova, T. E., Vernoslova, E. A., Urzhumtsev, A. G., and Podjarny, A. D. On the Ab-Initio solution of the phase problem for macromolecules at very low resolution: the few atoms model method. *Acta Cryst.* D., in press.
25. Jones, T. A. (1992) a, yaap, asap, @#*? A set of averaging programs. *Collaborative Computational Project Number 4: Proceedings of the Study Weekend "Molecular replacement,"* 91–105.
26. Turk, D. (1992) Weiterenwicklung eines Programs für Molekülgraphik und Elektronendichte-Manipulation und seine Anwendung auf verschiedene Protein-Strukturaufklärungen. PhD Thesis, University of München.
27. Vellieux, F. M. D., Hajdu, J., Verlinde, C. L. M. J., Groendijk, H., Read, R. J., Greenhough, T. J., Campbell, J. W., Kalk, K. H., Littlechild, J. H., Watson, H. C., and Hol, W. G. J. (1993) Structure of glycosomal glyceraldehyde-3-phosphate dehydrogenase from *Trypanosoma brucei* determined from Laue data. *Proc. Natl. Acad. Sci. USA* **90,** 2355–2359.
28. Tête-Favier, F., Rondeau, J.-M., Podjarny, A., and Moras, D. (1993) Structure determination of aldose reductase: joys and traps of local symmetry averaging. *Acta Cryst.* **D49,** 246–256.
29. Rossmann, M. G. and Blow, D. M. (1962) The detection of sub-units within the crystallographic asymmetric unit. *Acta Cryst.* **15,** 24–31.
30. Johnson, J. E. (1978) Averaging of electron density maps. *Acta Cryst.* **B34,** 576,577.
31. Rossmann, M. G., Blow, D. M., Harding, M. M., and Coller, E. (1964) The relative positions of independent molecules within the same asymmetric unit. *Acta Cryst.* **17,** 338–342.
32. Jones, E. Y., Walker, N. P. C., and Stuart, D. (1991) Methodology employed for the structure determination of tumour necrosis factor, a case of high non-crystallographic symmetry. *Acta Cryst.* **A47,** 753–770.
33. Rees, B. (1990) RMOL program and manual. Internal report, IBMC, Strasbourg, France.

34. Blow, D. M. and Crick, F. H. C. (1959) The treatment of errors in the isomorphous replacement method. *Acta Cryst.* **12,** 794–802.
35. Urzhumtsev, A. G. and Lunin, V. Yu., unpublished.
36. Sim, G. A. (1959) The distribution of phase angles for structures containing heavy atoms II. A modification of the normal heavy atom method for noncentrosymmetrical structures. *Acta Cryst.* **12,** 813–815.
37. Read, R. J. (1986) Improved Fourier coefficients for maps using phases from partial structures with errors. *Acta Cryst.* **A42,** 140–149.
38. Srinivasan, R. (1966) Weighting functions for use in the early stages of structure analysis when a part of the structure is known. *Acta Cryst.* **20,** 143,144.
39. Rice, D. W. (1981) The use of phase combination in the refinement of phosphoglycerate kinase at 2.5 Å resolution. *Acta Cryst.* **A37,** 491–500.
40. Stuart, D. and Artymiuk, P. (1985) The use of phase combination in crystallographic refinement: the choice of amplitude coefficients in combined syntheses. *Acta Cryst.* **A40,** 713–716.
41. Zelwer, C. (1988) The isomorphous pseudo-derivative technique for phase refinement by density modification. *Acta Cryst.* **A44,** 485–495.
42. Shiono, M. and Woolfson, M. M. (1992) Direct-space methods in phase extension and phase determination. I. Low-density elimination. *Acta Cryst.* **A48,** 451–456.
43. Cannillo, E., Oberti, R., and Ungaretti, L. (1983) Phase extension and refinement by density modification in protein crystallography. *Acta Cryst.* **A39,** 68–74.
44. Argos, P., Ford, G. C., and Rossmann, M. G. (1975) An application of the molecular replacement technique in direct space to a known protein structure. *Acta Cryst.* **A31,** 499–506.
45. Rayment, I. (1983) Molecular replacement method at low resolution: optimum strategy and intrinsic limitations as determined by calculations on icosahedric virus model. *Acta Cryst.* **A39,** 102–116.
46. Rossmann, M. G. (1990) The molecular replacement method. *Acta Cryst.* **A46,** 73–82.
47. Podjarny, A. D., Schevitz, R. W., and Sigler, P. B. (1981) Phasing low-resolution macromolecular structure factors by matricial direct methods. *Acta Cryst.* **A37,** 662–668.
48. Urzhumtsev, A. G. (1991) Low-resolution phases: influence on SIR syntheses and retrieval with double-step filtration. *Acta Cryst.* **A47,** 794–801.
49. Podjarny, A. D. and Yonath, A. (1977) Use of matrix direct methods for low-resolution phase extension for tRNA. *Acta Cryst.* **A33,** 655–661.
50. Luzzati, V., Mariani, P., and Delacroix, H. (1988) Cubic phases of lipid-containing system. Structure analysis and biological implications. *J. Mol. Biol.* **204,** 165–189.
51. Harrison, R. W. (1988) Histogram specification as a method of density modification. *J. Appl. Cryst.* **21,** 949–952.
52. Collins, D. M., Brice, M. D., La Cour, T. F. M., and Legg, M. G. (1976) Fourier phase refinement and extension by modification of electron density maps, in *Crystallographic Computing Techniques* (Ahmed, F. R., Huml, F., and Sedlacek, B., eds.), Munskgaard, Copenhagen, pp. 330–335.
53. Urzhumtsev, A. G. (1985) The use of local averaging to analyse macromolecular images in the electron density maps. Preprint, USSR Academy of Sciences, Pushchino, USSR.

54. Wilson, C. and Agard, D. A. (1993) PRISM: Automated crystallographic phase refinement by iterative skeletonization. *Acta Cryst.* **A49,** 97–104.
55. Baker, D., Bystroff, C., Fletterick, R. J., and Agard, D. A. (1993) PRISM: Topologically constrained phase refinement for macromolecular crystallography. *Acta Cryst.* **D49,** 429–439.
56. Bystroff, C., Baker, D., Fletterick, R. J., and Agard, D. A. (1993) PRISM: Application to the solution of two protein structures. *Acta Cryst.* **D49,** 440–448.
57. Cura, V., Podjarny, A. D., and Moras, D. (1992) Heavy atom refinement against solvent-flattened phases. *Acta Cryst.* **A48,** 756–764.
58. Rould, M. A., Perona, J. T., Söll, D., and Steitz, T. A. (1989) Structure of *E. Coli* glutaminyl tRNA synthetase complexed with $tRNA^{Gln}$ and ATP at 2.8 Å resolution. *Science* **246,** 1135–1142.
59. Ruff, M., Krishnaswamy, S., Boeglin, M., Polterszman, A., Mitschler, A., Podjarny, A., Rees, B., Thierry, J. C., and Moras, D. (1991) Class II aminoacyl transfer synthetases: crystal structure of yeast aspartyl tRNA synthetase complexed with $tRNA^{Asp}$. *Science* **252,** 1682–1689.
60. Rondeau, J. M., Tête-Favier, F., Podjarny, A. D., Reymann, J. M., Barth, P., Biellmann, J. F., and Moras, D. (1992) Novel NADPH-binding domain revealed by the crystal structure of aldose reductase. *Nature* **355,** 469–472.
61. Sayre, D and Toupin, R. (1975) Major increase in speed of least-squares phase refinement. *Acta Cryst.* **A31,** S20.
62. Navaza, J., Castellano, E. E., and Tsoucaris, G. (1983) Constrained density modifications by variational techniques. *Acta Cryst.* **A39,** 622–631.
63. Lunin, V. Yu. (1985) Use of Fast Differentiation algorithm for phase refinement in protein crystallography. *Acta Cryst.* **A41,** 551–556.
64. Navaza, J. (1986) The use of non-local constraints in maximum-entropy electron density reconstruction. *Acta Cryst.* **A42,** 212–223.
65. Zhang, K. Y. J. (1993) SQUASH- combining constraints for macromolecular phase refinement and extension. *Acta Cryst.* **D49,** 213–222.
66. Cowtan, K. D. and Main, P. (1993) Improvement of macromolecular electron-density maps by the simultaneous application of real and reciprocal space constraints. *Acta Cryst.* **D49,** 148–157.
67. Lunin, V. Yu., Urzhumtsev, A. G., and Skovoroda, T. P. (1990) Direct low-resolution phasing from electron-density histograms in protein crystallography. *Acta Cryst.* **A46,** 540–544.
68. Lunin, V. Yu. and Vernoslova, E. A. (1991) Frequency-restrained structure-factor refinement. II. Comparison of methods. *Acta Cryst.* **A47,** 238–243.
69. CCP4 (1979) The SERC (UK) Collaborative Computational Project Number 4, a suite of programs for protein crystallography, distributed from Daresbury Laboratory, Warrington WA4 4AD, UK.
70. Rossman, M. G., McKenna, R., Tong, L., Xia, D., Dai, J., Wu, H., Choi, H.-K., and Lynch, R. E. (1992) Molecular replacement real-space averaging. *J. Appl. Cryst.* **25,** 166–180.
71. Lunina, N. L. (1992) DSF/DMP. Internal report, IMPB, Pushchino, Moscow Region, Russia.

72. Bricogne, G. (1976) Methods and programs for direct space exploitation of geometric redundancies. *Acta Cryst.* **A32,** 832–847.
73. Rees, B., unpublished.
74. Leslie, A. G. W. (1987) A reciprocal-space method for calculating a molecular envelope using the algorithm of B. C. Wang. *Acta Cryst.* **A43,** 134–136.
75. Urzhumtsev, A. G., Lunin, V. Yu., and Luzyanina, T. B. (1989) Bounding a molecule in a noisy synthesis. *Acta Cryst.* **A45,** 34–39.
76. Reynolds, B., Remington, S. J., Weaver, L. H., Fisher, R. G., Anderson, W. F., Ammon, H. L., and Matthews, B. W. (1985) Structure of a serine protease from rat mast cells determined from twinned crystals by isomorphous and molecular replacement *Acta Cryst.* **B41,** 139–147.
77. Rees, B., Bilwes, A., Samama, J.-P., and Moras, D. (1990) Cardiotoxin $V^{II}4$ from *Naja mossambica*. The refined crystal structure. *J. Mol. Biol.* **214,** 281–297.
78. Rees, D. C. (1983) Largest likely values for R-factors calculated after phase refinement by non-crystallographic symmetry averaging. *Acta Cryst.* **A39,** 916–920.
79. Brünger, A. (1992) Free R value: a novel statistical quantity for assessing the accuracy of crystal structures. *Nature* **355,** 472–474.
80. Brünger, A. (1993) Assessment of phase accuracy by cross validation; the free R value. Methods and applications. *Acta Cryst.* **D49,** 24–36.
81. Artymiuk, P. J. and Blake, C. C. F. (1981) Refinement of human lysozyme at 1.5Å resolution. Analysis of non-bonded and hydrogen-bond interactions. *J. Mol. Biol.* **152,** 737–762.
82. Bhat, T. N. and Cohen, G. H. (1984) Omit-map: an electron density map suitable for the examination of errors in a macromolecular model. *J. Appl. Cryst.* **17,** 244–248.
83. Lunin, V. Yu. (1992) Personal communication.
84. Baker, D., Krukowski, A. E., and Agard, D. A. (1993) Uniqueness and *ab initio* phase problem in macromolecular crystallography. *Acta Cryst.* **D49,** 186–192.
85. Tulinsky, A. (1985) Phase refinement/extension by density modification. *Methods Enzymol.* **115,** 77–89.
86. Podjarny, A. D. (1989) Improving protein phases in real space. *Collaborative Computational Project Number 4: Proceedings of the Study Weekend "Improving Protein Phases,"* pp. 65–72.
87. Dodson, E. (1989) Improving electron density maps by density modification. *Collaborative Computational Project Number 4: Proceedings of the Study Weekend "Improving Protein Phases,"* pp. 73–87.
88. Vellieux, F. M. D., Groendijk, H., Huintema, F., Swarte, M. B. A., Drenth, J., and Hol, W. G. J. (1989) The use of solvent-flattening procedures in the crystal structure determination of quinoprotein methylamine dehydrogenase. *Collaborative Computational Project Number 4: Proceedings of the Study Weekend "Improving Protein Phases,"* pp. 88–99.

CHAPTER 9

Refinement of Protein and Nucleic Acid Structures

Eric Westhof and Philippe Dumas

1. Introduction

Refinement constitutes the last of the three main steps in the crystallographic establishment of a molecular structure, which are first, crystal growth and data collection; second, phase determination and calculation of electron density maps; and third, model building and refinement. This final part is necessary because the structural models arrived at after the first two steps are approximate and usually contain errors in the tracing of the macromolecular chain. During the crystallographic refinement process, the macromolecular model is changed so that the agreement between the measured diffraction intensities and those calculated improves. This improved agreement between observed and calculated structure factors leads to better phases and, concomitantly, to improved electron density maps. The refinement process is monitored by the conventional crystallographic R factor, defined by the equation:

$$R = (\sum_{hkl} \left| |F^{hkl}_{obs}| - |F^{hkl}_{calc}| \right| / \sum_{hkl} |F^{hkl}_{obs}| \qquad (1)$$

In this expression, the sums should include all the observed reflections. An R value around 60% for a biological macromolecular crystal indicates that the model does not fit the data better than a random distribution of atoms. Before starting the refinement process, R factors around 40–50% are commonly obtained. Depending on the size of the macromolecule, the resolution, and the quality of the data set, the refinement process converges to R factors between 10 and 25%. The latter value is nowadays considered to be insufficient.

From: *Methods in Molecular Biology, Vol. 56: Crystallographic Methods and Protocols*
Edited by: C. Jones, B. Mulloy, and M. Sanderson Humana Press Inc., Totowa, NJ

The errors contained in the starting model can be of different types and severities *(1)*: incorrect orientations of secondary-structure elements, wrongly built connections between secondary-structure elements, locally wrong sequence, or conformational errors (either in the macromolecular backbone or in the side chains). With the first two types of errors, the refinement usually stalls and the R factor does not improve beyond 25–27%. A revision of the model with partial rebuilding is then necessary to continue the refinement process. Since refinement is a gradual and iterative process, the other errors can usually be corrected by careful study of difference electron density maps followed by manual fitting.

The relationships between an atomic model and its associated diffraction pattern are well known *(2,3)*. The calculated intensities depend on the spatial, thermal, and occupancy parameters of each atom. The refinement techniques are based on the least-squares methods, which iteratively adjust the parameters of the atoms constituting the asymmetric unit of the crystal in order to minimize the sum of the squares of the properly weighted differences between observed and calculated structure factors. Whatever the algorithm used, the radius of convergence of such methods is small, about one-fourth of the highest resolution (or smallest spacing between diffracting planes) used in the calculation. If the initial parameters are outside that radius, the refinement may, and most often will, converge into false minima (usually with distorted conformation and geometry). Therefore, human intervention is necessary, and manual corrections are made at the graphics terminal. This refitting process is extremely slow, time-consuming, and requires a good knowledge of stereochemistry. Consequently, techniques of simulated annealing using molecular dynamic algorithms are now the rule, despite their high computing costs. Such methods are discussed elsewhere in this book (Chapter 10).

The choice of the refinement algorithm depends mainly on the ratio of observations to parameters. The number of observations is determined by the diffracting power, the crystalline order, and the resolution or spacing to which reflections could be measured. Except for small proteins or some DNA oligomers, biological macromolecules rarely diffract to spacings <1.7–2.0 Å (1 Å = 0.1 nm). Usual resolution is 2.4–2.5 Å for proteins and 2.8–3.0 Å for nucleic acids. The number of reflections is inversely proportional to the cube of the resolution (thus, for lysozyme, there are 290 independent data at 6 Å, but 120,000 at 1 Å resolution). Since at least four parameters per atom are required, the ratio of data to parameters is often low (e.g., for a DNA hexamer diffracting at 1.4 Å, that ratio is around 4, whereas for a tRNA crystal diffracting at 3.0 Å, it is around 0.7). Thus, full-matrix least-squares methods are not appropriate for biological macromolecules and additional information needs to be added to the observed data in order to define the atomic parameters uniquely. This fact clearly appeared to be necessary a long time ago for the refinement of smaller molecules *(4)*. This external knowledge consists of geometrical values (distances, angles, planarity, and so on) and stereochemical conditions (chiralities, torsion angles, and so on), both obtained from small molecule crystallography and stored in dictionaries accessible to the refinement programs. At low values of data-to-parameter ratios, constrained or rigid-body least-squares algorithms are used, whereas at higher values, restrained least-squares algorithms that vary all atomic parameters are very efficient.

The number of degrees of freedom is reduced in constrained refinement by fixing the geometrical and stereochemical parameters of chosen groups, leaving as variables the six translational and rotational degrees of freedom, whereas the links between groups are restrained to standard values. In restrained refinement, the reduction in degrees of freedom is accomplished by forcing each geometrical and stereochemical parameter to be within a specified standard deviation (σ) close to a standard or dictionary value. The ideal values, as well as the σ's, are evaluated on the basis of crystal structures of the components refined with full-matrix, least-squares methods.

In short, the minimized function is the sum of several terms, each one of the same mathematical form:

$$\phi = \sum_i [1/\sigma(i)^2]\,[\mathrm{Ideal}(i) - \mathrm{Model}(i)]^2 \qquad (2)$$

2. Programs

Various computer programs are available for refining biological macromolecules, proteins, or nucleic acids either alone or in the presence of small ligands. Computer programs that are able to refine complexes of protein and nucleic acids are not currently as well documented or available. The following list gives the names with their addresses of the scientists you should contact in order to obtain the program(s) you wish to use. This list is based on a literature search for the most frequently used programs.

1. Constrained least-squares for proteins and nucleic acids: CORELS *(5)*
   Joel L. Sussman
   Department of Structural Chemistry

Weizmann Institute of Science
Rehovot, Israel

2. Restrained least-squares for proteins: PROTIN-PROLSQ *(6)*
   Wayne A. Hendrickson
   Department of Biochemistry and Molecular Biophysics
   Columbia University
   630 West 168th Street
   New York, NY 10032
3. Restrained least-squares for nucleic acids: NUCLIN-NUCLSQ *(7a,b)*
   Eric Westhof
   Institut de Biologie Moléculaire et Cellulaire
   Centre National de la Recherche Scientifique
   15, rue R. Descartes
   F-67084 Strasbourg-Cedex, France
4. Jack-Levitt energy refinement: EREF *(8,9)*
   Johann Deisenhofer
   Department of Biochemistry
   University of Texas Southwestern Medical Center
   5323 Harry Hines Blvd.
   Dallas, TX 75235
5. Refinement by simulated annealing: X-PLOR *(10,11)*
   Axel T. Brünger
   Department of Molecular Biophysics and Biochemistry
   Yale University, New Haven, CT 06511
6. Refinement by simulated annealing: GROMOS *(12)*
   Wilfred F. Van Gunsteren
   Department of Chemistry
   University of Groningen
   Nijenborgh 16
   9747 AG Groningen, The Netherlands
7. Macromolecular building: FRODO *(13)* and O *(14)*
   T. A. Jones
   Department of Molecular Biology
   Box 590, Biomedical Center
   S-751 24, Uppsala, Sweden
8. Macromolecular building: FRODO for the Evans & Sutherland (E&S) PS300 *(14)*
   F. A. Quiocho
   Rice University
   Department of Biochemistry
   PO Box 1892 Houston, TEXAS 77251

9. Macromolecular building: TOM-FRODO for Silicon Graphics *(15)*
C. Cambillau
Laboratoire CCMB Faculté Nord
Boulevard P. Dramard
F-13326 Marseille Cedex 15
France

3. Reciprocal Space Methods

With the presently available algorithms and programs, one must admit that nowadays there is no unique recipe for the refinement of a biological macromolecule. As stated by W. A. Hendrickson, "each problem is idiosyncratic." The aim is to obtain a better agreement between observed and calculated intensities while simultaneously maintaining correct and meaningful geometry and stereochemistry. As long as those criteria are met, Machiavellian principles are applicable, i.e., the end justifies the means. This statement stems from the fact that there is a lack of theoretical knowledge about the influence of the relative weights on the various terms during the refinement process. As a main guiding rule, the user should go from the known to the unknown parts, i.e., first the well-defined regions should be properly defined before tackling uncertain regions (for example, loops with high-temperature factors or complexed ligands). In addition, different approaches should be undertaken depending on the quality of the electron density map and the size and nature of the macromolecule. Figure 1 displays a typical summary of a well-driven refinement process.

1. The gross features should be settled by rigid-body or constrained least-squares together with the overall scale factor bringing the observed and calculated structure factors on the same scale and the mean temperature factor (Debye-Waller parameter or B factor). Use intensity statistics if possible. Plot the ratio of the sum of observed structure factors over the sum of calculated structure factors (divided in resolution shells) as a function of resolution. The resolution range of the data should be restricted to the low-angle part.
2. If the resolution allows for it, switch to restrained least-squares refinement. Start with strong restraints on geometry and stereochemistry and weak ones on the structure factor term. Monitor changes in the scale factor and the mean B factor. If the refinement process behaves well, i.e., the geometry and the stereochemistry do not deteriorate, put stronger weights on the structure factor term. First, choose the weight so that the low resolution has a higher contribution than the high resolution. Then, a weight

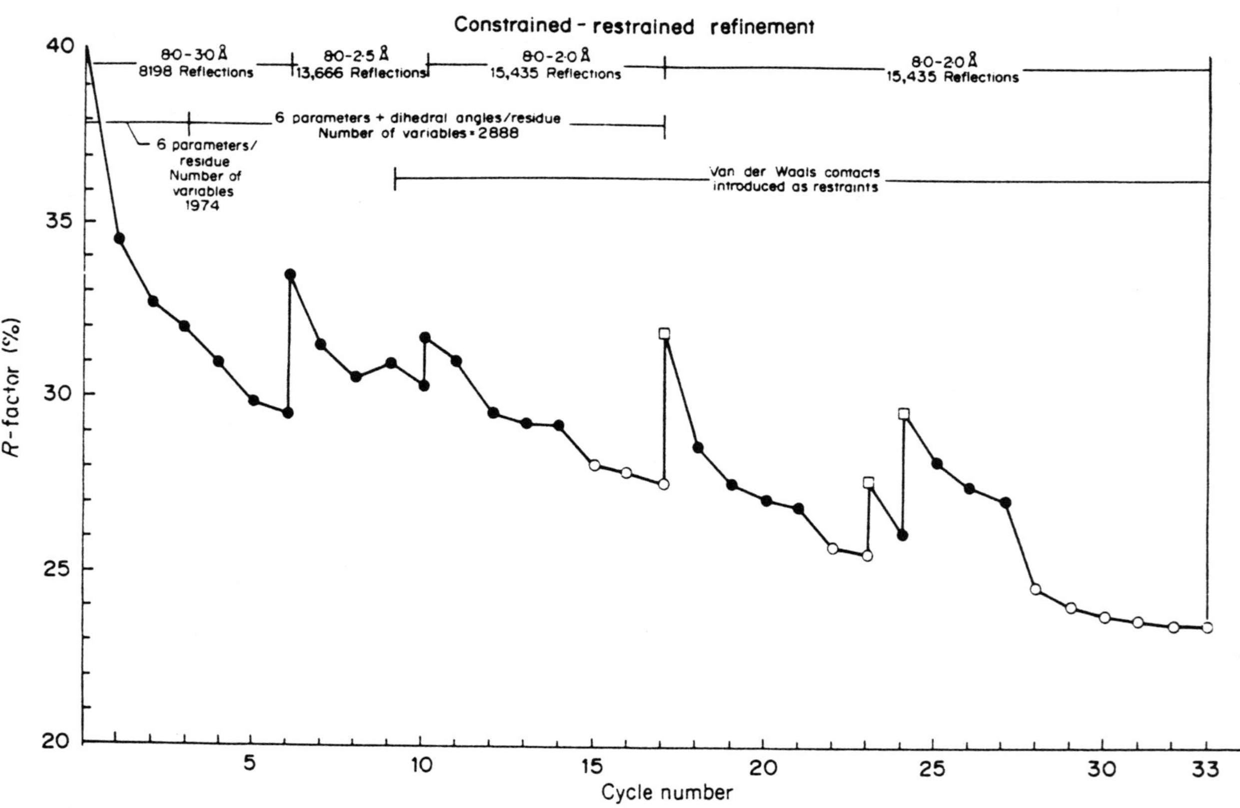
A
Constrained - restrained refinement
8·0-3·0 Å
8198 Reflections
8·0-2·5 Å
13,666 Reflections
8·0-2·0 Å
15,435 Reflections
8·0-2·0 Å
15,435 Reflections
6 parameters + dihedral angles/residue
Number of variables = 2888
6 parameters/
residue
Number of
variables
1974
Van der Waals contacts
introduced as restraints
R-factor (%)
40
35
30
25
20
5
10
15
20
25
30
33
Cycle number

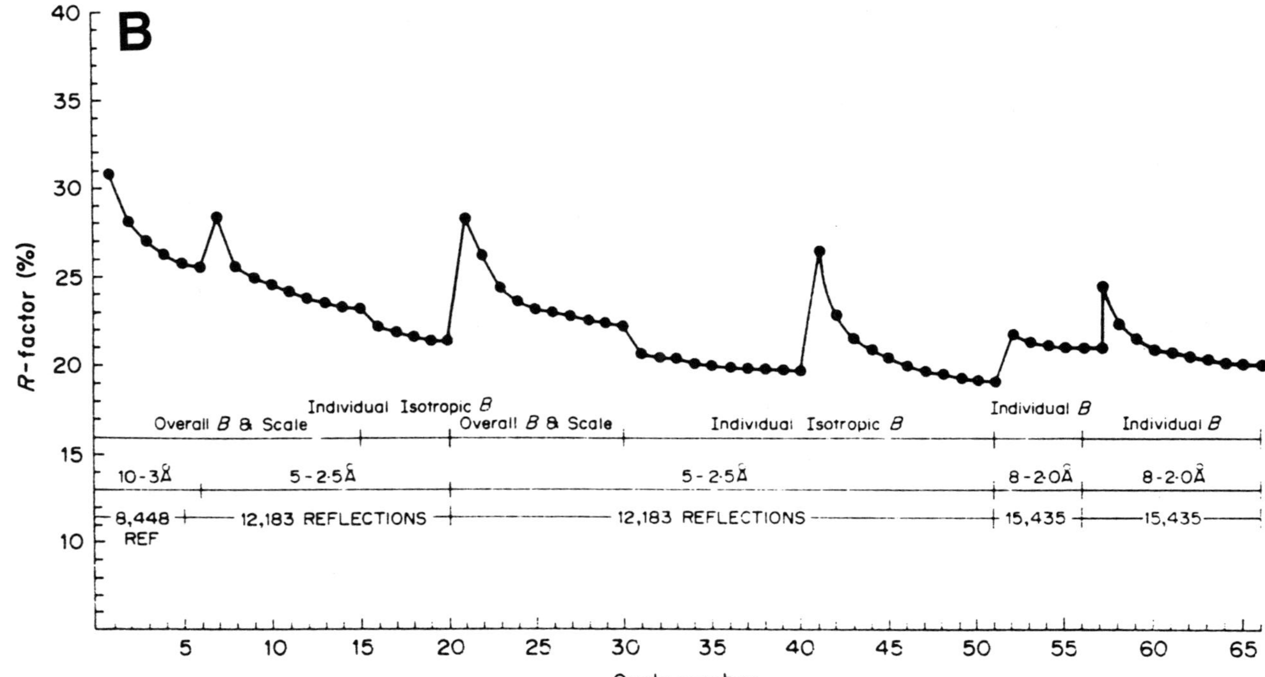

Fig. 1. Summary of the refinement of dogfish M4 apo-lactate dehydrogenase *(37)*. **(A)** Constrained-restrained refinement with CORELS *(5)*. (□) stereochemistry refinement only; (○) coordinates and B-factor refinement; (●) coordinates refinement only. **(B)** Restrained refinement with PROLSQ *(6)*. Manual intervention or inclusion of higher resolution data lead to the spikes of the otherwise smooth curve. Reprinted from ref. *37* with permission.

varying with resolution can be applied. Plot frequently the R factor (divided in resolution shells) as a function of the resolution *(16)*. The data of low-resolution range, say above 10 Å (but this limit depends on the crystal), are usually dominated by the interstitial bulk solvent and should be excluded if it is not considered theoretically during the structure factor calculation *(17)*.

3. Fourier maps (direct, with F_{obs} as coefficients or difference, with $[nF_{obs}-(n-1)F_{calc}]$ as coefficients usually with $n = 3$) should be calculated periodically and the fitting of the model in the electron density checked on a graphical device. This step is rarely unnecessary or done too frequently. The most widely used program for visualizing the electron density and manipulating the macromolecular model is FRODO (*see* Section 4.).
4. Contrary to common practice, temperature factors should be introduced and refined early in the refinement process. Start with a temperature factor per residue, and then, if enough data are available, introduce atomic temperature factors. The early introduction of temperature factors in the refinement is supported both by theoretical considerations and numerical tests *(18)*. Restraints on temperature factors should be soft, and large departures from the mean or the group values (side chain vs backbone atoms) must be monitored. Unrealistically strong temperature factor restraints will stall the refinement or, if the geometric restraints are soft, destroy the molecule. This has been observed in the case of a B-DNA oligomer *(19)* and has been established also by refinement against electron density obtained from molecular dynamics simulation *(20,21)* (Fig. 2).
5. In proteins, as well as in nucleic acid oligomers, especially at medium to high resolution (<2.0 Å), it is soon necessary to search for solvent molecules in difference Fourier maps and to include them in the refinement. At such resolution, the noninclusion of solvent molecules can lead to severe distortion in the sugar-phosphate backbone of the nucleic acid *(22)* (Fig. 3). Obviously, the solvent molecules should be monitored at later refinement steps in order to make sure that they do reappear in the difference maps.
6. The refinement is driven by progressively decreasing the σ of the structure factor term, which is chosen commonly at half the value of the mean difference between F_{obs} and F_{calc}. As the refinement progresses, this difference becomes smaller. In principle, one can stop the refinement when the value of the difference between F_{obs} and F_{calc} is within the σ value on observed structure factors (obtained, for example, from counting statistics and after various data-processing techniques of the measurements). In practice, this is rarely achieved with macromolecules. The restraints on geometrical and stereochemical terms should not be relaxed during the refinement; on the contrary, with good data and resolution, they can be enforced.

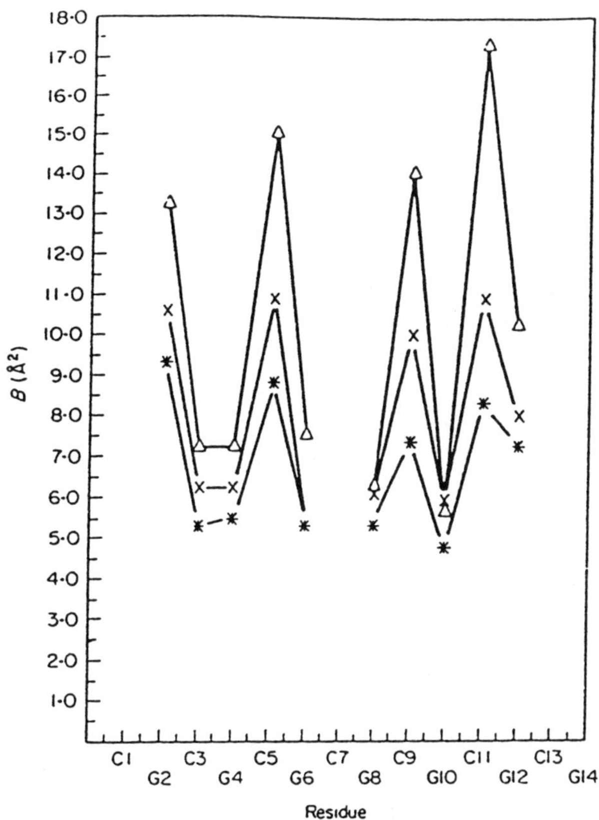

Fig. 2. Temperature factors for the phosphate groups of d(5BrCG5BrCG5BrCG) at 300 K (21). (△) Evaluated using the mean-square atomic fluctuations calculated directly from the molecular dynamics simulations; (x) obtained after refinement without B factor restraints; (*) obtained after refinement with strong B factor restraints. This plot illustrates the fact that soft restraints for B factors should be used in refinement. Reprinted from ref. *21* with permission.

7. The programs devised for X-ray refinement can be used without the structure factor terms for modeling studies. Stronger geometrical and stereochemical restraints (smaller σ values) should be applied progressively during regularization of model structures, especially on bond lengths and van der Waals contacts.
8. The σ values controlling each geometrical restraint at the end of the refinement should reflect the distribution of that particular parameter in well-refined crystal structures (*see* Table 1). However, at the beginning or during the refinement other values can be adopted, so that all terms contrib-

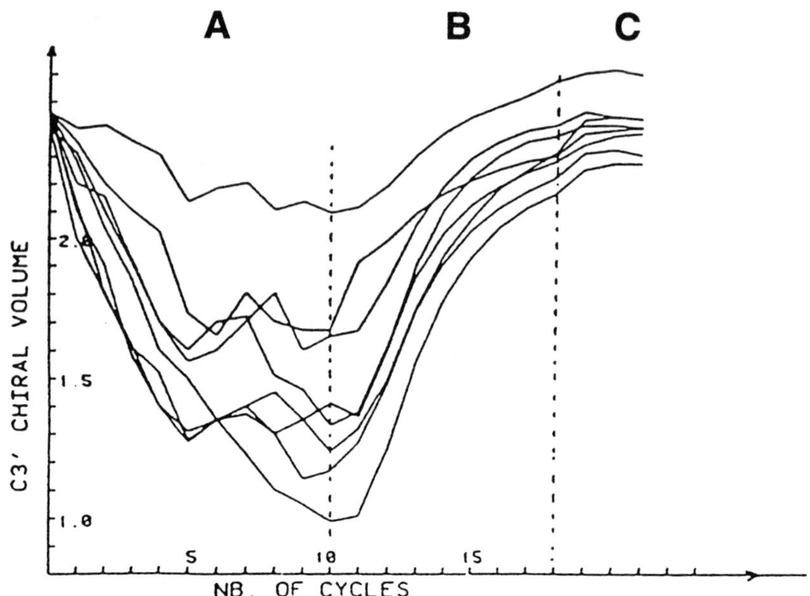

Fig. 3. Variations of the chiral volumes of the C3' carbon atoms of the deoxyribose sugars of the oligomer d(GG5BrUA5BrUACC) *(22)*. Chiral volumes are used to maintain the proper chirality at chiral atoms and are calculated as the triple product of the three vectors converging to the chiral atom. Stage A: refinement without solvent molecules; stage B: refinement with 40 solvent molecules; stage C: final converging steps with 68 solvent molecules. This is a striking example of how the neglect of solvent molecules can distort nucleic acid stereochemistry. Reprinted from ref. *22* with permission.

ute about equally to the overall sum function being minimized. Too severe restraints early in the refinement can lead to ill-conditioning with oscillatory behavior, deterioration of the model, or of the R factor, but do not forget "That which is not restricted will take its liberties" (Wayne Hendrickson).

9. Torsion angle values and bad contacts should be frequently listed. In nucleic acids, aberrant values for some torsion angles (especially, the torsion about the C5'–O5' or C3'–O3' bonds) can reflect a wrong construction. For proteins, Ramachandran plots should be periodically computed to spot nonstandard conformations. The use of good restraints by highlighting the deviant residues or regions helps identify those parts of the model presenting antagonism between geometry and electron density. Consequently, the number of distances (planes, contacts, and so on) deviating by more than two σ should be frequently checked and monitored.

Table 1
σ Values Commonly Used in Restrained Refinement[a]

Restraint	σ
Bond length	0.02 Å
Bond angle (1–3 distance)	0.03 Å
Intraplanar distance	0.05 Å
Hydrogen bond	0.05/0.04 Å
Deviation from planarity	0.02/0.01 Å
Chiral volume	0.15/0.04 Å
Pseudorotation phase angle	none/9.0°
Pseudorotation amplitude	none/3.0°
Single torsion contact	0.50/0.25 Å
Multiple-torsion contact	0.50/0.25 Å
Possible hydrogen bond	0.50 Å
Isotropic thermal factors[b]	1.00/10.0 Å
	1.50/12.5 Å
	2.00/12.5 Å
Noncrystallographic symmetry[c]	0.05–5.00 Å
	0.5–10.00 Å

[a]Adapted from ref. *6* for proteins and from ref. *7* for nucleic acids. If different, first number refers to proteins and second number to nucleic acids.

[b]First number is for atoms connected by a bond length, second number for atoms connected by a bond angle, third number for P–O bonds or side-chain atoms defining an angle, or for atoms in hydrogen bonding or in phosphate angles.

[c]First line is for positional restraints and the second line for thermal restraints. The range spans the tight to the weak restraint classes.

10. The refinement of solvent structure is a delicate procedure *(23,24)*. Clearly, solvent molecules are mobile and only the highly occupied sites can be safely identified, but the coordinates obtained are those of a position in space frequently occupied by a solvent molecule. However, those positions are not always either precisely defined or fully occupied. The spatial delocalization will be reflected by the B factor and the residency time by a fourth parameter, the occupancy factor. Those two parameters are highly correlated and should not be varied simultaneously. Instead cycles should be made with temperature factor refinement and fixed occupancies in alternation with cycles with fixed temperature factors and variable occupancies. Sites with low occupancies or very high temperature factors should be re-examined and possibly eliminated. The geometrical

parameters of the solvent sites should obey hydrogen bonding and coordination rules.

11. Contacts between symmetrically related molecules are not normally restrained. This, in itself, gives an idea of the errors in the refinement. However, this can be done in PROLSQ by the use of the appropriate routine and in NUCLSQ by the use of the noncrystallographic symmetry routine (in which case, the residues involved should be duplicated with fixed half occupancies).

4. Model Building Methods

The aim of this section is to consider a few aspects of the model-building problem. It is certainly not out of the scope of this chapter to recall that building models for biomolecules originates from the outstanding milestones that are the deciphering of the α-helix structure for proteins and of the double-helix structure for nucleic acids. In the field of macromolecular crystallography, one must recall the Richard's box device *(25)* that has been used for years and allowed the construction of tens of structures with Kendrew's models. Though venerable, the Richard's box is now out of date, and we will focus only on the computational/graphics aspect of the question. However, it would certainly be a mistake for a beginner to think that some practice with the skeleton "hardware" models is a waste of time! Also to be noted is the use of minimaps drawn on stacked transparent sheets. This practice remains valuable for the interpretation of rather noisy initial multiple isomorphous replacement (MIR) maps, especially for delineating the contours of the macromolecule. With the presently available softwares, the initial picking of the C-α atomic coordinates is, however, better made directly on a graphic station.

To our knowledge, several graphics programs have been written as specific tools for building a molecule in an electron density map, but only one, FRODO (*13*, and references therein), proliferated. FRODO was originated by Jones and gave rise to multiple "mutants." The initial "selective pressure" was mainly the rapidly changing hardware and software environment. FRODO for the Multi Picture System of E&S *(26)*, FRODO for the PS300 series of E&S *(14)*, and FRODO for the Silicon Graphics *(15)* successively appeared. Also to be noted is that other versions of the program appeared in different places *(27)*. Another "selective pressure" rapidly emerged with the need for more powerful tools. The leading group in the field is unambiguously that of Jones at Uppsala. Their work tends towards an increasing automatization of the building process *(28)* and is now incorporated into a new program, named "O" *(29)*, which cannot be considered simply as a "mutant" of FRODO, but more as a new species.

5. Model Building and Stereochemical Regularization Using FRODO/"O"

We will first examine the "automated procedure" (for proteins only), and then the different tricks for the still laborious method of model building by hand (for proteins and nucleic acids). The basis for the automated procedure in "FRODO" and "O" from Uppsala is the skeletonization of the electron density map *(30)* with the program "BONES." It goes without saying that the resulting ridge lines are an aid for the interpretation of a reasonably good starting map, not a way of replacing a bad set of phases by a good one! BONES gives a more or less continuous backbone, with tentative side chains. The important thing is that the resulting graphical object, namely the set of BONES atoms, can be submitted to interactive modifications after critical examination. For example, putative side chains can become pieces of backbone and vice versa, connectivity can be modified, wrong atoms can be suppressed, and so on. During this step, looking at more than one electron density map may be useful. It seems quite common indeed that maps from different sources, e.g., a pure MIR map and an MIR + Solvent flattened map, can be complementary in many places (especially close to the borders of the mask).

When a reasonable part of a tentative chain tracing is usable, one can go to the second step, which makes use of a structural database *(28)*. The database is built from a representative subset of well-refined protein structures present in the Protein Data Bank (PDB). It is stored as distance matrices of the Cα's. In practice, the user selects the option .DGNL (for DiaGoNaL) from the menu and then defines a relatively short segment, typically five residues long, in the backbone. Then, the program automatically searches and very rapidly finds a number of segments from the database fitting best with the trace of Cα atoms chosen. The user can choose the one which gives the best agreement with the surrounding electron density map. For more details on the method and its power, one should consult refs. *29* and *30*.

Though very efficient, this automated procedure, as well as all refinement programs subsequently used, do not fully replace manual interventions. *A priori*, building by hand means that one uses all degrees of freedom to put each residue at its correct location with a proper conformation. A common method consists of moving each residue separately, or even pieces of residues, as rigid bodies. One then relies on the regularization procedure (REFI) to "glue the pieces together" *(13)*. An alternative way is possible; with some experience, it is possible to model build using exclusively variations in torsion angles, which are the natural degrees of freedom of polymers. This method preserves the correct stereochemical links between residues, and REFI is used only when the fragment being built joins an already built fragment. Undoubtedly, this technique is not very simple and requires some practice before being used with success. The major difficulty is that one must learn to anticipate the effect of successive rotations. One can give as a guide a few general rules, valid for both proteins and nucleic acids, and then specific rules for each family. These rules can depend on the level of development of the software and should not always be considered definite.

1. General rules:
   a. Provided you are not building a piece of the chain from scratch, keep in background (for example, with the MOL option in FRODO) an image of the previous model as a positive or negative guide.
   b. Try to progress in the natural direction, i.e., that defined by the chain numbering. This is merely the result of the present weakness of most software, especially for automatically extending the chain from a given residue.
   c. Try as often as possible to answer the question: what movement do I want to apply to the current residue and then which bond is likely to produce this movement?
   d. A pure translational movement can be achieved by a crankshaft-like motion obtained by two successive rotations of opposite angles around two parallel bonds (*see* below for practical details). The use of a small rotation about a remote bond is also an effective way of producing a motion of a mainly translational nature.
   e. Do not try to attain an acceptable fit at once, but rather through successive approximations by going back and forth.
2. Specific rules for proteins and nucleic acids: The rules concerning the free torsional angles in proteins are simple and well known (for details, *see 31*):
   a. The angles Cα-N and Cα-C surrounding each Cα are only limited by steric considerations (Ramachandran's plot).
   b. The planar peptide links are almost always in *trans*.

   The rules for nucleic acids are:
   a. Angles around the P–O bonds are essentially free.

   b. The angles about the C–O bonds are restricted around the *trans* conformation, but considerably less than for the peptide link.
   c. The angle about C4'–C5' depends on the sugar pucker (especially if it is a ribose), but altogether prefers the +60° orientation.
   d. The sugar conformations are essentially clustered around two preferential ones: the C2'-*endo* and C3'-*endo*. Both conformations are possible within helices for DNA, but for RNA the C2'-*endo* is possible only at the 3'-end of a helix or in loops.
3. Practical considerations: These considerations strongly depend on the present state of development of FRODO (independently of the version) and/or on the specific version you are using:
   a. Definition of the torsional angles for proteins: Start anywhere along the backbone and hit up to nine bonded atoms. At most, four torsional angles Cα-N and Cα-C among the six defined torsions will be really useful for building since two correspond to peptide bonds. Do not pass the cycle of a proline! Opposite rotations for the angles Cα-N and Cα-C of two successive residues, on both sides of their peptide link, will produce a crankshaft motion.
   b. Definition of the torsional angles for nucleic acids: **Always** start at C4' of one sugar and hit the eight atoms up to C3' of the next residue and hit .YES on the menu (again do not pass the cycle of a sugar). Opposite rotations for the angles about P–O5' and C4'–C5' will produce a crankshaft motion if C5'–O5' is not too far from 180°. For the angle around C5'–O5', consider departures from 180° only after exploration of all other angles.
   c. With nucleic acids to flip the *N*th residue from C3'-*endo* to C2'-*endo* (or conversely), do the following sequence (valid for all versions of FRODO):
      i. SAM: option REPLACE to change the residue type from X to XE, or conversely, with X = A,T,U,G, or C. (X = C3'-*endo*, XE = C2'-*endo*).
      ii. Define the RZON as N-1, N.
      iii. REFI: option EXTEND to build the residue N with the correct conformation defined in the dictionary (DSN4) by Sussmann *(5)*.

   In the mutant version of FRODO from the authors' laboratory in Strasbourg, it is possible to vary continuously the phase of pseudorotation and monitor its variation on the screen in real time *(32)*.

6. Notes

1. The importance of a good starting model should not be underestimated, since serious faults in model building are not mended by available automatic refinement procedures. A model "refractory" to refinement may require obtaining measurements of additional isomorphous derivatives *(33)*.

2. Errors and accuracy of X-ray structures have been discussed by Janin *(34)*, who concludes that the positions of backbone atoms are reproducible to 0.5-Å precision.
3. Very recently, new automatic procedures for generating the coordinates of backbone and side-chain atoms, given the amino acid sequence and the trace for the Cα atoms, have been described *(35,36)*.
4. The problem of tracking down possible errors in atomic models constitutes one of the crystallographer's major concerns *(1,34)*. Although the R factor is essential to "ring the alarm bell" (25% for high-resolution data cannot be considered good enough), good R factors can still be obtained with significant errors in a structure. Thus, other criteria have been proposed to help decide whether or not a refinement procedure has properly converged. A useful procedure is the real space per-residue R factor, which identifies regions where the sequence is out-of-register with the density *(29)*. Another simple and extremely efficient procedure has been devised by Brünger *(38)*. It is based on the assessment of any modification in the model by the calculation of an R value based on a randomly chosen subset of the data (ca. 10%) not included in the refinement. This "free R" value is significantly higher than a usual R value, but is a good indicator of possible overrefinement, a pitfall as dangerous as underrefinement. Finally, we would like to advocate another method that is based on experimental data independent of all those used for building and refining the atomic model. Nowadays, with the use of 2D detectors, collecting accurate data has become common. It is thus quite possible to take into consideration the anomalous differences between Friedel mates in the native data (*see*, for example, 3). Such a signal is essentially owing to the sulfur atoms of cysteines and methionines. When used together with the model phases (to which a systematic shift of 90° has been applied), these data allow calculation of an anomalous difference map that should reveal the positions of the anomalous scatterers *(3)*. This has been first shown successfully for the recA protein *(39)*. This is a very stringent test for two reasons. First, it is based on completely independent experimental data. Second, any mistake in the folding, or even any shift by one residue along the chain, will erase the agreement between the highest peaks of the anomalous difference map and the sulfur positions. Obviously, this holds true only for the domains of a protein where sulfur atoms are present.

References

1. Branden, C.-I., and Jones, T. A. (1990) Between objectivity and subjectivity. *Nature* **343,** 687–689.
2. Dunitz, J. D. (1979) *X-Ray Analysis and the Structure of Organic Molecules.* Cornell University Press, Ithaca.

3. Blundell, T. L. and Johnson, L. (1976) Protein Crystallography. Academic, London.
4. Waser, J. (1963) Least-squares refinement with subsidiary conditions. *Acta Cryst.* **16,** 1091–1094.
5. Sussman, J. L. (1985) Constrained-restrained least-squares (CORELS) refinement of proteins and nucleic acids. *Methods Enzymol.* **115,** 271–303.
6. Hendrickson, W. A. (1985) Stereochemically restrained refinement of macromolecular structures. *Methods Enzymol.* **115,** 252–270.
7a. Westhof, E., Dumas, P., and Moras, D. (1985) Crystallographic refinement of yeast aspartic acid transfer RNA. *J. Mol. Biol.* **184,** 119–145.
7b. Westhof, E., Dumas, P., and Moras, D. (1988) Restrained refinement of two crystalline forms of yeast aspartic acid and phenylalanine transfer RNA crystals. *Acta Cryst.* **A44,** 112–123.
8. Jack, A., and Levitt, M. (1978) Refinement of large structures by simultaneous minimization of energy and R factor. *Acta Cryst.* **A34,** 931–935.
9. Deisenhofer, J., Remington, S. J., and Steigemann, W. (1985) Experiences with various techniques for the refinement of protein structures. *Methods Enzymol.* **115,** 303–323.
10. Brünger, A. T., Kuriyan, J., and Karplus, M. (1987) Crystallographic R factor refinement by molecular dynamics. *Science* **235,** 458–460.
11. Brünger, A. T. (1988) Crystallographic refinement by simulated annealing. *J. Mol. Biol.* **203,** 803–816.
12. Fujinaga, M., Gros, P., and Van Gunsteren, W. F. (1989) Testing the method of crystallographic refinement using molecular dynamics. *J. Appl. Cryst.* **22,** 1–8.
13. Jones, T. A. (1985) Interactive computer graphics: FRODO. *Methods Enzymol.* **115,** 157–171.
14. Pflugrath, J. W., Saper, M. A., and Quiocho, F. A. (1984) in *Methods and Applications in Crystallographic Computing* (Hall, S. and Ashida, T., eds.), Oxford University Press, London, pp. 404–407.
15. Cambillau, C. and Horjales, E. (1987) TOM: a FRODO subpackage for protein-ligand fitting with interactive energy minimization. *J. Mol. Graphics* **5,** 174–177.
16. Luzzati, V. (1952) Traitement statistique des erreurs dans la determination des structures cristallines. *Acta Cryst.* **5,** 802–810.
17. Rees, D., Bilwes, A., Samama, J. P., and Moras, D. (1990) Cardiotoxin $V^{II}4$ from *Naja mossambica*. The refined crystal structure. *J. Mol. Biol.* **214,** 281–297.
18. Read, R. (1990) Structure-factor probabilities for related structures. *Acta Cryst.* **A46,** 900–912.
19. Westhof, E. (1987) Re-refinement of the B-dodecamer d(CGCGAATTCGCG) with a comparative analysis of the solvent in it and in the Z-hexamer d(5BrCG5BrCG5BrCG) *J. Biomol. Struct. Dyn.* **5,** 581–600.
20. Yu, H., Karplus, M., and Hendrickson, W. A. (1985) Restraints in temperature factor refinement for macromolecules: an evaluation by molecular dynamics. *Acta Cryst.* **B41,** 191–201.
21. Westhof, E., Chevrier, B., Gallion, S. L., Weiner, P. K., and Levy, R. M. (1986) Temperature-dependent molecular dynamics and restrained X-ray refinement simulations of a Z-DNA hexamer. *J. Mol. Biol.* **191,** 699–712.

22. Kennard, O., Cruse, W. T. B., Nachman, J., Prangé, T., Shakked, Z., and Rabinovich, D. (1986) Ordered water structure in an A-DNA octamer at 1.7 Å resolution. *J. Biomol. Struct. Dynamics* **3,** 623–647.
23. Savage, H. and Wlodawer, A. (1986) Determination of water structure around biomolecules using X-Ray and neutron diffraction methods. *Methods Enzymol.* **127,** 162–183.
24. Westhof, E. (1988) Water: An integral part of nucleic acid structure. *Ann. Rev. Biophys. Biophys.* **17,** 125–144.
25. Richards, F. M. (1968) The matching of physical models to 3-dimensional electron density maps: a simple optical device. *J. Mol. Biol.* **37,** 225–230.
26. Bush, B. L. (1984) Interactive modeling of enzyme-inhibitor complexes at Merck macromolecular modeling graphics facility. *Comput. Chem.* **8,** 1–6.
27. FRODO from MRC (Cambridge, GB) by Phil. Evans. FRODO from IBMC (Strasbourg, France) with some nucleic acids specificity by B. Amerein, M. Bergdoll, P. Dumas, R. Ripp. FRODO from EMBL (Heidelberg, RFA) by C. Carlsson, H. Bosshard.
28. Jones, T. A. and Thirup, S. (1986) Using known substructures in protein model building and crystallography. *EMBO J.* **5,** 819–822.
29. Jones, T. A., Zou, J. Y., Cowan, S. W., and Kjeldgaard, M. (1991) Improved methods for the building of protein models in electron density maps and the location of errors in these models. *Acta Cryst.* **A47,** 110–119.
30. Greer, J. (1974) 3D pattern recognition: an approach to automated interpretation of electron density maps of proteins. *J. Mol. Biol.* **82,** 279–302.
31. Richardson, J. (1981) The anatomy and taxonomy of protein structure. *Adv. Prot. Chem.* **34,** 167–339.
32. Amerein, B., Ripp, R., and Dumas, P. (1987) PUCK: a real-time modification of sugar pucker on a PS300. *J. Mol. Graphics* **5,** 184–189.
33. Kim, Y., Grable, J. C., Love, R., Greene, P. J., and Rosenberg, J. M. (1990) Refinement of *Eco*RI endonuclease crystal structure: a revised protein chain tracing. *Science* **249,** 1307–1309.
34. Janin, J. (1990) Errors in three dimensions. *Biochimie* **72,** 705–709.
35. Holm, L. and Sander, C. (1991) Database algorithm for generating protein backbone and side-chain co-ordinates from a C-alpha trace. Application to model building and detection of co-ordinate errors. *J. Mol. Biol.* **218,** 183–194.
36. Tuffery, P., Etchebest, C., Hazout, S., and Lavery, R. (1991) A new approach to the rapid determination of protein side chain conformations. *J. Biomol. Struct. Dynamics* **8,** 1267–1289.
37. Abad-Zapatero, C., Griffith, J. P., Sussmann, J. L., and Rossmann, M. G. (1987) Refined crystal structure of dogfish M4 apo-lactate dehydrogenase. *J. Mol. Biol.* **198,** 445–467.
38. Brünger, A. (1992) Free R value: a novel statistical quantity for assessing the accuracy of crystal structures. *Nature* **355,** 472–475.
39. Story, R. M., Weber, I. T., and Steitz, T. A (1992) The structure of the *E. coli* recA protein monomer and polymer. *Nature* **355,** 318–325.

CHAPTER 10

Recent Developments for Crystallographic Refinement of Macromolecules

Axel T. Brünger

1. Introduction

In the past decade macromolecular crystallography has undergone major advances in crystallization, data collection by synchrotron X-ray sources and area detectors, and data analysis by high performance computers and new computational techniques. In addition, recombinant gene technology in many cases allows the expression of large amounts of protein. This has resulted in an unprecedented increase in the number of protein crystal structures elucidated. Despite these successes, the fundamental problem in X-ray crystallography, the phase problem, remains unchanged. From a monochromatic diffraction experiment of a single crystal, it is possible to obtain the amplitudes, but not the phases of the reflections. Construction of the electron density by Fourier transformation requires both components of the complex structure factors. Phase information has to be obtained through experimental procedures, most commonly multiple isomorphous replacement, or knowledge-based procedures referred to as Patterson search or molecular replacement. Phase information obtained through these techniques is usually of limited accuracy and resolution, making it often difficult to interpret electron density maps in certain regions of the molecule. Furthermore, macromolecular crystals mostly diffract to less than atomic resolution, causing the process of fitting an atomic model to the observed intensities to be

From: *Methods in Molecular Biology, Vol. 56: Crystallographic Methods and Protocols*
Edited by: C. Jones, B. Mulloy, and M. Sanderson Humana Press Inc., Totowa, NJ

underdetermined. Therefore, the observed data have to be augmented with knowledge about local geometry and nonbonded interactions.

The goal of crystallographic refinement is to achieve a best fit between observed and calculated data of an atomic model, while producing good local geometry and nonbonded interactions. A well-refined model is a prerequisite for the interpretation of the structure in terms of its function and interaction with other molecules.

Despite geometric restraints, it is possible to overfit the diffraction data: an incorrect model can be refined to fairly good conventional R values, as several publications of (partially) incorrect crystal structures have shown *(1)*. Cross-validation, a tool of modern statistics, can be used to assess the quality of a model that is fitted against observed data. The cross-validated or free R value is a better measure than the conventional R value to distinguish between correct and incorrect atomic models *(2,3)*. Furthermore, the free R value can be used to optimize the model's phase accuracy.

This chapter will review recent developments for crystallographic refinement of macromolecules, and will describe the target function for positional refinement, discuss methods to optimize the target function, describe cross-validation, summarize criteria for the quality of refined crystal structures, and describe efforts to incorporate thermal motion and bulk solvent into refinement.

2. Target Function for Refinement

Positional refinement can be formulated to find the global minimum of the target:

$$E = E_{chem} + w_{x\text{-}ray}E_{x\text{-}ray} \quad (1)$$

where E_{chem} comprises empirical information about chemical interactions, $E_{x\text{-}ray}$ describes the difference between observed and computed diffraction data, and $w_{x\text{-}ray}$ is an appropriately chosen weight *(4,5)*. The most common form of $E_{x\text{-}ray}$ consists of the crystallographic residual defined as the sum over the squared differences between the observed $[|F_{obs}(\mathbf{h})|]$ and calculated $[|F_{calc}(\mathbf{h})|]$ structure factor amplitudes:

$$E_{x\text{-}ray} = \sum_{\mathbf{h}} [|\mathbf{F}_{obs}(\mathbf{h})| - k|\mathbf{F}_{calc}(\mathbf{h})|]^2 \quad (2)$$

The scale factor k is chosen to minimize the residual. The computation of the structure factors $\mathbf{F}_{calc}$ is accomplished efficiently by Fast Fourier transformation of the electron density *(6–8)*.

Phase information can be incorporated into $E_{x\text{-ray}}$ in order to improve the ratio of observables to parameters. Since the experimental phase information for macromolecules is usually not very accurate, the errors of the phase observations have to be taken into account. Ideally, this would be accomplished by computing a phase probability distribution for each reflection and to refine the model phases against these distributions *(9)*. A simple representation of the phase probability distribution consists of a double-well potential around the phase centroid where the width of the double-well is determined by the individual figure of merit *(10)*. Another approach is taken by Arnold and Rossmann *(11)* who do not separate the amplitude and phase information, but rather restrain the real and imaginary parts A, B of the structure factor simultaneously.

E_{chem} is a function of all atomic positions of the system describing covalent interactions (bond lengths, bond angles, dihedral torsion angles, chiral centers, planarity of aromatic rings) as well as nonbonded (Van der Waals, hydrogen bonding, and electrostatic) interactions.

$$E_{chem} = \sum_{bonds} k_b(r - r_0)^2 + \sum_{angles} k_\theta(\theta - \theta_0)^2 + \sum_{dihedrals} k_\phi \cos(n\phi + d) + \sum_{chiral,planar} k_w(\omega - \omega_0)^2 + \sum_{atom-pairs} (ar^{-12} + br^{-6} + cr^{-1}) \quad (3)$$

Empirical energy functions were originally developed for energy minimization and molecular dynamics studies of macromolecular structure and function (*see* ref. *12* for a review).

Crystallographic refinement is not very sensitive to the accuracy of the empirical energy function. In fact, the electrostatic term in Eq. (3) can be omitted without introducing large uncertainties. Therefore, a geometric energy function is often used that consists of restraints for covalent bonds lengths, bond angles, chirality, planarity, and nonbonded repulsion in accordance with prior knowledge of these terms obtained from small molecule crystallography *(5,13)*. The differences between a geometric energy function and an empirical energy function mainly affect regions that are not well determined by the experimental information. However, the geometry statistics (deviations of bond lengths and angles from ideality) are clearly dependent on the parameters of the energy function.

Additional restraints based on non-crystallographic symmetry or prior chemical knowledge may be required if the resolution of the crystal structure is limited. For example, base pair hydrogen bonding, base planarity, and sugar puckering of oliogonucleotide structures often need to be restrained to values observed in high-resolution structures.

3. Refinement Methods

The high-dimensionality of the parameter space of the atomic model (typically three times the number of atoms) introduces many local minima of the target function, and thus, gradient descent methods, such as conjugate gradient minimization or least-squares methods *(14)*, normally do not achieve shifts of atomic positions large enough to refine the structure fully. Electron density maps computed by a combination of native crystal amplitudes and experimentally observed phases are sometimes insufficient to allow a complete and unambiguous tracing of the macromolecule. Furthermore, electron density maps for macromolecules are usually obtained at lower than atomic resolution and, thus, are prone to human errors when interpreting the maps. Thus, initial atomic models are likely to contain (partially) incorrect regions and require refinement with a large radius of convergence. The average phase difference between the initial and a refined atomic model can be as large as 54° *(10)*.

Several algorithms to refine macromolecular crystal structures have been developed over the past 20 yr *(15)*. These algorithms can be generally classified into constrained or restrained least-squares optimization *(5,16,17)*, conjugate gradient minimization *(4,18)*, and simulated annealing (SA) refinement *(19)*. Restrained least-squares refinement techniques were reviewed by Hendrickson *(5)*. An improved algorithm for conjugate gradient minimization was recently described by Tronrud *(20)* that employs information about curvature in order to speed convergence and to enable simultaneous refinement of positions and B factors. The following sections will focus on refinement by simulated annealing, because it has the largest radius of convergence.

3.1. Refinement by Simulated Annealing

Annealing denotes a physical process in which a solid in a heat bath is heated up by increasing the temperature of the heat bath to a value at which all particles of the solid randomly arrange themselves in the liquid phase and then cooled by slowly lowering the temperature of the heat bath. In this way, all particles arrange themselves in the lowest energy ground state of the solid, provided the maximum temperature is sufficiently high and the cooling is carried out sufficiently slowly. By formally identifying the target of an optimization problem with the "energy" of the system, "simulated" annealing can be carried out. Compared to gradient descent methods where search directions are restricted to be downhill, simulated annealing achieves more optimal solutions by allowing uphill directions as well *(21)*. The likelihood of going uphill is determined by a control parameter referred to as "temperature": the higher the temperature, the more likely the optimization can overcome barriers. It should be noted that the temperature of simulated annealing normally has no physical meaning and merely determines the likelihood of overcoming barriers of the target function.

Simulated annealing can be applied to combinatorial optimization problems in which the parameters of the system assume a set of discrete values as well as to continuous optimization problems in which the parameters can assume a continuous range of values. There is usually no guarantee that the solution found by simulated annealing is the global minimum. It can be shown, however, that asymptotically the algorithm must find the global minimum *(22)*.

Simulated annealing requires the definition of a target function, a generation mechanism to create a Boltzmann distribution at a given temperature T, and an annealing schedule, that is, a sequence of temperatures $T_1 > T_2 > \ldots > T_l$ at which the Boltzmann distribution is computed *(22)*. Implementations of the generation mechanism differ in the way they generate a transition or "move" from one set of parameters to another which is consistent with the Boltzmann distribution at a given temperature. The two most widely used generation mechanisms are Monte Carlo *(23)* and molecular dynamics *(24)* simulations where the former can be applied to both discrete and continuous optimization problems, but the latter is restricted to continuous problems. The moves can proceed in Cartesian coordinate space or in reduced-dimensional coordinate spaces *(81)*, such as torsion-angle space or rigid-body coordinate space. The molecular dynamics implementation has the advantage of restricting moves of the system to physically reasonable pathways, whereas the inertia of the system enables transitions between small minima. Therefore, molecular dynamics rather than Monte Carlo was employed for crystallographic refinement by simulated annealing *(19)*.

Molecular dynamics consist of the numerical solution of the familiar Newton's equations of motion:

$$m_i\,(\partial^2 r_{i,\mu}/\partial t^2) = -(\partial E/\partial r_{i,\mu}) \tag{4}$$

The quantities $r_{i,\mu}$ and m_i are the coordinates and masses of atom i, respectively, and E is the potential energy. In the context of simulated annealing, E denotes the target function of the optimization problem, which may contain "physical" energies, such as covalent and nonbonded energy terms, as well as "unphysical" energies that restrain the system to observed data. The solution of the partial differential equations (Eq. [4]) is normally carried out numerically by finite-difference methods, such as the Verlet algorithm *(24)*. The initial velocities are usually assigned from a Maxwellian distribution at the appropriate temperature.

For the implementation of simulated annealing, it is necessary to be able to control the temperature during molecular dynamics. The three most commonly used methods are: velocity scaling, Langevin dynamics and coupling to a heat bath *(25)*. The maintenance of a particular temperature can also be viewed as enhancing the search behavior of the optimization process. Suppose the molecular dynamics trajectory reaches a minimum. Without temperature control, the "particles" would gain momentum and spend very little time in the region of the minimum, which would work against the goal of finding a minimum in the first place. The temperature control removes excessive kinetic energy and, thus, ensures a slower exploration of the minimum.

The use of molecular-dynamics-like trajectories was suggested earlier for general global optimization problems by Griewank *(26)*. Here, non-Newtonian molecular dynamics was employed in order to search conformational space more efficiently. Unfortunately, the algorithm involves a large computational effort. Van Schaik et al. *(27)* recently proposed a computationally less expensive variation of this approach. The potential energy of the system is coupled to an energy bath, i.e., the system explores a potential energy surface. The reference level of the bath is then slowly decreased over the course of the molecular dynamics run. It remains to be seen if this method can be successfully applied to crystallographic refinement.

The success and efficiency of simulated annealing critically depend on the choice of the annealing schedule *(22,28)*, that is, the sequence of and numerical values $T_1 > T_2 > \ldots > T_l$ for the temperature. Scaling the temperature T by a factor s is equivalent to scaling the target E by $1/s$ and scaling the time variable by $[1/(s)^{1/2}]$ *(13)*. The equivalence of temperature control and scaling of E suggest a generalization of annealing schedules where, in addition to the overall scaling of E, relative scale factors

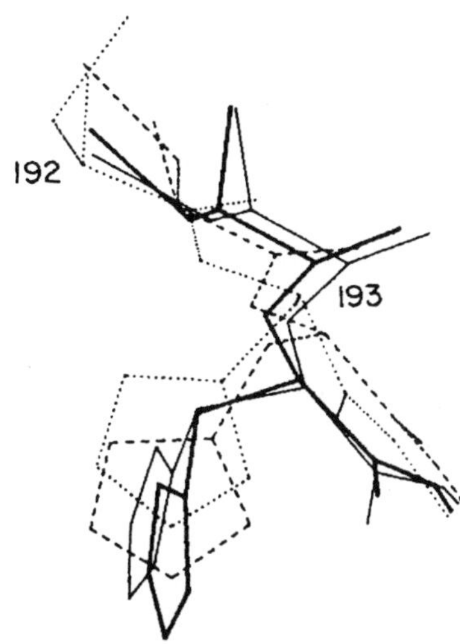

Fig. 1. The segment consisting of residues Cys-192 and His-193 of the 2.8-Å resolution structure of aspartate aminotransferase *(10)*. Superimposed are the initial structure (dotted lines) obtained by fitting the atomic model to a multiple isomorphous replacement map, the structure obtained after several cycles of rebuilding and restrained least-squares refinement (thick lines), the structure obtained after SA refinement (thin lines), and the structure obtained after conjugate gradient minimization (dashed lines).

between components of the target E are introduced, i.e., simulated annealing is carried out with a variable target function.

Kirkpatrick et al. *(21)* suggested reducing the cooling rate at phase transitions, since the system is in a critical state where fast cooling might trap the system in a *meta*-stable state. A simple but not necessarily optimal annealing schedule is "slow cooling." Here the temperature is decreased monotonically at a rate slow enough to ensure that the system reaches equilibrium at each stage of the annealing schedule. Slow cooling is the best compromise for crystallographic refinement in terms of generality and efficiency *(29)*.

SA refinement starting from initial models produces a significantly improved R value and geometry compared to least-squares optimization or conjugate gradient minimization without model building *(10,19,30,31)*. Atoms can move by more than 2 Å, and side-chain conformations can be changed. Figure 1 shows a representative case where SA refinement has essentially converged to a manually refined structure of the enzyme aspartate aminotransferase refined by Brünger *(10)*. The imidazole ring of the histidine side chain has undergone a 90° rotation around the χ_1 bond during SA refinement. This rotation was accompanied by significant structural changes of the backbone atoms. This resulted in convergence of the SA-refined structure to the manually refined structure. The conformational changes were not accomplished by conjugate gradient minimization without rebuilding. Rigid-body-like corrections of up to 10° through SA refinement were observed by Gros et al. *(32)*.

SA refinement is most useful when the initial model is relatively crude. Given an already well-refined model, SA refinement offers little advantage over conventional methods, with the possible exception of providing information about the accuracy of the refined structure *(10)*. However, when only a crude model is available, SA refinement is able to reduce greatly the amount of human intervention. The initial model can be as crude as one that is obtained by automatic building based on C^{α} positions alone *(33)*.

3.2. Automation of Refinement

Simulated annealing has greatly improved the efficiency of crystallographic refinement. However, this method is still insufficient to refine a crystal structure automatically. First, an initial model needs to be manually built before refinement can proceed by minimization of Eq. (1). Second, the model needs to be periodically refit to electron density maps with interactive computer graphics. Improved graphics programs are now available for assisting the rebuilding process using data bases of small fragments of known protein structures *(33)*. However, these programs still require a good deal of human intervention and intuition. In a sense, the human fitting process complements the numerical minimization of Eq. (1). Recent attempts at further automation of density map fitting have been reported by Read and Moult *(34)* and Lamzin and Wilson *(35)*. Fortier et al. *(36)* suggest using artificial intelligence methods to guide the process.

4. Assessing Phase Accuracy by Cross-validation

The most common measure for the agreement between a refined model and the observed diffraction data is the R value, defined as:

$$R = [\Sigma_{\mathbf{h}}||\mathbf{F}_{obs}(\mathbf{h})| - k|\mathbf{F}_{calc}(\mathbf{h})||/ \Sigma_{\mathbf{h}}|\mathbf{F}_{obs}(\mathbf{h})|] \tag{5}$$

The R value is closely related to the crystallographic residual $E_{x\text{-}ray}$ (Eq. [2]), which has more convenient mathematical properties. It can be shown that $E_{x\text{-ray}}$ is a linear function of the negative logarithm of the likelihood of the atomic model assuming that all observations are independent and normally distributed *(14)*. Crystallographic refinement consists of minimizing $E_{x\text{-ray}}$ and thus maximizing the likelihood of the atomic model. $E_{x\text{-ray}}$ can be made arbitrarily small by increasing the number of model parameters and subsequent refinement. The theory of linear hypothesis tests has been employed in order to decide whether the addition of parameters or the imposition of fixed relationships between parameters results in a significant improvement or a significant decline in the agreement between atomic model and diffraction data *(37)*. This theory strictly applies to the situation where the restraints can be expressed as holonomic boundary conditions, e.g., fixed bond lengths, and thus does not apply to nonlinear restraints, such as E_{chem} (Eq. [3]).

In close analogy to testing statistical models by cross-validation, Brünger *(2,3)* proposed the free R_T^{free} statistic that measures the agreement between the atomic model and the diffraction data for a "test" set of reflections that is omitted during refinement. The test was obtained by random selection from a unique set of observed reflections. The size of the test set was chosen to be 10% of the observed diffraction data, which was a compromise between minimizing statistical fluctuations of R_T^{free} and avoiding a significant effect on the atomic model. A high correlation between R_T^{free} and the phase accuracy of the atomic model was observed, independent of the number of model parameters and restraints.

In a different context, partitioning of observed reflections into test and working sets has been used for the multisolution strategy by combined maximization of entropy and likelihood *(38,39)*. Karle *(40)* proposed using "rolling" working sets to aid the convergence behavior of least-squares minimization. Refined omit maps can be viewed as the real-space analog to cross-validation: part(s) of the model is omitted and then the remaining model is refined (cf Section 5.). The structure determination of virus structures often proceeds with subsets of data *(11)*.

The principle of cross-validation states that the model must not be refined against the test set. Thus, if a crystal structure has been refined with all diffraction data included, a procedure is required to remove the refined model's "memory" of the test set. To accomplish this, one needs a refinement method with a large radius of convergence, preferably SA refinement (Section 3.1.). R_T^{free} is then defined as the R value computed for the test set.

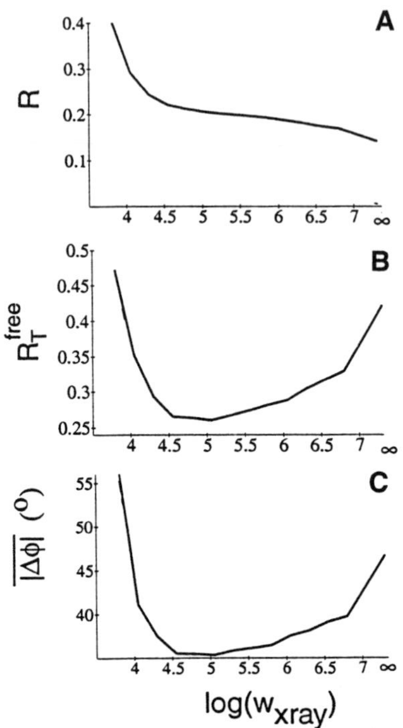

Fig. 2. Cross-validation for SA refinements of penicillopepsin *(48)* at 6–1.8 resolution as a function of $w_{x\text{-ray}}$ (Eq. [1]). Details of the penicillopepsin model and refinement procedure are the same as in Fig. 3, except that B-factors were restrained during refinement. $w_{x\text{-ray}} = \infty$ represents the completely unrestrained case. The test set was obtained by a 10% random selection. **(A)** R as a function of $\log(w_{x\text{-ray}})$. **(B)** R_T^{free} as a function of $\log(w_{x\text{-ray}})$. **(C)** $\Delta\phi$ as a function of $\log(w_{x\text{-ray}})$. $\Delta\phi$ is the figure-of-merit weighted mean phase difference between model phases and the most probable multiple isomorphous replacement phases at 6–2.8 Å resolution.

The R_T^{free} method can be used to optimize the overall weighting between diffraction data and chemical restraints in crystallographic refinements (Eq. [1]). If $w_{x\text{-ray}}$ is chosen too small, too much emphasis is put on the geometry as provided in the parameter set of the refinement program, which results in a poor fit to the diffraction data. If $w_{x\text{-ray}}$ chosen is too large, the structure will become overfitted: although the conventional R value is very small, the geometry of the structure becomes severely distorted. Optimizing R_T^{free} is as a function of $w_{x\text{-ray}}$ is an objective method to determine $w_{x\text{-ray}}$ *(2,3)*. As shown in Fig. 2, there is a high correlation between R_T^{free} and the model's phase accuracy. A similar correlation was observed when determining the weight for restrained thermal B factor refinement *(2)*. Optimal relative weighting among bond length, bond angle, dihedral angle, and van der Waals restraints was obtained for the parameters by Engh and Huber *(41,42)*. It was found that the distribution of bond lengths and bond angles as it is found in the Cambridge Structural Database *(43)* is in fact optimal for the penicillopepsin crystal structure at 1.8-Å resolution. The deviations of the geometry from ideality were surprisingly small (0.008 Å and 1° for bond lengths and bond angles, respectively).

R_T^{free} qualitatively assesses the model's phase accuracy without the requirement that phases of the crystal structure be known. This feature of R_T^{free} is not restricted to refinement. It applies to any optimization procedure in crystallography where certain fixed parameters need to be adjusted. For example, Baker et al. *(44)* applied this procedure to derive optimal parameters for skeletonization in order to maximize phase accuracy of a density modification procedure.

In general, R_T^{free} will be higher than R, since the test set has been omitted in the refinement process. The difference between R and R_T^{free} can be caused by noise in the data, incompleteness of the atomic model, or an unfavorable observable to parameter ratio. Reported values for R_T^{free} vary between the low 20s to the middle 30s for macromolecular crystal structures (for example, *45–47*). A structure of plant ribulose-1,5-biphosphate carboxylase oxygenase (RuBisCO) with a major error in the chaintrace produced an R_T^{free} value of 47% *(2)*.

4.1. The Uniqueness of the Phase Problem

Brünger *(2)* refined the positions of a liquid consisting of equal atoms using diffraction data of penicillopepsin at 1.8-Å resolution collected by James and Sielecki *(48)*. Atomicity was ensured by employing a repulsive potential that allowed the atoms to get within covalent bonding distance, but otherwise no chemical restraints were applied. Two configurations were refined, and the agreement with the diffraction data assessed (Fig. 3). One configuration consisted of the nonhydrogen atom positions of the crystal structure. The other configuration consisted of the same number of atoms, randomly distributed in an asymmetric unit of the crystal. After SA refinement against the crystallographic residual (Eq. [2] in Section 2.), both configurations produced essentially the same low R value (Fig. 3). However, the random configuration contained no

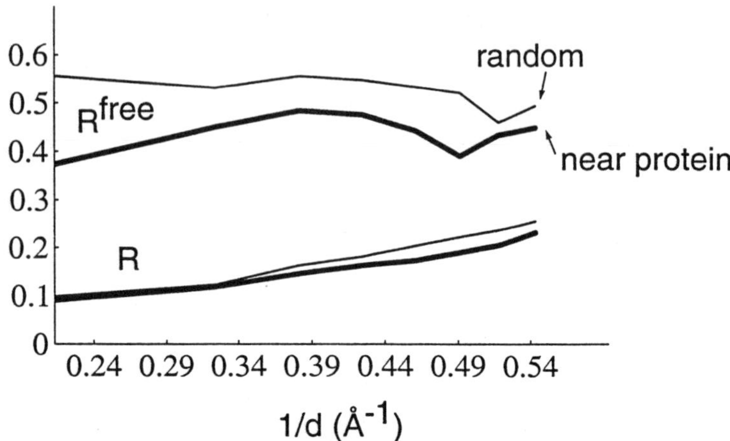

Fig. 3. R value distributions for refinements of equal atoms against the diffraction data of penicillopepsin *(48)*; 2365 atoms with a reduced van der Waals radii of 1.57 Å were randomly placed in the asymmetric unit of the crystal or placed at the nonhydrogen positions of the penicillopepsin structure. SA refinements were carried at 6–1.8 Å resolution with about 10% of the data randomly omitted for evaluating R_T^{free} (Section 4.). Each SA refinement consisted of a slow-cooling protocol *(29)* using the program X-PLOR *(41)* starting at 1000 K, overall B-factor refinement, and individual unrestrained B-factor refinement. Shown are R and R_T^{free} distributions as a function of reciprocal resolution ($1/d$, where d is the resolution in Å). Thick lines correspond to the configuration close to the penicillopepsin crystal structure.

phase information (90° phase error), whereas the other configuration had a phase error of around 48°. Furthermore, a large number of other configurations could be generated with equally good R values as that of the crystal structure but with essentially random phase errors *(3)*. Thus, a target function based on the R value is degenerate, and it cannot produce a unique solution of the phase problem. Baker et al. *(49)* came to the same conclusion, and showed that inclusion of solvent boundaries, maximization of the entropy of the liquid model, or enforcement of positivity of the electron density cannot break this degeneracy.

Thus far, these results do not rule out the existence of a target function that breaks the degeneracy. Brünger *(2,3)* showed that the free R value (R_T^{free}) (cf Section 4.) reduces the number of false solutions. In fact, a random configuration produces a significantly higher R_T^{free} value than that of the correct configuration (Fig. 3). In this case, the discriminatory power of

R_T^{free} was most pronounced for the low-resolution reflections (Fig. 3). Similar results were obtained for a number of other random configurations *(3)*.

The refinement of the random configuration and the crystal structure in Fig. 3 was carried out against a working set consisting of 90% of the observed diffraction data in order to enable the computation of R_T^{free}. What happens when the random configuration is extensively refined against the complete data set prior to the R_T^{free} test? One obtains R values of 16.3 and 15.7% and R_T^{free} values of 47.3 and 42.3% for the random and the correct configuration, respectively (Brünger, unpublished results). Thus, although less pronounced than in Fig. 3, R_T^{free} is still larger for the random configuration. The R_T^{free} difference is larger than the R difference both in absolute and in relative terms regardless of whether or not the random configuration has been refined against the complete data set prior to the R_T^{free} test.

Based on these results, one is tempted to define R_T^{free} as a target function to score the configurations of the liquid. By choosing this target function, the number of incorrect configurations might be significantly reduced. A difficulty with this target function is that R_T^{free} cannot be expressed as an analytic function of the atomic coordinates. The target function could only be optimized by conformational searches or trial and error, but not by optimization methods involving analytical derivatives. In this sense R_T^{free} bears resemblence to the maximum likelihood criterion described by Bricogne and Gilmore *(39)* although in the latter case an approximate analytic formulation exists. Another difficulty is that bias toward the test set needs to be minimized in order to compute R_T^{free} (cf Section 4.). This can only be achieved by extensive refinements against the working set. However, the atoms of the liquid-like model tend to move during refinement even when starting with the crystal structure coordinates. This is because of an adverse observable-to-parameter ratio for unrestrained refinement at 1.8-Å resolution. For example, when using gradient descent minimization the root mean-square (r.m.s.), difference between the refined model and the crystal structure is 0.6 Å; SA refinement further increases this difference. Thus, in the process of removing the bias toward the test set, the quality of the model seriously degrades. This is in contrast to the applications of R_T^{free} discussed in the previous section where chemical restraints alleviate this problem.

The R_T^{free} test reduces the degeneracy of the phase problem, but it is probably not sufficient to break it. The incorporation of knowledge about the covalent bonding structure and the atomic distributions in macromolecules remains an important challenge in any *ab initio* phasing approach. Global conformational searches of a detailed atomic model with covalent bonding and nonbonded constraints or restraints are not feasible for macromolecules at the present time. The problem bears some relationship to the Levinthal paradox of protein folding *(50)*: On folding, the polypeptide chain finds its distinct minimum on the free energy surface within an extremely short period of time, compared with the time predicted for a global conformational search mechanism. In a similar vein, global conformational searches of macromolecules against a crystallographic target function appear impossible with currently available computational resources.

5. Quality of X-Ray Crystal Structures

The quality of X-ray crystal structures can be assessed by the fit to the experimental data and by the agreement with prior knowledge of molecular geometry, structures, and atomic distributions. Tools to assess the fit to the data include the real space R factor *(33)*, distribution of refined B factors, convergence of multiple refinements, and fit and continuity of density maps. The interpretation of density maps can be affected by "model bias." This model bias can obscure the detection of errors in atomic models if sufficient phase information is unavailable. In fact, during the past decade several cases of incorrect or partly incorrect atomic models have been reported where model bias may have played a role *(1)*.

An example of model bias is shown in Fig. 4, which shows the Trp 91 side chain of an antidinitrophenyl-spin-label murine monoclonal antibody (MAb) (AN02) complexed with its hapten at 2.9-Å resolution *(51,52)*. During the course of refinement of AN02, electron density maps of the binding pocket suggested two possible orientations of the side chain of Trp 91 (gray and black lines). Initially, the tryptophan side chain was placed in the incorrect orientation (gray lines), and the complete model was refined. The resulting $(2F_o - F_c)$ density map was ambiguous despite the fact that the trytophan side chain and neighboring atoms were omitted in the map calculation and Read's *(53)* weighting procedure was used (Fig. 4a). Apparently, overfitting of the complete model had caused a "memory" for the incorrectly placed atoms in the rest of the atomic model. Hodel et al. *(54)* investigated the reduction of model bias by various omit map techniques with or without refinement. It was concluded

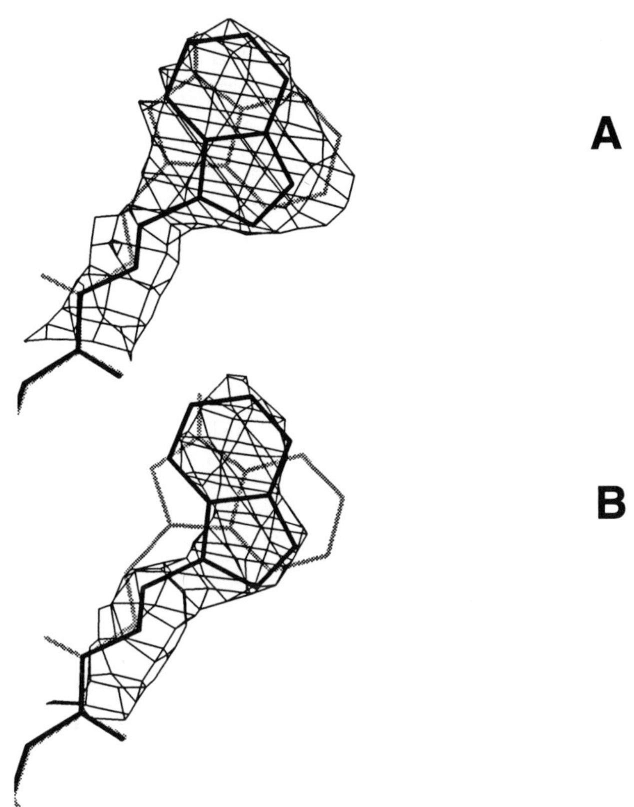

Fig. 4. The effects of omit map techniques around Trp 91 located in the light chain of the AN02 Fab fragment *(52,54)*. The correct conformation of Trp 91 is shown in black, and the initial incorrect conformation is shown in gray. All maps are of the σ_A-weighted $2F_o - F_c$ type *(53)* at 2.8 Å resolution. Residues 89–97 in the light chain, 93–102 from the heavy chain, and the hapten molecule were omitted for the map calculations and refinements. **(A)** Ordinary omit map of the initial incorrect structure shown at a contour level of 1.2σ. **(B)** SA omit map of the initial structure shown at a contour level of 1.2σ. The partial structure was SA refined using a slow-cooling protocol *(29)* with a starting temperature of 3000 K at 8.0–2.8 Å resolution.

that extensive refinement of the partial model without the tryptophan side chain and its neighboring atoms was required to reduce model bias. SA refinement produced the best results, although conjugate gradient minimization was sufficient in other cases. An example for an SA refined omit map is shown in Fig. 4b, where model bias toward the incorrect conformation is removed, and the density map now clearly shows the correct conformation (black lines).

Examples for geometric criteria for the quality of crystal structures are the agreement with ideal covalent geometry obtained from small molecule crystal structures for bond lengths, bond angles, and torsion angles, and the absence of close nonbonded contacts. Conformational energies obtained from an empirical energy function are not useful at present, since they can be affected by the inaccuracy of the empirical energy function. Criteria for the quality for three-dimensional structures have also been derived from statistical analyses of known protein structures. A good structure should minimize the number of polypeptide backbone torsion angles in disallowed regions *(55)*, maximize the number of hydrogen bonds *(56)*, minimize the free energy of solvation *(57)*, which can distinguish between certain correct and incorrect folds of the protein *(58,59)*, and satisfy contact maps *(60)* or profiles of environmental classes derived from the database of known protein structures *(61,62)*.

The quality of an atomic model depends on the availability of good experimental data. Perhaps the most important criterion for the correctness of an atomic model is that it must agree with biochemical or biophysical data other than that obtained from the X-ray diffraction experiment.

6. Models for Thermal Motion and Bulk Solvent

Up to now, even the best protein crystal structures produce relatively large final R values of 10–20% as compared to the noise in the data, which is typically estimated to be around 5%. This is probably because of an inadequate description of thermal motion, disorder, bulk solvent, or some other kind of systematic error.

Water constitutes a large portion of the volume in water-soluble macromolecule crystals *(63)*. The macromolecule is surrounded by a tightly bound layer of water molecules. The remaining bulk water is disordered, and its contribution in refinement procedures is usually neglected or approximated by a “flat” solvent model. Structure factors of the flat solvent model have been computed using Babinet’s principle *(64,65)* or by construction of a solvent mask *(66)*. This produces discrepancies between observed and calculated structure factors especially at low and medium resolution. A more detailed description of the solvent has been achieved by dividing the solvent volume into shells extending outward from the surface of the protein and refinement of two parameters for each shell, the solvent-scattering density, and an isotropic temperature factor *(67)*. Two hydration layers in myoglobin crystals were observed. In order to obtain more detailed information about the three-dimensional structure of the bulk solvent, an iterative density modification method has been proposed by Badger and Caspar *(68)*. Both a real-space and a reciprocal restraint were applied where the former consisted of the requirement that the protein density should remain unchanged, and the latter consisted of the requirement that observed and calculated structure factors should agree as closely as possible. Fluctuations in solvent density in cubic insulin crystals were observed, demonstrating nonrandom arrangements of water molecules extending several layers from the first solvation shell of the protein.

The flexibility of macromolecular structure produces thermal motion and disorder that should be accounted for during refinement *(69)*. The customary use of isotropic temperature factors and a single molecular conformation provides only a poor description for these phenomena. Owing to the adverse parameter to observable ratio typical for macromolecular diffraction data, anisotropic temperature factors normally cannot be used. Thus, the goal is to describe thermal motion and disorder with a minimal set of additional parameters.

Conformational disorder has been studied by including two independent structures ("twins") with isotropic temperature factors in least-squares optimization *(70)*. For several test cases, a reduction in the R value was achieved by slight displacement of the corresponding atoms in the twin structure. The displacements were correlated with anisotropic temperature factors in the case of crambin where 0.95-Å resolution data allows anisotropic temperature factor refinement. By repeated application, the method automatically identified side chains with two or more alternative conformations.

Another approach makes use of a minimal set of low-frequency normal modes to model temperature factors *(71–73)*. Refinement is carried out for amplitude coefficients for the normal modes *(71)* or, more generally, for the variances and covariances of the normal modes *(72)*. In the case of a crystal structure of bovine pancreatic trypsin inhibitor, the refinement of the amplitude coefficients for only 19 normal modes derived from a molecular mechanics computation was sufficient to reproduce a description obtained by 892 isotropic temperature factors. The rigid body translational and vibrational modes provided the largest contribution to the temperature factors. This is corroborated by the observation that the

TLS model of Schomaker and Trueblood *(74)* for the oscillations of rigid protein molecules is successful in qualitatively reproducing the isotropic temperature factor profiles of a wide variety of proteins ranging in size from lysozyme to influenza virus hemagglutinin *(75)*. For more detailed descriptions, a segmented TLS refinement has been suggested by Howlin et al. *(76)*. This TLS refinement methodology might benefit from improvements reported by He and Craven *(77)*.

Diffraction data reflect a time average over many possible conformations of the crystallized molecule. Gros et al. *(78)* have suggested using a time-average restraint that incorporates previously encountered conformations of the macromolecule during a long molecular dynamics simulation. This method yields an ensemble of structures in which many possible thermal motions are allowed, including anisotropic and anharmonic motions.

All approaches discussed in this section are aimed at reducing the R value, but there is no guarantee that these methods actually produce a significant improvement of the information content of the model. Computation of R_T^{free} now provides an objective test to determine if a particular model actually increases the information content. Results indicate that none of the approaches discussed in this section fully bridge the gap between protein and small molecule R values *(79–81)*.

References

1. Brändén, C. I. and Jones, A. (1990) Between objectivity and subjectivity. *Nature* **343,** 687–689.
2. Brünger, A. T. (1992) The Free *R* value: a novel statistical quantity for assessing the accuracy of crystal structures. *Nature* **355,** 472–474.
3. Brünger, A. T. (1993) Assessment of phase accuracy by cross validation: The free *R* value. Methods and applications. *Acta Cryst.* **D49,** 24–36.
4. Jack, A. and Levitt, M. (1978) Refinement of large structures by simultaneous minimization of energy and *R* factor. *Acta Cryst.* **A34,** 931–935.
5. Hendrickson, W. A. (1985) Stereochemically restrained refinement of macromolecular structures. *Meth. Enzymol.* **115,** 252–270.
6. Ten Eyck, L. F. (1973) Crystallographic fast fourier transforms. *Acta Cryst.* **A29,** 183–191.
7. Ten Eyck, L. F. (1977) Efficient structure-factor calculation for large molecules by the fast fourier transform. *Acta Cryst.* **A33,** 486–492.
8. Brünger, A. T. (1989) A memory-efficient fast fourier transformation algorithm for crystallographic refinement on supercomputer *Acta Cryst.* **A45,** 42–50.
9. Hendrickson, W. A. and Lattman, E. E. (1970) Representation of phase probability distributions for simplified combination of independent phase information. *Acta Cryst.* **B26,** 136–143.

10. Brünger, A. T. (1988) Crystallographic refinement by simulated annealing: Application to a 2.8 Å resolution structure of aspartate aminotransferase. *J. Mol. Biol.* **203,** 803–816.
11. Arnold, E. and Rossmann, M. G. (1988) The use of molecular-replacement phases for the refinement of the human rhinovirus structure. *Acta Cryst.* **A44,** 270–282.
12. Karplus, M. and Petsko, G. A. (1990) Molecular-dynamics simulations in biology. *Nature* **347,** 631–639.
13. Brünger, A. T. (1991) Simulated annealing in crystallography. *Ann. Rev. Phys. Chem.* **42,** 197–223.
14. Press, W. H., Flannery, B. P., Teukolosky, S. A., and Vetterling, W. T. (eds.) (1986) *Numerical Recipes.* Cambridge University Press, Cambridge, pp. 498–546.
15. Stout, G. H. and Jensen, L. H. (eds.) (1989) *X-ray Structure Determination, A Practical Guide.* John Wiley, New York, pp. 341–419.
16. Sussman, J. L., Holbrook, S. R., Church, G. M., and Kim, S. H. (1977) Structure-factor least-squares refinement procedure for macromolecular structure using constrained and restrained parameters. *Acta Cryst.* **A33,** 800–804.
17. Konnert, J. H. and Hendrickson, W. A. (1980) A restrained-parameter thermal-factor refinement procedure. *Acta Cryst.* **A36,** 344–349.
18. Tronrud, D. E., Ten Eyck, L .F., and Matthews, B. W. (1987) An efficient general-purpose least-squares refinement program for macromolecular structures. *Acta Cryst.* **A43,** 489–500.
19. Brünger, A. T., Kuriyan, J., and Karplus, M. (1987) Crystallographic R factor refinement by molecular dynamics. *Science* **235,** 458–460.
20. Tronrud, D. E. (1992) Conjugate-direction minimization: an improved method for the refinement of macromolecules. *Acta Cryst.* **A48,** 912–916.
21. Kirkpatrick, S., Gelatt, C. D., Jr., and Vecchi, M. P. (1983) Optimization by simulated annealing. *Science* **220,** 671–680.
22. Laarhoven, P. J. M. and Aarts, E. H. L. (eds.) (1987) *Simulated Annealing: Theory and Applications.* D. Reidel Publishing, Dordrecht, pp. 187.
23. Metropolis, N., Rosenbluth, M., Rosenbluth, A., Teller, A., and Teller, E. (1953) Equation of state calculations by fast computing machines. *J. Chem. Phys.* **21,** 1087–1092.
24. Verlet, L. (1967) Computer "experiments" on classical fluids. I. Thermodynamical properties of Lennard-Jones molecules. *Phys. Rev.* **159,** 98–105.
25. Berendsen, H. J. C., Postma, J. P. M., van Gunsteren, W. F., DiNola, A., and Haak, J. R. (1984) Molecular dynamics with coupling to an external bath. *J. Chem. Phys.* **81,** 3684–3690.
26. Griewank, A. O. (1981) Generalized descent for global optimization. *J. Optimization Theory and Applications* **34,** 11–39.
27. van Schaik, R. C., van Gunsteren, W. F., and Berendsen, H. J. C. (1992) Conformational search by potential energy annealing: Algorithm and application to cyclosporin A. *J. Comp.-Aided Mol. Design* **6,** 97–112.
28. Bounds, D. G. (1987) New optimization methods from physics and biology. *Nature (Lond.)* **329,** 215–219.
29. Brünger, A. T., Krukowski, A., and Erickson, J. (1990) Slow-cooling protocols for crystallographic refinement by simulated annealing. *Acta Cryst.* **A46,** 585–593.

30. Fujinaga, M., Gros, P., and van Gunsteren, W. F. (1989) Testing the method of crystallographic refinement using molecular dynamics. *J. Appl. Cryst.* **22,** 1–8.
31. Kuriyan, J., Brünger, A. T., Karplus, M., and Hendrickson, W. A. (1989) X-ray refinement of protein structures by simulated annealing: Test of the method on myohemerythrin. *Acta Cryst.* **A45,** 396–409.
32. Gros, P., Betzel, Ch., Dauter, Z., Wilson, K. S., and Hol, W. G. J. (1989) Molecular dynamics refinement of a thermitase-eglin-c complex at 1.98 Å resolution and comparison of two crystal forms that differ in calcium content. *J. Mol. Biol.* **210,** 347–367.
33. Jones, T. A., Zou, J.-Y., Cowan, S. W., and Kjeldgaard, M. (1991) Improved methods for building protein models in electron density maps and the location of errors in these models. *Acta Cryst.* **A47,** 110–119.
34. Read, R. J. and Moult, J. (1992) Fitting electron density by systematic search. *Acta Cryst.* **A48,** 104–113.
35. Lamzin, V. S. and Wilson, K. S. (1993) Automated refinement of protein models. *Acta Cryst.* **D49,** 129–147.
36. Fortier, S., Castleden, J., Glasgow, J., Conklin, D., Walmsley, C., Leherte, L., and Allen, F. H. (1993) Molecular scene analysis: the integration of direct methods and artificial intelligence strategies for solving protein crystal structures. *Acta Cryst.* **D49,** 168–178.
37. Hamilton, W. C. (1965) Significance tests on the crystallographic R factor. *Acta Cryst.* **18,** 501–510.
38. Bricogne, G. (1984) Maximum entropy and the foundation of direct methods. *Acta Cryst.* **A40,** 410–445.
39. Bricogne, G. and Gilmore, C. J. (1990) A multisolution method of phase determination by combined maximization of entropy and likelihood. I. Theory, algorithms and strategy. *Acta Cryst.* **A46,** 284–297.
40. Karle, J. (1991) Direct calculation of atomic coordinates from diffraction intensities: Space group *P*1. *Proc. Natl. Acad. Sci. USA* **88,** 10,099–10,103.
41. Brünger, A. T. (1992) *X-PLOR. A System for X-ray Crystallography and NMR.* Yale University Press, New Haven.
42. Engh, R. A. and Huber, R. (1991) Accurate bond and angle parameters for X-ray structure refinement. *Acta Cryst.* **A47,** 392–400.
43. Allen, F. H., Kennard, O., and Taylor, R. (1983) Systematic analysis of structural data as a research technique in organic chemistry. *Accounts of Chemical Research* **16,** 146–153.
44. Baker, D., Bystroff, C., Fletterick, R. J., and Agard, D. A. (1993) PRISM: Topologically constrained phase refinement for macromolecular crystallography. *Acta Cryst.* **D49,** 429–439.
45. Swaminathan, S., Furey, W., Pletcher, J., and Sax, M. (1992) Crystal structure of staphylococcal enterotoxin B. a superantigen. *Nature* **359,** 801–806.
46. Musacchio, A., Noble, M., Pauptit, R., Wierenga, R., and Saraste, M. (1992) Crystal structure of a Src-homology 3 (SH3) domain. *Nature* **359,** 851–855.
47. Guo, H.-C., Jardetzky, T. S., Garrett, T. P. J., Lane, S. W., Strominger, J. L., and Wiley, D. C. (1992) Different length peptides bind to HLA-Aw68 similarly at their ends but bulge out in the middle. *Nature* **360,** 364–366.

48. James, M. N. G. and Sielecki, A. R. (1983) Structure and refinement of penicillopepsin at 1.8 Å resolution. *J. Mol. Biol.* **163,** 299–361.
49. Baker, D., Krukowski, A. E., and Agard, D. A. (1993a) Uniqueness and the *ab initio* phase problem in macromolecular crystallography. *Acta Cryst.* **D49,** 186–192.
50. Levinthal, Z. (1968) Are there pathways for protein folding? *J. Chem. Phys.* **65,** 44,45.
51. Leahy, D. J., Hynes, T. R., McConnell, H. M., and Fox, R. O. (1988) Crystallization of an anti-tempo-dinitrophenyl monoclonal antibody fab fragment with and without bound hapten. *J. Mol. Biol.* **203,** 829,830.
52. Brünger, A. T., Leavy, D. J., Hynes, T. R., and Fox, R. O. (1991) The 2.9 Å resolution structure of an anti-dinitrophenyl-spin-label monoclonal antibody fab fragment with bound hapten. *J. Mol. Biol.* **221,** 239–256.
53. Read, R. J. (1990) Structure-factor probabilities for related structures. *Acta Cryst.* **A46,** 900–912.
54. Hodel, A., Kim, S.-H., and Brünger, A. T. (1992) Model bias in macromolecular crystal structures. *Acta Cryst.* **48,** 851–859.
55. Ramachandran, G. N. and Sasisekharan, V. (1968) Conformation of polypeptides and proteins. *Advan. Protein Chem.* **23,** 283–438.
56. Morris, A. L., MacArthur, M. W., Hutchinson, E. G., and Thornton, J. M. (1992) Stereochemical quality of protein structure coordinates. *Proteins* **12,** 345–364.
57. Eisenberg, D. and McLachlan, A. D. (1986) Solvation energy in protein folding and binding. *Nature* **319,** 199–203.
58. Novotný, J., Rashin, A. A., and Bruccoleri, R. E. (1988) Criteria that discriminate between native proteins and incorrectly folded models. *Proteins* **4,** 19–30.
59. Chiche, L., Gregoret, L. M., Cohen, F. E., and Kollman, P. A. (1990) Protein model structure evaluation using the solvation free energy of folding. *Proc. Natl. Acad. Sci. USA* **87,** 3240–3243.
60. Vriend, G. and Sander, C. (1993) Quality control of protein models: Directional atomic contact analysis. *J. Appl. Cryst.* **26,** 47–60.
61. Bowie, J. U., Lüthy, R., and Eisenberg, D. (1991) A method to identify protein sequences that fold into a known three-dimensional structure. *Science* **253,** 164–170.
62. Lüthy, R., Bowie, J. U., and Eisenberg, D. Assessment of protein models with three-dimensional profiles. (1992) *Nature* **356,** 83–85.
63. Matthews, B. W. (1968) Solvent content of protein crystals. *J. Mol. Biol.* **33,** 491–497.
64. Fraser, R. D. B., Macrae, T. P., and Suzuki, E. (1978) An improved method for calculating the contribution of solvent to the X-ray diffraction pattern of biological molecules. *J. Appl. Cryst.* **11,** 693,694.
65. Moews, P. C. and Kretsinger, R. H. (1975) Refinement of the structure of the carp muscle calcium-binding parvalbumin by model building and difference fourier analysis. *J. Mol. Biol.* **91,** 201–228.
66. Phillips, S. E. W. (1980) Structure and refinement of oxymyoglobin at 1.6 Å resolution. *J. Mol. Biol.* **142,** 531–554.
67. Cheng, X. and Schoenborn, B. P. (1990) Hydration in protein crystals. A neutron diffraction analysis of carbonmonoxymyoglobin. *Acta Cryst.* **B46,** 195–208.

68. Badger, J. and Caspar, D. L. D. (1991) Water structure in cubic insulin crystals. *Proc. Natl. Acad. Sci. USA* **88,** 622–626.
69. Kuriyan, J., Petsko, G. A., Levy, R. M., and Karplus, M. (1986) Effect of anisotropy and anharmonicity on protein crystallographic refinement. *J. Mol. Biol.* **190,** 227–254.
70. Kuriyan, J., Ösapay, K., Burley, S. K., Brünger, A. T., Hendrickson, W. A., and Karplus, M. (1991) Exploration of disorder in protein structures by X-ray restrained molecular dynamics. *Proteins* **10,** 340–358.
71. Diamond, R. (1990) On the use of normal modes in thermal parameter refinement: Theory and application to the bovine pancreatic trypsin inhibitor. *Acta Cryst.* **A46,** 425–435.
72. Kidera, A. and Go, N. (1990) Refinement of protein dynamic structure: Normal mode refinement. *Proc. Natl. Acad. Sci. USA* **87,** 3178–3722.
73. Kidera, A., Inaka, K., Matsushima, M., and Gō, N. (1992) Normal mode refinement: Crystallographic refinement of protein dynamic structure applied to human lysozyme. *Biopolymers* **32,** 315–319.
74. Schomaker, V., and Trueblood, K. N. (1968) On the rigid-body motion of molecules in crystals. *Acta Cryst.* **B24,** 63–76.
75. Kuriyan, J. and Weis, W. I. (1991) Rigid protein motion as a model for crystallographic temperature factors. *Proc. Natl. Acad. Sci. USA* **88,** 2773–2777.
76. Howlin, B., Moss, D. S., and Harris, G. W. (1989) Segmented anisotropic refinement of bovine ribonuclease A by the application of the rigid-body TLS model. *Acta Cryst.* **A45,** 851–861.
77. He, X.-M. and Craven, B. M. (1993) Internal vibrations of a molecule consisting of rigid segments. I. Non-interacting internal vibrations. *Acta Cryst.* **A49,** 10–22.
78. Gros, P., van Gunsteren, W. F., and Hol, W. G. J. (1990) Inclusion of thermal motion in crystallographic structures by restrained molecular dynamics. *Science* **249,** 1149–1152.
79. Burling, F. T. and Brünger, A. T. (1994) Thermal motion and conformational disorder in protein crystal structures: comparison of multi-conformer and time-averaging models. *Israel Journal of Chemistry* **34,** 165–175.
80. Jiang, J.-S. and Brünger, A. T. (1994) Protein hydration observed by x-ray diffraction: solvation properties of penicillopepsin and neuraminidase crystal structures. *J. Mol. Biol.* **243,** 100–115.
81. Rice, L. M. and Brünger, A. T. (1994) Torsion angle dynamics: reduced variable conformational sampling enhances crystallographic structure refinement. *Proteins: Structure, Function, and Genetics* **19,** 277–290.

CHAPTER 11

The Crystallization and Structure Analysis of Oligonucleotide Sequences

Stephen Neidle

1. Introduction

Until 1976, the study of nucleic acid structure was exclusively the domain of fiber diffractionists. Between the original Watson-Crick structure in 1953 and this date, there was considerable activity in refining the original B-form model of DNA and extending the approach to other polymorphs and a number of synthetic, repetitious polynucleotides, all of which were based on data from fiber-diffraction samples. These studies reached their zenith with the development and use of a "linked-atom," least-squares refinement procedure for the optimization of mono-or dinucleotide repeat units against the relatively sparse para-crystalline diffraction data from ordered fibers. It is a tribute to the sophistication of these analyses, in spite of the inherent limitations of fiber data, that the refined "canonical" A- and B-DNA double helices are still major reference points for many studies *(1)*.

The overriding limitation of structural studies of fibrous polynucleotides is their inherent inability to address questions of structure and sequence at the individual nucleotide level, as opposed to the averaged mono- or dinucleotide repeating units defined and refined by the polynucleotide modeling procedures. This limitation is not relevant to single-crystal studies, which are unequivocally able to determine structural features at all points along a sequence, without having to define artificial averaged residues. Since 1976, there has been a progressive

From: *Methods in Molecular Biology, Vol. 56: Crystallographic Methods and Protocols*
Edited by: C. Jones, B. Mulloy, and M. Sanderson Humana Press Inc., Totowa, NJ

increase in the number of single-crystal analyses of short (<16 nucleotide residues in length) oligonucleotides, following the determination of the structures of dinucleotide monophosphate duplexes $(ApU)_2$ and $(GpC)_2$ *(2,3)*. The first real oligonucleotide determination was that of the Z-DNA structure in the $d(CGCGCG)_2$ duplex *(4,5)*. This has been followed by a large number of analyses of duplex helices, which have shown:

1. That the earlier fiber structure is correct in general terms;
2. Ways in which a number of drug molecules interact with DNA;
3. Details of a number of types of base mispairing; and
4. A wealth of information on detailed sequence-dependent and other features, such as hydration. Almost all of these are DNA oligonucleotide sequences, with relatively few RNA sequences.

Single-crystal studies have thus provided unequivocal confirmation of Watson-Crick base pairing and of the double-helix itself, with structures having been determined *ab initio* and without recourse to any assumptions from models. A number of these studies have revealed novel features of DNA structure and conformation. Much knowledge has now been obtained on sequence-dependent structural and conformational features; however, it is increasingly clear that several independent determinations (i.e., in differing crystal lattice and flanking sequence contexts) of a particular sequence run need to be obtained in order for these findings to be generalized into sequence–structure relationships with a high degree of confidence *(6)*. As yet, only some of the possible di- or tetra-nucleotide combinations have been observed in sufficient sequence contexts *(7,8)*. Under some circumstances, crystal-packing forces can induce local distortions. However, with care, it is possible to dissect out their effects and even to appreciate that one can obtain information on deformability from such regions *(9)*. Early attempts *(10)* to formulate general rules governing sequence-dependent structural features were based on just a single structure (the Dickerson-Drew dodecamer $d[CGCGAATTCGCG]_2$), and suffer from obvious limitations.

This chapter does not attempt to review the details of oligonucleotide structures. Instead the interested reader is referred to several recent reviews *(1,7,11)*. Rather, we survey the scope, extent, and problems of oligonucleotide crystallization and structure determination. Details of unit cell dimensions, space groups, resolution of diffraction data, and so forth, have been extensively tabulated elsewhere *(7)* and are not duplicated here. Similarly, routine procedures common to other areas of macromolecular crystallization and crystallography are not covered. The refinement of oligonucleotide structures is discussed in detail elsewhere in this volume (Chapter 9).

2. Oligonucleotide Crystallization

2.1. Synthesis and Purification

The development of efficient large-scale methods for DNA oligonucleotide synthesis has been vital to the development of structural studies, both NMR and crystallography. X-ray crystallography requires relatively large quantities of material for crystallization trials. A minimum of 5–10 mg is typical. High purity is essential–impurities from the synthetic chemistry, which are irrelevant for molecular biological applications, are in general inimical to successful crystallization. Side reactions, such as base deamination, can occur, and their products are usually only separable from the required sequence by HPLC.

Synthetic methods involve three stages: the preparation of suitably protected nucleotide monomers of the common bases (and increasingly of modified ones as well), their coupling in the order defined by the sequence required, and finally the deprotection of the final oligomer. A variety of protecting groups have been developed, all of which ensure that the sensitive hydroxyl and other substituents on a nucleotide are not themselves reacted during coupling stages. Original development of these stages employed phosphotriester chemistry in solution *(12)*; such solution-phase synthesis has the advantage of large-scale production being possible. However, a skilled chemist is required for this time-consuming approach, in contrast to automated machine synthesis. This latter method is now in common use in many laboratories; modern DNA synthesizers, with microprocessor control, provide very considerable ease and speed of use. They have been responsible for the current wide availability of synthetic oligonucleotides. Cyanoethyl phosphoramidite chemistry *(13)* is almost universally employed in synthesizers, with coupling times of 12–30 min/nucleotide and the monomers being purchased in fully-protected form. A single overnight synthetic run at the 10-μmol level can produce approx 10 mg of pure oligomer after purification. The machine approach can produce oligomers at this scale of up to 30 bp in length, whereas solution-phase coupling is very inefficient beyond about 10–12 bases. Recent advances in base-protection chemistry have led to significant decreases in the thus far slow deprotection step, reducing it from 8 h to 1. As yet, automated RNA synthesis has not quite reached an equivalent stage of reliability *(14)*, hence the lack of RNA structures in spite of the obvious biological importance of having structural data on features, such as short stem loops. Longer RNA sequences, such as in ribozymes, are best obtained by in vitro transcription with T7 bacteriophage RNA polymerase *(15,16)*.

Modified DNA oligomers can in principle be synthesized by machine, provided that the appropriate protected modified nucleoside is available. In practice, only a restricted range is commercially available, and therefore, others of interest may have to be synthesized and protected *de novo*. Several starting blocks with methylated bases are available, including O^6-methylguanine. Nucleosides that are base-halogenated with either bromine or iodine may be of special use as heavy-atom markers for isomorphous replacement or anomalous scattering phasing; they are available commercially in protected form. The machine chemistry for routine laboratory synthesis of some backbone modifications (methylphosphonates and phosphothioates) is also available owing to their interest as antisense agents with nuclease resistance, although to date few structural studies have been reported on them.

Extensive purification of synthesized oligomers, to at least 95% purity, is essential in order to optimize the chances of obtaining high-quality crystals. Procedures used are usually reverse-phase, high-pressure liquid chromatography (to both purify and judge purity on the basis of a single sharp peak) or gel-filtration chromatography. These procedures remove unwanted blocking reagents, precursors, and any truncated sequences. Superior purification may be obtained by an initial pass through HPLC of the 5'-dimethoxytrityl end-protected oligomer, followed by its removal with glacial acetic acid and further HPLC. Several manufacturers supply purification cartridges packed with gel-filtration beads; with these, impure oligonucleotides are directly passed through by means of a syringe, thus greatly speeding up the purification process, although on a small scale. These methods are optimal for oligonucleotides up to 20 bp in length. Beyond this limit, gel electrophoretic methods are probably more effective, with separated bands being easily cut out and pure material excised.

Ease and extent of purification are often dependent on the sequence itself. Those containing runs of guanines are notoriously difficult to handle, with solubility often being a problem. This most probably arises from interstrand aggregation of guanines into, for example, four-stranded helical bundles.

The extent of purity of an oligonucleotide is often indicated by its physical appearance. Pure material should be white in color and floccular rather than yellow and gel-like. Correctness of the sequence can be checked by sequencing methods; one procedure uses enzymatic digestion with snake venom phosphodiesterase followed by two-dimensional electrophoresis on DEAE-cellulose. The differences in mobilities between different nucleosides is such that the sequence can usually be directly read off the plate *(12)*. Another method end-labels the oligomer with ^{32}P-ATP, followed by running on a gel against a standard.

2.2. General Aspects of Crystallization

All DNA structures reported to date are in the duplex form; several also have looped-out bases and one has a hairpin loop. The overwhelming majority of structures have self-complementary sequences. In large part, this has reflected a desire to minimize cost and effort, in that one rather than two sequences are required. However, especially when studying biologically relevant sequences and such phenomena as mispairing, non-self-complementary sequences should be the systems of choice.

Crystals of oligonucleotides are often difficult to produce in the quality required for structure analysis. It is a typical experience that success is limited to 10–20% of sequences examined. Examination of the literature also shows that apparently only certain lengths of oligomer are amenable to crystallization in forms that diffract to better than 2.5-Å resolution, probably on account of crystal-packing factors. Often, although several sequences of a given length may crystallize well, slight changes in sequence can result in total failure to crystallize. A general rule of thumb appears to be that both 5' and 3' terminal base pairs should be CG ones to minimize fraying and thus stabilize the sequence from melting. Exceptions to this rule are when the sequence is very GC-rich, so that single AT base pairs at the ends can be tolerated. In addition, high-CG content, in general, will improve prospects for crystallizing at ambient temperatures and possibly for higher resolution data (since thermal motion is reduced). It is always advisable prior to expending effort on crystallizing to obtain a melting curve for the oligomer being studied so that (1) one can ensure that crystallization attempts are made well below the mid-point of the helix-to-coil transition and (2) that there are no hairpins in the structure, shown by multiple transition points. Annealing a sequence in order to remove possible hairpins prior to crystallization, is always to be recommended. In general, both short sequences and decamer and longer sequences with a 5'-end cytosine are often crystallizable, but those with a 5' guanosine can only be crystallized with considerable difficulty. Two examples from the author's laboratory illustrate this point: the sequences dGCATGC and dGAAACGTTTC both produce crystals under a variety of conditions, the latter in particular resulting in exceptionally large and well-formed bipyramidal crystals. Yet neither diffract to better than ca. 6 Å, even with high-intensity X-ray sources. Octanucleotide duplexes crystallizing in the A form are the principal exceptions to this rule, with the overwhelming majority of structures reported in this class having a 5'-end guanosine residue *(7)*.

2.3. Crystallization Conditions

Successful oligonucleotide crystallizations have used a remarkably narrow range of conditions, in striking contrast to proteins, where a wide variety of precipitants, counterions, and so on, are routinely employed *(17)*. This in part reflects the uniformity of polyanionic surface presented by an oligonucleotide compared to the enormous range of charge and hydrophobicity patterns on the surface of proteins. The near-universal emphasis on just one type of oligonucleotide crystallizing approach, using 2-methyl-3,4-pentanediol (MPD), also reflects the success achieved by it in the early 1970s with the crystallization and subsequent structure analysis of yeast $tRNA^{Phe}$. Prior to this success, many tRNA species had been crystallized under a variety of conditions, but they were invariably poor X-ray diffractors, with data not extending beyond about 6 Å.

Tables 1–6 detail some of the crystallization conditions reported in the literature. Those with incomplete or ambiguous descriptions have been excluded, and, in some cases, the temperature used has been inferred from that of the data-collection experiments. There has been little reporting to date of usage of robotic methods, crystal screening with wide-ranging conditions, or of factorial approaches *(17)*.

Oligonucleotide crystallization experiments most commonly utilize MPD as precipitant in a closed vapor equilibration system. Some success has been reported with isopropyl alcohol, but there are very few reports of high-quality crystals being obtained with other agents, such as polyethylene glycol, or salts, such as ammonium sulfate. Polyethylene glycol, which is often the agent of choice for proteins, appears to be most useful in

(text continued on p. 279)

Table 1
A-DNA[a]

Sequence	[DNA]	$[Mg^{2+}]$	[Spermine]	$[MPD]_{droplet}$	$[MPD]_{reservoir}$	pH	Temperature	Time for Xtl	Ref.
GGCCGGCC	1.2	3	0.6	0	30	7	—	—	*(18)*
GGATGGGAG + CTCCCATCC	0.2	12	0.8	0	6–20	6.5	Ambient	5 mo	*(19)*
$GG^{Br}UA^{Br}UACC$	0.4	0.28	0.4	15	40	7	—	—	*(20)*
GTACGTAC	2	10	2	0	40	7	Ambient	few wk	*(21)*
CTCTAGAG	1.5	25	1	7	50	6.8	18°C	3–4 wk	*(22)*
GTACGTAC	2	15	8	5	30	6	20°C	3–4 wk	*(23,24)*
GGGCGCCC	2	5	0.15	5	30, 50	7	19°C	—	*(25)*
$r[U(UA)_6A]$	4	400	0	35	—	6.5	35°C	—	*(26)*
r[GCG]d[TATACGC]	1.5	15	8	0	40	6.0	4°C	1 wk	*(27)*

[a]Typical oligonucleotide crystallization where MPD has been used as precipitant. All DNA and Salt Concentrations are in m*M*. MPD Concentrations (in %) are the initial values, at the outset of the crystallization experiments. Time for Xtl is the time quoted for crystals to appear.

Table 2
B-DNA[a]

Sequence	[DNA]	[Mg^{2+}]	[Spermine]	$[MPD]_{droplet}$	$[MPD]_{reservoir}$	pH	Temperature	Time for Xtl	Ref.
CGCGAATTCGCG	0.6	6	0.2	10–35	—	7	7°C	—	*(28)*
CGCAAAAATGCG + CGCATTTTTGCG	1.1	44	0	0	40	7	10°C	—	*(29)*
CGCAAAAAAGCG + CGCTTTTTTGCG	0.2	10	0.5	5	30–45	7	4°C	—	*(30)*
ACCGGCCACA + TGGCCGCGGTGT	1.0	18	1.2	0	45	6	4°C	2–3 d	*(31)*
CCAGGC[Me]CTGG	2	50	0	0	40	7.5	4°C	3 wk	*(32)*
CGCGAAAACGCG + + CGCGTT TTCGCG	0.5	2.9	8.6	2	50	5	4°C + seeding	1 wk	*(33)*
CGTGAATTCGCG	1.22	15	2.0	10	22	7	4°C	1 wk	*(34)*
CGCATATATGCG	0.5	22	0.4	10	44	—	—	—	*(35)*

[a]Typical oligonucleotide crystallization conditions where MPD has been used as precipitant. All DNA and salt concentrations are in m*M*. MPD concentrations (in %) are the initial values, at the outset of the crystallization experiments.

Table 3
Z-DNA[a]

Sequence	[DNA]	$[Mg^{2+}]$	[Spermine]	$[MPD]_{droplet}$	$[MPD]_{reservoir}$	pH	Temperature	Time for Xtl	Ref.
CGCGCG	1.4	120	0	0	10	7	Ambient	3–14 d	*(36)*
^{Br}CGAT^{Br}CG	2.3	25	0	0	25	7	Ambient	3 wk	*(37)*
(^{Me5}CGCG)$_3$	2.0	4	3	0	10	7	Ambient	2 wk	*(38)*
CACGTG	2	35	0	3.5	70	6	Ambient	4 mo	*(39)*
CGr(CG)CG	1	2.5 (+60 m*M* NaCl)	2.0	1.5	10	7	Ambient	7–10 d	*(40)*
^{Me5}CGUA^{Me5}CG	4	15	0	8.5	30	7	Ambient	3 d	*(41)*
CGCG^{Mo4}CG	2	12	4.7	9.5	40	6.5	—	—	*(42)*
^{Me5}CG^{Me5}CG^{Me5}CG	2	4	3	2	10	7	Ambient	—	*(43)*

[a]Typical oligonucleotide crystallisation conditions where MPD has been used as precipitant. All DNA and salt concentrations are in m*M*. Concentrations (in %) are the initial values at the outset of the crystallization experiments.

Table 4
DNA Mismatches and Bulges[a]

Sequence	[DNA]	$[Mg^{2+}]$	[Spermine]	$[MPD]_{droplet}$	$[MPD]_{reservoir}$	pH	Temperature	Time for Xtl	Ref.
CGCGAAATTTACGCG	0.3	12	0.33	6	15	7.0	4°C	1 wk	*(44)*
$CGC^{Me6}GCG$	2	5	0.5	10–20	30	7.0	Ambient	8–12 mo	*(45)*
CGCAAGCTGGCG	0.75	25	2.5	0	32	7.2	5°C	10–14 d	*(46)*
GGGTGCCC	1.5	4.5	0.1	10	30	7.0	4°C	—	*(47)*

[a]Typical oligonucleotide crystallisation conditions where MPD has been used as precipitant. All DNA and salt concentrations are in m*M*. MPD concentrations (in %) are the initial values at the outset of the crystallisation experiments.

Table 5
Drug–DNA Minor Groove Complexes[a]

Sequence	[DNA]	[Drug]	$[Mg^{2+}]$	[Spermine]	$[MPD]_{droplet}$	$[MPD]_{reservoir}$	pH	Temperature	Time for Xtl	Ref.
CGCGAATTCGCG	0.3	Netropsin 0.6	6.3	0.3	10	32.5	—	4°C	—	*(48)*
CGCGAATTCGCG	1.2	Hoechst 33258 1.2	8	1	12	50	6.5	Ambient	2 mo	*(49)*
CGCGAATTCGCG	0.21	Hoechst 33258 0.21	5.38	0.21	20	35	7.0	—	2 mo	*(50)*
CGCGATATCGCG	1.0	Hoechst 33258 2.0	12.5	4.5	20	50	7.5	Ambient	4 wk	*(51)*
CGCGATATCGCG	1.0	Netropsin 2.0	7.0	4.5	5	50	6.5	Ambient	—	*(52)*
CGCAAATTTGCG	1.0	Distamycin 1.1	8.0	1.0	10	50	6.5	Ambient	—	*(53)*
CGCGAATTCGCG	3.0	Berenil 2.0	30.0	0	20	50	7.0	5°C	—	*(54)*
CGCGAATTCGCG	0.28	DAPI 0.23	5.38	0.22	20	40	7.0	Ambient	Weeks	*(55)*

[a]Typical oligonucleotide crystallization conditions where MPD has been used as precipitant. All DNA and salt concentrations are in m*M*. MPD concentrations (in %) are the initial values, at the outset of the crystallisation experiments.

Table 6

Drug–DNA Intercalation Complexes[a]

Sequence	[DNA]	[Drug]	$[Mg^{2+}]$	[Spermine]	$[MPD]_{droplet}$	$[MPD]_{reservoir}$	pH	Temperature	Time for Xtl	Ref.
CGTACG	2.0	Daunomycin 2.0	15.0	10	5	30	6.5	Ambient	1 wk	*(56,57)*
CGATCG	1.5	Daunomycin 1.5	20.0	3.5	9	50	6.5	20°C	Weeks	*(58)*
CGCG	0.7	Ditercalinium 0.2	0.8	0.3	6	30	6.0	Ambient	1 wk	*(59)*
GCGTACGC	1.2	Triostin A 1.1	8.0 (+10 m*M* NaCl)	1.0	12	30	4.5	Ambient	Several months	*(60)*
CGTACG	2.0	Triostin A 1.5	10.0	0	10	35	7.0	—	1 wk	*(61)*

[a]Typical oligonucleotide crystallisation conditions where MPD has been used as precipitant. All DNA and salt concentrations are in m*M*. MPD concentrations (in %) are the initial values at the outset of the crystallization experiments.

the crystallization of drug–oligonucleotide complexes where the drug is water-insoluble. Examples are nogalamycin *(62)* and actinomycin *(63)*.

The polycation spermine is often employed—its role is presumed to shield the polyanionic helices from each other and so promote crystal packing. There are, however, several well-documented reports—*see* Tables 1–6—where the presence of spermine has not been found to be essential. Magnesium ions are also required, probably to bridge and stabilize phosphate groups. The system is buffered at approx neutral pH with sodium cacodylate or tris buffer—major deviations from a value of ca. pH 7.0 can lead to base protonation or even acid-catalyzed strand scission in extreme cases. Phosphate buffer is not recommended, as magnesium phosphate can crystallize easily out in the concentrations and temperatures used. Spermine can sometimes play an active structural role, especially in Z-DNA, where for example, it has been in the minor groove of a Z-DNA duplex *(64)*.

Crystallization techniques used reflect the often small amounts of oligomer available. Several different microcrystallization setups are in use. Those that employ vapor diffusion are the most common, with the most popular systems being:

1. Glass depression plates;
2. Plastic petri dishes with a 4 × 4 matrix of drops containing the crystallizing solution *(65)*;
3. Hanging droplets.

   For all of these, surfaces should be siliconized to minimize the spreading of droplets. A typical droplet volume is 10–15 µL. This would contain oligomer at a concentration of, typically, 1 m*M*, which is sufficient for at least six good-sized (0.1 × 0.2 × 0.5 mm) crystals of a decamer. Microdialysis, a technique frequently used for protein crystallization, has only rarely been successfully reported for oligonucleotides *(32)*.

Other important factors include:

1. Time: Crystals of many dodecanucleotide sequences can appear within a few days of setting up, whereas other types of sequence may take many weeks or even months. For example, those of the O^6-methylated Z-form $(CG)_3$ hexamer duplex were reported to have taken a year *(45)*, and then to have only produced three crystals. Seeding has been occasionally used with success. Oligonucleotide crystals frequently deteriorate with age when kept in their mother liquor, with crystal surfaces clouding over and crazing, and diffraction quality being greatly diminished.

2. Temperature: It is axiomatic that sequences with low helix→coil transition→temperatures (T_m) will require cold-room conditions for successful crystallization. However, since this temperature is actually the midpoint of the transition, there will be a significant population of single-stranded species well below this temperature. It is therefore advisable initially to attempt all crystallizations in the cold. A notable exception to this rule is the sequence $rU(UA)_6A$, with a temperature of 35°C being needed for successful production of crystals diffracting to high resolution *(26,66)*. As with all macromolecular crystallizations, temperature stability is important.
3. Relative concentrations of oligonucleotide, magnesium ion, and precipitant: The magnesium ion is usually in considerable molar excess, although there does not appear to be any rule about its molar ratio to the DNA. Aqueous solubilities of different sequences can vary over a wide range, with those containing runs of guanines being the least soluble. Base and backbone modifications can also significantly decrease oligomer solubility.
4. Most drugs that have been complexed with DNA sequences are cationic and freely soluble in water to 10 m*M* concentration or more. Exceptions are Hoechst 33258 (only sparingly soluble), the anthracycline nogalamycin, and the bis-intercalators of the echinomycin class, such as triostin A. All of these except for Hoechst 33258 are uncharged and essentially insoluble in aqueous solution. These drugs can be dissolved in 1:1 chloroform/methanol (triostin A) or methanol (echinomycin and nogalamycin). Once in a droplet with oligonucleotide, the volatile organic solvent quickly evaporates, and the drug becomes solubilized by DNA complexation. Drug–oligonucleotide complexes are generally less water-soluble than the DNA itself, so precipitation is frequently observed at quite low MPD levels. Crystals of complexes often grow out of the precipitate, provided it is not too extensive.
5. The volume of the enclosed system, especially with respect to the surface area of the reservoir of higher concentration precipitant, is important both for the time taken for crystallization and for the amount of water lost from the droplet in order to achieve equilibration in the enclosure. Clearly, too large a volume of container is to be avoided. This factor has to be taken into account when trying to reproduce conditions from other laboratories.
6. The gradient in concentration between droplet and reservoir is a critical factor that governs speed of precipitation and/or crystallization. Too steep a gradient typically results in premature precipitation or a mass of small crystals before true equilibrium has been reached.

2.4. Correlations in Crystallizing Conditions

Crystallization of a newly synthesized oligonucleotide sequence normally starts with recourse to an examination of literature conditions

Table 7
Correlation Coefficients Calculated by Means of Linear Regression Analysis

	For concentrations given in Tables 1 and 2, for A and B DNAs together	For concentrations given in Table 5, for DNA-minor groove drug complexes
[DNA] vs [Mg^{2+}]	0.29	0.94
[DNA] vs [spermine]	0.14	0.03
[DNA] vs $[MPD]_{final}$	0.09	0.65
[Mg^{2+}] vs [spermine]	0.27	0.12
[Mg^{2+}] vs $[MPD]_{final}$	0.09	0.63
[Spermine] vs $[MPD]_{final}$	0.08	0.20

(Tables 1–6). Choice of oligomer concentration is often dictated by the quantity of material available. Can choices of concentrations for the other components in a crystallization experiment be made in a systematic manner? This question can be approached by a comparative examination of the data contained in Tables 1–6. Linear regression analysis has been used in order to determine whether there are correlations between pairs of the major variable concentrations of DNA, spermine, magnesium ion, and final MPD concentration in the droplet that have successfully produced crystals. It has been assumed that this MPD concentration is reached at equilibrium, and has been calculated as the simple average of initial droplet and reservoir concentrations.

Table 7 details correlation coefficients for (1) the A- and B-form oligonucleotide conditions together in Tables 1 and 2, and (2) the DNA–minor groove complexes in Table 5. It is apparent that none of the variables in the first group are correlated with any degree of significance. This is shown in the random scatter of points in the [DNA]:[MPD] concentration plot (Fig. 1). On the other hand, there are several significant correlations for the drug–DNA data. The magnesium ion:[DNA] correlation of 0.94 is highly statistically significant, even with the relatively small number of observations used (Fig. 2). The correlation between [DNA] and [MPD] levels (Fig. 3) is still statistically significant, although at not the same level. It is at the same level as the magnesium ion:[MPD] one (Fig. 4), although the correlation coefficient for this is increased to 0.70 when the aberrant value at high [Mg^{2+}] is removed. On the other hand, spermine levels are quite uncorrelated with any of the other variables. The correlation coefficient between [Mg^{2+}] and [MPD] for the

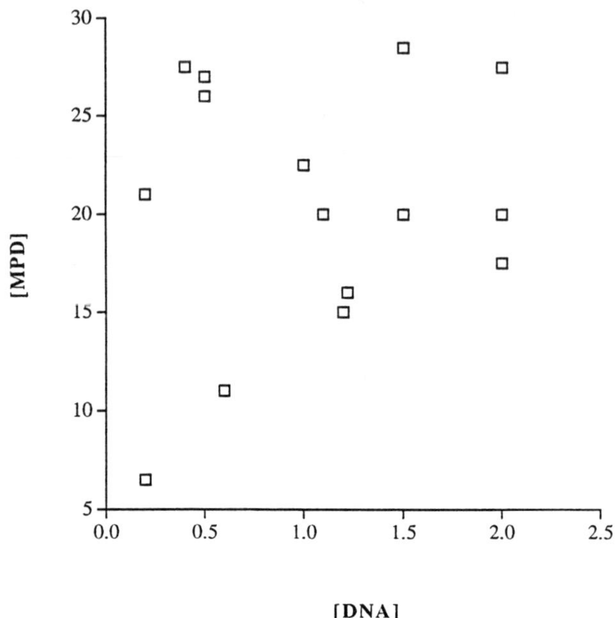

Fig. 1. Plot of DNA against MPD concentrations, for the A- and B-type oligonucleotide structures listed in Tables 1 and 2. Concentrations are in m*M* for DNA and % v:v for MPD. Values for A-type oligomers have filled-in circles.

Z-DNA oligomers (Table 3) is 0.80 (Fig. 5), which also has a weak correlation between [MPD] and [DNA].

These observations suggest that for a set of related sequences in length and sequence type, the concentration of the MPD-precipitating agent required for crystallization of a given oligonucleotide can be approximately predicted, as can magnesium levels. It is noteworthy that the magnesium:DNA molar ratio for the drug:DNA minor groove complexes averages 1:10–12, in accord with what one would expect for dodecanucleotides. This suggests that such a ratio, of one Mg^{2+}/nucleotide, may be a generally useful one. On the other hand, spermine concentrations do not show any discernible pattern—it may be significant that a number of oligomers have been crystallized in the absence of spermine, and only rarely has it been located in electron density maps. It may be that the presence of spermine is only fulfilling a secondary need, to provide a general counterion atmosphere in the crystallizing solution. The absence of correlation for the A and B oligomer conditions in Tables 1 and 2 may

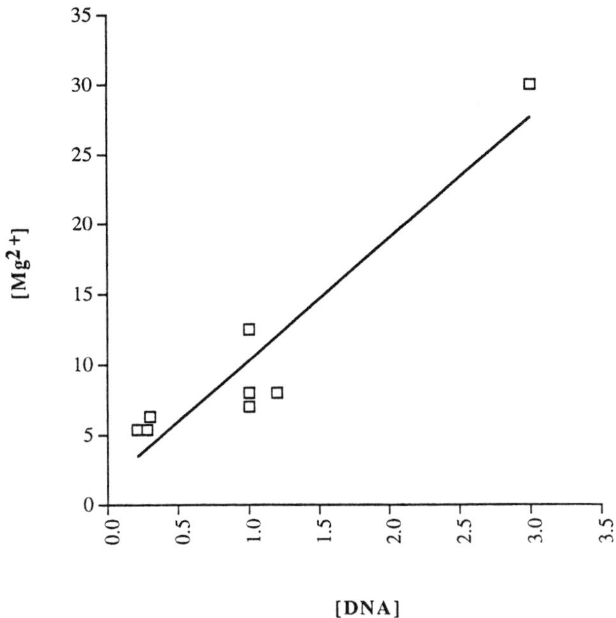

Fig. 2. Plot of DNA against Mg^{2+} concentrations (in m*M*) for the minor groove-oligonucleotide structures listed in Table 5.

be a reflection of the much greater disparity in other variables compared to the drug complexes in Table 5, such as sequence, temperature, and crystallization method.

3. Oligonucleotide Crystal Forms

It is widely accepted that crystal-packing considerations can force particular lengths of oligonucleotide duplex to adopt one helical type rather than another *(7)*. This is best illustrated by the octanucleotide family, of over 30 structures, all of which have been found in the A-form even though their crystallization conditions are often indistinguishable from those used to crystallize B-form oligonucleotides. Almost all octanucleotides crystallize in one of two space groups, the tetragonal $P4_32_12$ or the hexagonal $P6_1$. Several sequences, such as dGGGCGCCC *(25)*, have been found to crystallize in both forms, with consequent differences in hydration and conformation between them. The highest resolution reported for an A-DNA oligomer is 1.5 Å, although 2.0–2.2 Å is the norm. One octanucleotide sequence dGTACGTAC has uniquely been found to crys-

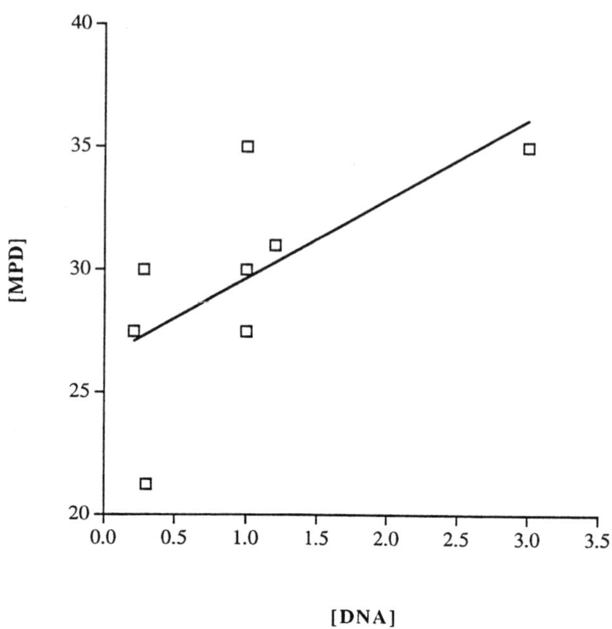

Fig. 3. Plot of DNA against MPD concentrations (in %) for the minor groove structures.

tallize in the orthorhombic space group $P2_12_12$ as well as the standard tetragonal form *(23,24)*.

The largest single group of oligonucleotide structures are B-form dodecamers, and are variants of the seminal "Dickerson-Drew" sequence of structures reported to date. The group includes the minor-groove drug complexes and a number of base mismatched structures. Almost all crystallize virtually isomorphously in an orthorhombic $P2_12_12_1$ unit cell. A nonisomorphous B-form dodecamer, in space group C2, has been reported *(67)*, although with a similar packing motif. Even more extreme has been the finding *(68,69)*, of two dodecamer sequences, dCCGTACGTACGG and dGCGTACGTACGC, that crystallize in an A-DNA conformation. It appears that a 5'-end sequence of CGC is required in order for a dodecamer sequence to crystallize in the "standard" $P2_12_12_1$ form. The highest resolution reported is 1.9 Å, with 2.2–2.5 Å being common. By contrast, several B-form decamers have been found to crystallize to much higher resolutions *(8)* (1.3–1.5 Å). These duplexes crystallize in a wide variety of packing modes, in contrast to the

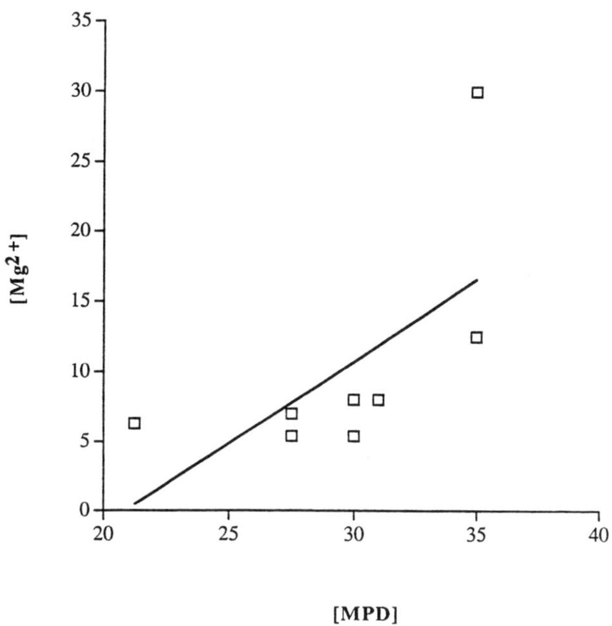

Fig. 4. Plot of MPD against Mg^{2+} concentrations for the minor groove structures.

dodecamers, with space groups $P2_12_12_1$, P6, R3, C2, $P3_22_1$, and $I2_12_12_1$ being reported *(8,32,70–74)*.

Z-form oligonucleotides generally diffract to still higher resolution, with dCGCGCG itself having data extending to 1.0 Å. The Z-DNA hexamers all crystallize in an isomorphous orthorhombic $P2_12_12_1$ cell.

The mono-intercalation drug complexes are almost all with hexanucleotide duplexes. They tend to crystallize in high-symmetry space groups and frequently diffract to high resolution.

4. Methods of Crystal Structure Analysis

The overwhelming majority of oligonucleotide structures have been solved by molecular replacement methods. This has been facilitated by the tendency of all oligomers of a given length to crystallize in the same space group and with closely similar unit-cell dimensions. The assumption that all such structures are isomorphous is, however, not necessarily correct, as was shown by the existence of two oppositely oriented "half-molecules" in the structure of an apparently normal dodeca-

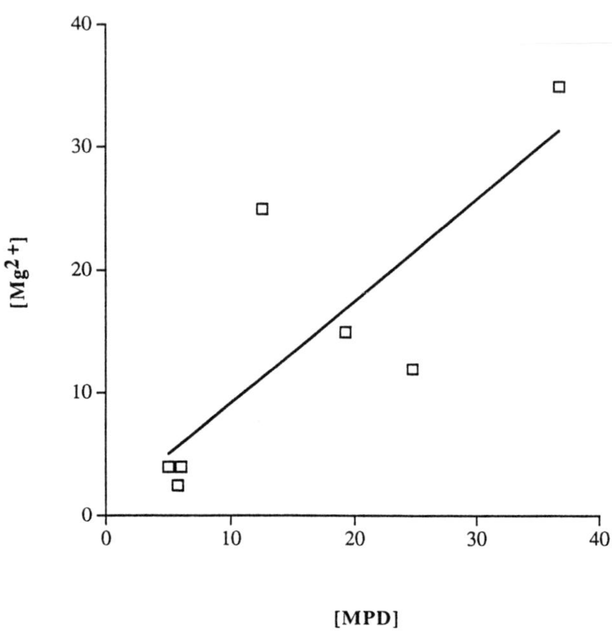

Fig. 5. Plot of MPD against Mg^{2+} concentrations for the Z-form oligonucleotide structures listed in Table 3.

nucleotide duplex, dCGCAAAAATGCG *(29)*. Nonetheless, most members of each oligonucleotide family are more truly isomorphous. This has been exploited to considerable effect in the case of the dodecamers, with such variations as sequence changes, base mispairing, base alkylation, and drug binding being analyzed, and much significant new information on them being obtained. These variations are invariable in the central 6–8 bp region, where crystal packing factors are not significant.

A number of structures have been solved by search procedures, notably using the ULTIMA *(75)* and MERLOT *(76)* computer programs. Oligonucleotide structures are particularly amenable to successful rotation searches, provided a good starting model is available. When this is so, the dominance of the intensity transform by base pairs generally provides an unambiguous and unique solution to a rotation function search. This same feature can cause problems with translation searches, as it can be difficult to distinguish solutions that differ by base-pair translations, especially when oligonucleotide helices are stacked end-to-end in the crystal lattice. Such a situation has been discussed for a B-DNA decamer

(32), when a solution was found that appeared to refine satisfactorily, to an R factor of 23%. However, examination of electron density maps and the detail of intermolecular distances indicated that the structure was in error. It is generally true that an incorrect structure does not pack satisfactory in its unit cell, that difference electron density maps show spurious features of significant and unexplained density, and that the R factor will not decrease below about 23–25% even with solvent molecules included. The structure reported for a 13-mer oligonucleotide duplex with a looped-out base *(78)*, initially refined to an R factor of 15%; however a subsequent careful re-examination has shown *(79)* that significant revisions to this structure are needed.

Direct methods have not been successfully used to solve an oligonucleotide structure. This is not surprising in view of the condition of uniform electron distribution not being obeyed in these structures. One structure has been reported as solved by the application of maximum entropy methods *(80,81)*, which use a variant of the standard sign relationships.

Only a small number of oligonucleotide structures have been solved by multiple isomorphous replacement *(82)* and can therefore be considered to be determined *ab initio*, independent of any prior structural model. These include the "Dickerson-Drew" B-form dodecamer dCGCGAATTCGCG and the Z-form hexamer dCGCGCG. In the former case, two derivatives were used, one with a covalently bound bromine atom and the other with cis-dichlorodiamino platinum (II) soaked into the native crystals. The dCGCGCG structure was solved with the aid of three heavy-atom derivatives, of Ba^{2+}, Co^{2+}, and Cu^{2+}, which were diffused into native crystals.

Acknowledgments

I am grateful to my colleagues and collaborators, especially Helen Berman and Mark Sanderson for much useful discussion on DNA conformation and structure determination, and Daniel Neidle for assistance with the correlation analyses. Studies on oligonucleotide structure in my laboratory are supported by the Cancer Research Campaign.

References

1. Neidle, S. (1994) *DNA Structure and Recognition.* Oxford University Press, Oxford, UK.
2. Seeman, N. C., Rosenberg, J. M., Suddath, F. L., Kim, J. J. P., and Rich, A. (1976) RNA double-helical fragments at atomic resolution: I. The crystal and molecular structure of sodium adenylyl-3',5'-uridine hexahydrate. *J. Mol. Biol.* **104,** 109–144.

3. Rosenberg, J. M., Seeman, N. C., Day, R. O., and Rich, A. (1976) RNA double-helical fragment at atomic resolution: II. The crystal structure of sodium guanylyl-3,5'-cytidine nonahydrate. *J. Mol. Biol.* **104,** 145–167.
4. Wang, A. H.-J., Quigley, G. J., Kolpak, F., Crawford, J. L., van Boom, J. H., van der Marel, G., and Rich, A. (1979) Molecular structure of a left-handed double helical DNA fragment at atomic resolution. *Nature* **282,** 680–686.
5. Gessner, R. V., Frederick, G. A., Quigley, G. J., Rich, A., and Wang, A. H. -J. (1989) The molecular structure of the left-handed Z-DNA double helix at 1. 0-Å atomic resolution. *J. Biol. Chem.* **264,** 7921–7935.
6. Shakked, Z. (1991) The influence of the environment on DNA structures determined by X-ray crystallography. *Curr. Opinion Structural Biol.* **1,** 446–451.
7. Dickerson, R. E. (1991) DNA structure from A to Z. *Methods Enzymol.* **211,** 67–111.
8. Yanagi, K., Privé, G. G., and Dickerson, R. E. (1991) Analysis of local helix geometry in three *B*-DNA decamers and eight dodecamers. *J. Mol. Biol.* **217,** 201–214.
9. Dickerson, R. E., Goodsell, D. S., and Neidle, S. (1994) "…the tyranny of the lattice…" *Proc. Natl. Acad. Sci. USA* **91,** 3579–3583.
10. Calladine, C. R. (1982) Mechanics of sequence-dependent stacking of bases in B-DNA. *J. Mol. Biol.* **161,** 343–352.
11. Kennard, O. and Hunter, W. N. (1989) Oligonucleotide structure: A decade of results from single-crystal X-ray diffraction studies. *Quart. Rev. Biophys.* **22,** 327–379.
12. Gait, M. J. (ed.) (1984) Oligonucleotide synthesis. IRL Press, Oxford.
13. Caruthers, M. H. (1985) Gene synthesis: DNA chemistry and its uses. *Science* **230,** 281–285.
14. Reese, C. B. (1989) The chemical synthesis of oligo- and poly-ribonucleotides, in *Nucleic Acids and Molecular Biology*, vol. 3 (Eckstein, F. and Lilley, D. M. J., eds.), Springer-Verlag, Berlin.
15. Gerwith, D. T. and Moore, P. B. (1988) Exploration of the L18 binding site on 5S RNA by deletion mutagenesis. *Nucleic Acids Res.* **16,** 10,717–10,732.
16. Szewczak, A. A., White, S. A., Gerwith, D. T., and Moore, P. B. (1990) On the use of T7 RNA polymerase transcripts for physical investigation. *Nucleic Acids Res.* **18,** 4139–4142.
17. Ducruix, A. and Giegé, R. (eds.) (1992) *Crystallization of Nucleic Acids and Proteins. A Practical Approach.* IRL Press at Oxford University Press, Oxford, UK.
18. Wang, A. H.-J., Fujii, S., van Boom, J. H., and Rich, A. (1982) Molecular structure of the octamer d(G-G-C-C-G-G-C-C): modified A-DNA. *Proc. Natl. Acad. Sci. USA* **79,** 3968–3972.
19. McCall, M., Brown, T., Hunter, W. N., and Kennard, O. (1986) The crystal structure of d(GGATGGGAG) forms an essential part of the binding site for transcription factor IIIA. *Nature* **322,** 661–664.
20. Kennard, O., Cruse, W. B. T., Nachman, J., Prange, T., Shakked, Z., and Rabinovich, D. (1986) Ordered water structure in an A-DNA octamer at 1.7Å resolution. *J. Biomol. Struct. Dyn.* **3,** 623–647.
21. Takusagawa, F. (1990) The crystal structure of d(GTACGTAC) at 2. 25 Å resolution: Are the A-DNA's always unwound approximately 10° at the C-G steps? *J. Biomol. Struct. Dyn.* **7,** 795–809.

22. Hunter, W. N., D'Estaintot, B. L., and Kennard, O. (1989) Structural variation in d(CTCTAGAG). Implications for protein–DNA interactions. *Biochemistry* **28,** 2444–2451.
23. Courseille, C., Dautant, A., Hospital, M., D'Estaintot, B. L., Precigoux, G., Molko, D., and Teoule, R. (1990) Crystal structure analysis of an A(DNA) octamer d(GTACGTAC). *Acta Crystallographica* **A46,** FC9–FC12.
24. Langlois d'Estaintot, B., Dautant, A., Courseille, C., and Precigoux, G. (1993) Orthorhombic crystal structure of the A-DNA octamer d(GTACGTAC). *Eur. J. Biochem.* **213,** 673–682.
25. Shakked, Z., Guerstein-Guzikevich, G., Eisenstein, M., Frolow, G., and Rabinovich, D. (1989) The conformation of the DNA double helix in the crystal is dependent on its environment. *Nature* **342,** 456–460.
26. Dock-Bregeon, A. C., Chevrier, B., Podjarny, A., Moras, D., deBear, J. S., Gough, G. R., Gilham, P. T., and Johnson, J. E. (1988) High resolution structure of the RNA duplex $[U(U\text{-}A)_6A]_2$. *Nature* **335,** 375–378.
27. Wang, A. H.-J., Fujii, S., van Boom, J. H., van der Marel, G., van Boeckel, S. A. A., and Rich, A. (1982) Molecular structure of r(GCG)d(TATACGC): a DNA-RNA hybrid helix joined to double helical DNA. *Nature* **299,** 601–604.
28. Kopka, M. L., Fratini, A. V., Drew, H. R., and Dickerson, R. E. (1983) Ordered water structure around a *B*-DNA dodecamer. A quantitative study. *J. Mol. Biol.* **163,** 129–146.
29. DiGabriele, A. D., Sanderson, M. R., and Steitz, T. (1989) Crystal lattice packing is important in determining the bend of a DNA dodecamer containing an adenine tract. *Proc. Natl. Acad. Sci. USA* **86,** 1816–1820.
30. Nelson, H. C. M., Finch, J. T., Luisi, B. F., and Klug, A. (1987) The structure of an oligo(dA).oligo(dT) tract and its biological implications. *Nature* **330,** 221–226.
31. Timsit, Y., Westhof, E., Fuchs, R. P. P., and Moras, D. (1989) Unusual helical packing in crystals of DNA bearing a mutation hot spot. *Nature* **341,** 459–462.
32. Heinemann, U. and Alings, C. (1991) The conformation of a B-DNA decamer is mainly determined by its sequence and not by crystal environment. *EMBO J.* **10,** 35–43.
33. Aymami, J., Coll, M., van der Marel, G., van Boom, J., Wang, A. H. J., and Rich, A. (1990) Molecular structure of nicked DNA: A substrate for DNA repair enzymes. *Proc. Natl. Acad. Sci. USA* **87,** 2526–2530.
34. Narendra, N., Ginell, S. L., Russu, I. M., and Berman, H. M. (1991) *Biochemistry* **30,** 4449–4455.
35. Yoon, C., Privé, C. Y., Goodsell, D. S., and Dickerson, R. E. (1988) Structure of an alternating-B DNA helix and its relationship to A-tract DNA. *Proc. Natl. Acad. Sci. USA* **85,** 6332–6336.
36. Gessner, R. V., Quigley, G. J., Wang, A. H.-J., van der Marel, G. A., van Boom, J., and Rich, A. (1985) Structural basis for stabilization of Z-DNA by cobalt hexaammine and magnesium cations. *Biochemistry* **24,** 237–240.
37. Wang, A. H.-J., Gessner, R. V., van der Marel, G. A., van Boom, J. H., and Rich, A. (1985) Crystal structure of Z-DNA without an alternating purine-pyrimidine sequence. *Proc. Natl. Acad. Sci. USA* **82,** 2611–3615.

38. Fujii, S., Wang, A. H.-J., van der Marel, G., van Boom, J. H., and Rich, A. (1982) Molecular structure of $(m^5dC\text{-}dG)_3$: the role of the methyl group on 5-methyl cytosine in stabilizing Z-DNA. *Nucleic Acids Res.* **10,** 7879–7892.
39. Coll, M., Fita, I., Lloveras, J., Subirana, J. A., Bardella, F., Huynh-Dinh, T., and Igolen, J. (1988) Structure of d(CACGTG), a Z-DNA hexamer containing AT base pairs. *Nucleic Acids Res.* **16,** 8695–8705.
40. Teng, M.-K., Liaw, Y.-C., van der Marel, G., van Boom, J. H., and Wang, A. H.-J. (1989) Effects of the 02' hydroxyl group on Z-DNA conformation: Structure of Z-RNA and (araC)-[Z-DNA]. *Biochemistry* **28,** 4923–4928.
41. Zhou, G. and Ho, P. S. (1990) Stabilization of Z-DNA by demethylation of thymine bases: 1.3Å single-crystal structure of d(m^5CGUAm^5CG). *Biochemistry* **29,** 7229–7236.
42. Van Meervelt, L., Moore, M. H., Lin, P. K. T., Brown, D. M., and Kennard, O. (1990) Molecular and crystal structure of d(CGCGmo^{4}CG): N^4-methoxycytosine. guanine basepairs in Z-DNA. *J. Mol. Biol.* **216,** 773–781.
43. Wang, A. H.-J., Hakoshima, T., van der Marel, G., van Boom, J. H., and Rich, A. (1984) AT base pairs are less stable than GC base pairs in Z-DNA: The crystal structure of d(m^5CGTAm^5CG). *Cell* **37,** 321–331.
44. Miller, M., Wlodawer, A., Appella, E., and Sussman, J. L. (1987) Crystallization of a DNA duplex 15-mer containing unpaired bases: d(CGCGAAATTTACGCG). *J. Mol. Biol.* **195,** 967,968.
45. Ginell, S. L., Kuzmich, S., Jones, R. A., and Berman, H. M. (1990) Crystal and molecular structure of a DNA duplex containing the carcinogenic lesion O^6-methylguanine. *Biochemistry* **29,** 10,461–10,465.
46. Webster, G. D., Sanderson, M. R., Skelly, J. V., Neidle, S., Swann, P. F., Li, B. F., and Tickle, I. J. (1990) Crystal structure and sequence-dependent conformation of the A. G mispaired oligonucleotide d(CGCAAGCTGGCG). *Proc. Natl. Acad. Sci. USA* **87,** 6693–6697.
47. Rabinovich, D., Haran, T., Eisenstein, M., and Shakked, Z. (1988) Structures of the mismatched duplex d(GGGTGCCC) and one of its Watson-Crick analogues d(GGGCGCCC). *J. Mol. Biol.* **200,** 151–161.
48. Kopka, M. L., Yoon, C., Goodsell, D., Pjura, P., and Dickerson, R. E. (1985) Binding of an antitumour drug to DNA: netropsin and C-G-C-G-A-A-T-T-BrC-G-C-G. *J. Mol. Biol.* **183,** 553–563.
49. Teng, M.-K., Usman, N., Frederick, C. A., and Wang, A. H. -J. (1988) The molecular structure of the complex of Hoechst 33258 and the DNA dodecamer d(CGCGAATTCGCG). *Nucleic Acids Res.* **16,** 2671–2690.
50. Pjura, P. E., Grzeskowiak, K., and Dickerson, R. E. (1987) Binding of Hoechst 33258 to the minor groove of B-DNA. *J. Mol. Biol.* **197,** 257–271.
51. Carrondo, M. A. F. de C. T., Coll, M., Aymami, J., Wang, A. H.-J., van der Marel, G., van Boom, J. H., and Rich, A. (1989) Binding of a Hoechst dye to d(CGCGATATCGCG) and its influence on the conformation of the DNA fragment. *Biochemistry* **28,** 7849–7859.
52. Coll, M., Aymami, J., van der Marel, G., van Boom, J. H., Rich, A., and Wang, A. H.-J. (1989) Molecular structure of the netropsin-d(CGCGATATCGCG) complex: DNA conformation in an alternating AT segment. *Biochemistry* **28,** 310–320.

53. Coll, M., Frederick, C. A., Wang, A. H.-J., and Rich, A. (1987) A bifurcated hydrogen-bonded conformation in the d(A.T) base pairs of the DNA dodecamer d(CGCAAATTTGCG) and its complex with distamycin. *Proc. Natl. Acad. Sci. USA* **84,** 8385–8389.
54. Brown, D. G., Sanderson, M. R., Skelly, J. V., Jenkins, T. C., Brown, T., Garman, E., Stuart, D. I., and Neidle, S. (1990) Crystal structure of a berenil-dodecanucleotide complex: the role of water in sequence-specific ligand binding. *EMBO J.* **9,** 1329–1334.
55. Larsen, T. A., Goodsell, D. S., Cascio, D., Grzeskowiak, K., and Dickerson, R. E. (1989) The structure of DAPI bound to DNA. *J. Biomol. Struct. Dyn.* **7,** 477–491.
56. Quigley, G. J., Wang, A. H.-J., Ughetto, G., van der Marel, G., van Boom, J. H., and Rich, A. (1980) Molecular structure of an anticancer drug–DNA complex: Daunomycin plus d(CpGpTpApCpG). *Proc. Natl. Acad. Sci. USA* **77,** 7204–7208.
57. Wang, A. H.-J., Ughetto, G., Quigley, G. J., and Rich, A. (1987) Interactions between an anthracycline antibiotic and DNA: Molecular structure of daunomycin complexed to d(CpGpTpApCpG) at 1.2-Å resolution. *Biochemistry* **26,** 1152–1163.
58. Moore, M. H., Hunter, W. N., d'Estaintot, B. L., and Kennard, O. (1989) DNA-Drug interactions. The crystal structure of d(CGATCG) complexed with daunomycin. *J. Mol. Biol.* **206,** 693–705.
59. Gao, Q., Williams, L. D., Egli, M., Rabinovich, D., Chen, S. -L., Quigley, G. J., and Rich, A. (1991) Drug-induced DNA repair: X-ray structure of a DNA-ditercalinium complex. *Proc. Natl. Acad. Sci. USA* **88,** 2422–2426.
60. Quigley, G. J., Ughetto, G., van der Marel, G. A., van Boom, J. H., Wang, A. H.-J., and Rich, A. (1986) Non-Watson-Crick G.C and A.T base pairs in a DNA-antibiotic complex. *Science* **232,** 1173–1304.
61. Wang, A. H. -J., Ughetto, G., Quigley, G. J., Hakoshima, T., van der Marel, G. A., van Boom, J. H., and Rich, A. (1984) The molecular structure of a DNA-triostin A complex. *Science* **225,** 1115–1121.
62. Gao, Y. -G., Liaw, Y. -C., Robinson, H., and Wang, A. H.-J. (1990) Binding of the antitumour drug nogalamycin and its derivatives to DNA: Structural comparison. *Biochemistry* **29,** 10,307–10,316.
63. Takusagawa, F., Goldstein, B. M., Youngster, S., Jones, R. A., and Berman, H. M. (1984) Crystallization and preliminary X-ray study of a complex between d(ATGCAT) and actinomycin D. *J. Biol. Chem.* **259,** 4714,4715.
64. Bancroft, D., Williams, L. D., Rich, A., and Egli, M. (1994) The low-temperature crystal structure of the pure spermine form of Z-DNA reveals binding of a spermine molecule in the minor groove. *Biochemistry* **33,** 1073–1086.
65. This technique was developed ca. 1972 by M. Spencer at King's College London, and extended to oligonucleotide use by the present author.
66. Dock-Bregeon, A. C., Chevrier, B., Podjarny, A., Johnson, J., de Bear, J. S., Gough, G. R., Gilham, P. T., and Moras, D. (1989) Crystallographic structure of an RNA helix: $[U(UA)_6)A]_2$. *J. Mol. Biol.* **209,** 459–474.
67. Leonard, G. A. and Hunter, W. N. (1993) Crystal and molecular structure of d(CGTAGATCTACG) at 2.25Å resolution. *J. Mol. Biol.* **234,** 198–208.
68. Bingman, C. A., Zon, G., and Sundaralingam, M. (1992) Crystal and molecular structure of the *A*-DNA dodecamer d(CCGTACGTACGG). *J. Mol. Biol.* **227,** 738–756.

69. Bingman, C., Jain, S., Zon, G., and Sundaralingam, M. (1992) Crystal and molecular structure of the alternating dodecamer d(GCGTACGTACGC) in the *A*-DNA form: comparison with the isomorphous non-alternating dodecamer d(CCGTACGTACGG). *Nucleic Acids Res.* **20,** 6637–6647.
70. Heinemann, D., Alings, C., and Bansal, M. (1992) Double helix conformation, groove dimensions and ligand binding potential of a G/C stretch in B-DNA. *EMBO J.* **11,** 1931–1939.
71. Goodsell, D. S., Grzeskowiak, K., and Dickerson, R. E. (1995) Crystal structure of C-T-C-T-C-G-A-G-A-G. Implications for the structure of the Holliday junction. *Biochemistry* **34,** 1022–1029.
72. Goodsell, D. S., Grzeskowiak, M. K., and Dickerson, R. E. (1994) The crystal structure of C-C-A-T-T-A-A-T-G-G. Implications for bending of B-DNA at T-A steps. *J. Mol. Biol.* **239,** 79–96.
73. Goodsell, D. S., Kopka, M. L., Cascio, D., and Dickerson, R. E. (1993) Crystal structure of CATGGCCATG and its implications for A-tract bending models. *Proc. Natl. Acad. Sci. USA* **90,** 2930–2934.
74. Lipanov, A., Kopka, M. L., Kacor-Grzeskowiak, M., Quintana, J., and Dickerson, R. E. (1993) Structure of the B-DNA decamer C-C-A-A-C-I-T-T-G-G in two different space groups: conformational flexibility of B-DNA. *Biochemistry* **32,** 1373–1389.
75. Rabinovich, D. and Shakked, Z. (1984) A new approach to structure determination of large molecules by multi-dimensional search methods. *Acta Crystallographica* **A40,** 195–200.
76. Fitzgerald, P. M. D. (1988) MERLOT, an integrated package of computer programs for the determination of crystal structures by molecular replacement. *J. Appl. Crystallography* **21,** 273–278.
77. Chattopadhyaya, R. and Chakrabarti, P. (1988) Solving DNA structures by MERLOT. *Acta Crystallographica* **B44,** 651–657.
78. Joshua-Tor, L., Rabinovich, D., Hope, H., Frolow, F., Appella, E., and Sussman, J. L. (1988) The three-dimensional structure of a DNA duplex containing looped-out bases. *Nature* **334,** 82–84.
79. Joshua-Tor, L., Frolow, F., Appella, E., Hope, H., Rabinovich, D., and Sussman, J. L. (1992) Three-dimensional structures of bulge-containing DNA fragments. *J. Mol. Biol.* **225,** 397–431.
80. Miller, M., Harrison, R. W., Wlodawer, A., Appella, E., and Sussman, J. L. (1988) Crystal structure of 15-mer DNA duplex containing unpaired bases. *Nature* **334,** 85,86.
81. Harrison, R. W. (1989) Minimization of cross entropy: a tool for solving crystal structures. *Acta Crystallographica* **A45,** 4–10.
82. Wing, R., Drew, H., Takano, T., Broka, C., Tanaka, S., Itakura, K., and Dickerson, R. E. (1980) Crystal structure analysis of a complete turn of B-DNA. *Nature* **287,** 755–758.

CHAPTER 12

Crystallography in the Study of Protein–DNA Interactions

David G. Brown and Paul S. Freemont

1. Introduction

Over the past few years, there has been a dramatic increase in the number of protein–DNA cocrystal structures solved at atomic resolution (Table 1; for reviews *see [74–76]*). This has allowed a detailed understanding of how specific proteins interact with DNA (e.g., prokaryotic repressors), but has not facilitated the formulation of a common protein–DNA recognition code, which although suggested earlier *(54)*, now appears to be too simplistic *(77)*. However, some general observations have emerged from these recent protein–DNA crystal structures, some of which are discussed below.

The structure of DNA when complexed to a protein can adopt a variety of distorted conformations. The structure of the nucleosome core particle, for example, shows the DNA as a superhelix formed by a series of kinks in the DNA *(78–80)*. Other examples of DNA kinking and bending can be found in the cocrystal structures of yeast TBP, *Eco*RI, CAP, and 434 repressor complexed to DNA where significant alterations in the dimensions of the major and minor grooves of DNA are observed *(4,20,33,49)*. Another example of changes in groove dimensions is well exhibited in the complex of the zinc-finger protein GLI bound to its operator site *(28)*. In a number of repressor-operator structures (Table 1), it has been shown that two α-helices (helix-turn-helix motif) are responsible for DNA recognition and binding. One of these helices (recognition helix) lies in the major groove of the operator site and makes specific

From: *Methods in Molecular Biology, Vol. 56: Crystallographic Methods and Protocols*
Edited by: C. Jones, B. Mulloy, and M. Sanderson Humana Press Inc., Totowa, NJ

Table 1
Crystallization Conditions for DNA–Protein Complexes

Protein	Res (Å)	Synthetic oligonucleotide (5'-3')	Refs	Space Group	T °C	P:DNA ratio	Precipitant	Buffer	pH	Additives
λ Repressor	2.5	TATATCACCAGTGGTAT TATAGTGGTCACCATAA	(1, 2, 3)	$P2_1$	20	1:2	PEG 400	15 mM BTP	7.0	1 mM NaN_3
λ – Cro Repressor	3.9	TATCACCGCGGGTGATA ATAGTGGCGCCCACTAT	(4, 5)	$P3_2$		1:2	NaCl	20 mM Na Cacodylate	6.9	
434 Repressor/OR1	2.5	TATACAAGAAAGTTTGTACT TATGTTCTTTCAAACATGAA	(6)	$P2_12_12_1$	4	2:1	PEG 3000			100 mM NaCl, 120 mM $MgCl_2$, 2 mM spermine
434 Repressor/OR2	2.5	ACAAACAAGATACATTGTAT GTTTGTTCTATGTAACATAT	(7)	$P2_12_12_1$	4	2:1	PEG 3000	100 mM MES	5.7	120 mM NaCl, 120 mM $MgCl_2$, 2 mM cobaltic hexamine
434 Repressor/OR3	2.5	TATACAAGAAAAACTGTACT TATGTTCTTTTTGACATGAA	(8)	$P2_12_12_1$	4	1:1	PEG 3000	100 mM MES	5.5	80 mM NaCl, 100 mM $MgCl_2$, 2mM spermine
434 Repressor/OL2	3.2	ACAATATATATTGT TGTTATATATAACA	(9, 10)	I422	4	2:1	$(NH_4)_2SO_4$	5 mM Na phosphate	4.7	
434 Cro Repressor/OL2	3.2	ACAATATATATTGT TGTTATATATAACA	(11, 12)	C2	20	2:1	$(NH_4)_2SO_4$	20 mM Tris-HCl	8.0	1 mM EDTA,1 mM DTT, 0.7 M NaCl
434 Cro Repressor/OR1	2.5	TATACAAGAAAGTTTGTACT TATGTTCTTTCAAACATGAA	(13)	$P2_1$	4	2:1	PEG 3500	100 mM MES	6.2	160 mM NaCl, 120 mM $MgCl_2$, 2 mM spermidine
Trp Repressor	2.4	TGTACTAGTTAACTAGTC CATGATCAATTGATCAGT	(14, 15)	$P2_1$	20	1:2	MPD	10 mM Na cacodylate	7.2	11 mM $CaCl_2$, 2 mM L-Trp
Trp repressor	2.4	TAGCGTACTAGTACGCT TCGCATGATCATGCGAT	(16)	C2	23	2.2:1	Na Formate	0.1 M Na Acetate	4.8	0.1 mM L-Trp
Lac Repressor	6.5	TTGTGAGCGCTCACAA AACACTCGCGAGTGTT	(17)	C2		1:2	$(NH_4)_2SO_4$	1.4 M Na Acetate	6.5	0.1 M N-(2-acetamido)- -2-iminoiacetic acid
Forkhead HNF-3	2.5	†GACTAAGTCAACC CTGATTCAGTTGG	(18)	$P3_1$	4	1:1	NH_4 Acetate	100 mM K Acetate	5.5	50 mM KCl, 2 mM $MgCl_2$, 20 mM DTT
Cap Repressor 30mer	3.0	†GCGAAAAGTGTGACAT•AT GCTTTTCACACTG	(19, 20)	$C222_1$		1:1.5	PEG 3350	50 mM MES	5 - 6	0.2 M NaCl, 0.1 M $CaCl_2$ 2 mM spermine, 0.3% N-octyl-β D-glucopyranosic
Matα2	2.7	ACATGTAATTCATTTACACGC GTACATTAAGTAAATGTGCGT	(21, 22)	$P2_1$		4:1	PEG 400	15 mM Tris-HCl	7.5	25 mM NaCl, 75 mM $CaCl_2$
engrailed homeodomain	2.8	TTTTGCCATGTAATTACCTAA AAACGGTACATTAATGGATTA	(23, 24)	C2			$(NH_4)OH$	30 mM Tris-HCl	6.7	
Zif268	2.1	AGCGTGGGCGT CGCACCCGCAT	(25)	$C222_1$		1:1	PEG 400	25 mM Tris-HCl	8.0	0.35-0.65 M NaCl 1.5 M equivalence $ZnCl_2$
Gal4	2.7	CCGGAGGACAGTCCTCCGG GGCCTCCTGTCAGGAGGCC	(26)	$P4_32_12$		1:1.5	MPD	12 mM Na Cacodylate	6.8	75 mM $CaCl_2$, 24 mM NaCl
TRAMTRAK	2.8	CTAATAAGGATAACGTCCG ATTATTCCTATTGCAGGCT	(27)	$P2_12_12_1$	20	1:1		20 mM MES	6.0	20-50 mM NaCl, 1.75-3.5 mM spermine
GLI	2.6	ACGTGGACCACCCAAGACGAA GCACCTGGTGGGTTCTGCTTT	(28)	$P2_12_12_1$	22	1:1.2	PEG 400	50 mM BTP-HCl	7.0	$CoCl_2$ 2 molar equivalent Co^{2+} pe finger, 60-100 mM $MgCl_2$
glucocorticoid receptor	2.9	CCAGAACATCGATGTTCTG GTCTTGTAGCTACAAGACC	(29)	$P2_12_12_1$	8	2:1	MPD	18 mM Na cacodylate	6.0	1.8 mM $MgSO_4$, 25 mM NaCl, 2 μM $ZnCl_2$, 1.8 mM spermine
estrogen receptor	2.4	CCAGGTCACAGTGACCTG GTCCAGTGTCACTGGACC	(30)	$P2_12_12_1$	20	2.25:1	MPD	20 mM MES	6.0	1.8 mM spermine, 2μM $ZnCl_2$
Met Repressor	2.8	TTAGACGTCTAGACGTCTA ATCTGCAGATCTGCAGATT	(31)	$P6_322$			MPD	10 mM Na cacodylate	7.0	6 mM NaCl, 1mg/ml SAM, 30 mM $CaCl_2$, 1.5 mM lanthanide chloride
Arc repressor	2.6	TATAGTAGAGTGCTTCTATCAT TATCATCTCACGAAGATAGTAA	(32)	$P2_1$	20	1:2	PEG 400	30 mM BTP	7.4	100 mM $MgCl_2$, 1 mM NaN_3

yTBP	2.5	G**TATATAAA**ACGGGTGG **CATATATTA**TGC	(33)	$P4_3$	22	1:2	PEG 8000	20 mM BTP	7.5	500 mM NaCl, 3.75% glycerol, 1% ethylene glycol
TBP2	2.2	GC**TATAAAAG**GGCA CG**ATATTTTC**CCGT	(34)	$P2_1$	4	1:1	NH_4 Acetate	250 mM MES	6.2	100 mM KCl, 4 mM $MgCl_2$, 14% glycerol, 15 mM DTT
Max	2.9	GTGTAGGC**CACGTG**ACCGGGTG CACATCCG**GTGCAC**TGGCCCAC	(35)	$P6_122$	4	2:1	PEG 1000	100 mM Na cacodylate	5.5	100 mM KCl, 2 mM $MgCl_2$, 5-10% glycerol
GCN4	2.9	TTCCT**ATGACTCAT**CCAGTT AGGA**TACTGAGTA**GGTCAAA	(36)	$P2_12_12_1$	22	1.5:1	PEG 400	25 mM MES	5.8	30 mM $MgCl_2$, 1 mM spermine
MyoD	2.8	TCAA**CAGCTGTT**GA AGTT**GTCGACAA**CT	(37)	$P2_12_12_1$	20	2:1	PEG 3500	100 mM Tris-HCl	8.5	100 mM Na Citrate, 20 mM $BaCl_2$
BPV E2	1.7	CCG**ACCG**ACGT**CGGT**CG GC**TGGC**TGCA**GCCA**GCC	(38)	R32	20	1:1.2	PEG 3350	28 mM Tris-HCl	7.5	75 mM NaCl, 12.5 mM KCl, 1.5 mM lanthanide chloride, 0.5 mM $CaCl_2$
POU	3.0	TGT**ATGCAAAT**AAGG CA**TACGTTTA**TTCCA	(39)	$C222_1$	20	2:1	PEG 3350	50 mM Na citrate	5.4	100-150 mM $(NH_4)_2SO_4$
HIV reverse transcriptase	3.5	**ATGGCGCCCGAACAGGGAC** **ACCGCGGGCTTGTCCCTG**	(40, 41)	$P3_112$/ $P3_212$	4	1:2	$(NH_4)_2SO_4$	100 mM Na cacodylate	5.6	75 mM NaCl
HhaI Methyl-transferase	2.8	TGATA**GC**^{F}**GC**TATC CTAT**CG CG**ATAGT	(42)	$R3_2$	16	1:1	$(NH_4)_2SO_4$	50 mM Na Citrate	5.6	25 mM NaCl, 0.25 mM EDTA
Hin recombinase	3.0	TG**TTTTTGATA**AGA **CAAAAACTAT**TCTA	(43)	$C222_1$	20	3:1	PEG 1500	50 mM Tris-HCl	7.0	20-50 mM NaCl, 20 mM $MgCl_2$
Dnase I	2.3	**GGTATACC** **CCATATGG**	(44)	$P2_12_12_1$	20	1:4	Na Citrate	5 mM HEPES	4 - 5	0.3M NaCl, 20mM EDTA
Dnase I	2.0	**GCGATCGC** **CGCTAGCG**	(45, 46)	$C222_1$	4	-	PEG 600			15 mM EDTA
Eco RI	2.8	†**GAATTC** **CTTAAG**	(47)	$P42_12$	4		MPD	15 mM KPO_4	7.4	0.075 mM DTT, 0.75 mM EDTA, 0.23 M NaCl
Eco RI	2.6	TCGC**GAATTC**GCG GCG**CTTAAG**CGCT	(48, 49, 50)	P321	4	2:9	PEG 400	40 mM BTP	7.4	15% Dioxane, 0.5M NH_4 Acetate
DNA Polymerase I	2.8	pdT4	(51)		18		$(NH_4)_2SO_4$	0.4 M Citrate	5 - 6	1 mM EDTA
DNA Polymerase I	3.8	AGACCGCCCGG GGCGGGCC	(51)							
Hha II	4.0	G**GAGTC**C C**CTCAG**G	(52)	C2	20	2:1	PEG 4000	10 mM K_2PO_4	7.2	0.2M $NaCl_2$, 1 mM EDTA, 0.1 mM DTT
Eco RV	3.0	GG**GATATC**CC CC**CTATAG**GG	(53, 54)	$C222_1$	20	1:1.7	PEG 4000	10 mM $NaPO_4$	7.0	80 mM NaCl
Eco RV	3.0	CGAGCTCG GCTCGAGC	(53, 54)	$P2_1$	20	1:5	PEG 400	20 mM MES	6.0	100 mM NaCl
NF-κB	2.3	**TGGGAATTCCC** **CCCTTAAGGGT**	(55)	C2	18	1.1:1	PEG 4000	25 mM HEPES	7.5	50 mM NH_4Cl, 5 mM DTT, 5 mM $CaCl_2$, 0.05 mM spermine, 0.05% β-octyl glucoside
NF-κB	2.6	AGATG**GGGAATCCC**CTAGA AGATC**CCCTAAGGG**GTAGA	(56)	$P4_12_12$		1:1.1	PEG 3000	50 mM Na Acetate	4.7	50 mM $MgCl_2$, 1 mM DTT, 2 mM spermine
c-Fos-c-Jun	3.0	TTCTCCTA**TGACTCA**TCCAT AGAGGAT**ACTGAGT**AGGTAA	(57)	$P2_12_12$	20	1:1.6	PEG 400	25 mM BTP	6.7	100-200 mM NH_4 Acetate, 150 mM NaCl, 27.5 mM $MgCl_2$, 1.0 mM spermine, 5-10 mM DTT
paired homeodomain	2.5	AAC**GTCACGGTTG**AC **GCAGTGCCAAC**TGTT	(58)	$P2_12_12_1$	20	1:1.3	PEG 1000	10 mM BTP	7.0	0.15-0.2 M NH_4 Acetate 5 mM DTT, 0.1 mM EDTA 10 mM $MgCl_2$
P53	2.2	ATAATT**GGGCAAGTCT**AGGAA ATTAA**CCCGTTCAGA**TCCTTT	(59)	C2	4	1:1	PEG 400	50 mM BTPI	6.4	10 mM DTT, 50 mM NaCl 100 mM MES

(continued)

Table 1 *(continued)*

γδ resolvase	3.0	†GCAG**TGTCCGATAATT**TATA**AA**· GTCA**CAGGCTATT**	(60)	$P2_12_12_1$		1:2	PEG 3350	50 mM MES	6.0	10 mM Tris, 0.5 mM EDTA, 0.2 M $(NH4)_2SO_4$ 2.5% ethylene glycol
HaeIII	2.8	ACCAGCA**GGC**^{F}**C**ACCAGTG GGTCGT**CC**^{M}**GG**TGGTCACT	(61, 62)	$P2_12_12_1$			PEG 3500	100 mM MES	6.5	120 mM NaCl, 1 mM DTT, 13% glycerol
E47	2.8	AA**CACCTG**GCT **TGTGGAC**CGAT	(63)	C2		1:1.4	PEG 4000	20 mM MES	6.5	50 mM $CaCl_2$,1mM spermine
USF	2.9	CACCCGGT**CACGTG**GCCTACA TGGGCCA**GTGCAC**CGGATGTG	(64)	$P2_12_12_1$	4	2:1	PEG 400	100 mM Na Acetate	4.7 5	100 mM KCl, 1.4 mM $MgCl_2$, 0.7 mM Cd(II) Acetate, 5 mM HEPES
PPR1	3.0	**TCGGCAATTGCCGA** **AGCCGTTAACGGCT**	(65)	I222		1:1.2	PEG 3500	20 mM Na cacodylate	6.8	1% MPD, 87.5 mM NaCl, 62.5 mM $CaCl_2$, 50 μM $Zn(Acetate)_2$
BamHI	2.2	TAT**GGATCC**ATA ATA**CCTAGG**TAT	(66, 67)	$P2_12_12_1$		1:2	PEG 8000	3.9 mM Tris	7.6	150 mM KCl, 3% glycerol, 6.1 mM KPO_4, 3.9 mM NaCl, 0.4 mM EDTA
PvuII	3.0	TGAC**CAGCTG**GTC CTG**GTCGAC**CAGT	(68, 69)	$P2_12_12_1$	16	2.7:1	PEG 4000	50 mM Na Acetate	4.5	0.3 mM EDTA
a1/MAT α2 homeodomain	2.7	TAC**ATGTAATTTATTACATCA** **GTACATTAAATAATGTAG**TAT	(70)	$P6_1/P6_5$	20	1:1:1.4	$CaCl_2$	100 mM HEPES	6.0	5-10 mM cobaltic hexamine
Even-skipped homeodomain	2.0	**TCAATTAAAT**TCAATTAAAT **GTTAATTTA**AGTTAATTTAA	(71)	$P2_1$	20	1.8:1	7-15% 4 M Na Acetate		6.1	KCl
Even-skipped homeodomain	3.5	**TTCAGCACCG**TTCAGCACCG **AGTCGTGGC**AAGTCGTGGCA	(71)	$P6_3$	20	1.8:1	PEG 8000	0.4 M K Acetate	6.2	3 mM DTT
T4 Endonuclease V	2.75	dodecamer	(72)	$P6_5$						
TFIIB-TBP	2.7	GGC**TATAAAAG**GGCTG CCG**ATATTTTC**CCGAC	(73)	$P2_12_12_1$	4		NH_4 Acetate	40 mM Tris-HCl	8.5	40 mM KCl, 5 mM $MgCl_2$, 5 mM $CaCl_2$, 10 mM DTT, 10 μM zinc acetate,10% glycerol, 2% ethylene glycol

† half site F Fluorinated base Bold text indicates recognition site

amino acid–DNA contacts through either side-chain or main-chain atoms. The DNA phosphate backbone and base edges of the DNA form the main protein–DNA contact points. However, in the Trp–repressor operator complex, specific protein–DNA contacts are mediated by solvent *(15)*. Since the initial identification of the helix-turn-helix motif, a number of other recognition motifs have been observed to interact with the DNA in cocrystal structures, such as the zinc-finger, β-zip, and b/HLH/Z motifs *(25,31,35)*. It has also been suggested that the flexibility of DNA, in relation to specific base sequences, may be part of the overall protein–DNA recognition mechanism *(6)*. This is clearly exhibited in the 434 repressor–operator interaction where the central base pairs are twisted allowing favorable protein–DNA interactions *(6)* and the cocrystal structure of TATA binding protein, which induces sharp kinks in the DNA and an opening of the minor groove allowing recognition by its antiparallel β-sheet motif *(34)*.

In summary, it appears that the protein–DNA interface is extremely complex, with for example several side chains able to recognize either one or two bases and other side chains providing favorable surfaces for DNA interaction. Therefore, the prediction and modeling of specific protein–DNA interactions, based on amino acid sequence motifs and even three-dimensional structures, is extremely difficult *(1)*. Thus, there remain a large number of DNA binding proteins where no structural information is yet available and about which little structural interpretation can be made.

The starting point for any X-ray crystallographic investigation of protein–DNA interaction is the growth of suitable protein–DNA cocrystals. In this chapter, therefore, the major focus will be on a description of methods and techniques required for cocrystallizing proteins complexed to DNA, an area of study that is accessible to most biochemists and molecular biologists. A description of the biochemical and crystallographic analyses of protein–DNA cocrystals will also be presented. However, the general crystallographic methods used to solve protein structures (described in the rest of this volume) will not be discussed here unless they specifically apply to the analysis and structure solution of protein–DNA cocrystals.

2. Materials

2.1. Cocrystallization

1. Suitable source or overexpression system for the protein under study.
2. Purified protein (10–100 mg).

3. Purified synthetic oligonucleotides (1/10-μmol scale using an automated DNA synthesizer).
4. 24-Well tissue-culture trays (1 cm diameter × 2 cm deep) (Linbro, Flow Laboratories Ltd.).
5. Siliconized, circular, microscope coverslips (22 mm) (Hampton Research Ltd.).
6. Standard binocular microscope (max. mag. 40×) with analyzer and polarizer.
7. Vacuum grease.
8. A variety of precipitating reagents (for a review, *see* refs. *81* and *82* and Chapter 2 in this volume).
9. Different temperature controlled environments (e.g., incubators, cold room).

2.2. X-ray Crystallographic Analysis

For those laboratories that do not possess specialized X-ray analysis equipment, a suitable collaboration with a protein crystallography laboratory should be undertaken. However, for completeness, a list of some of the specialized equipment required for the X-ray crystallographic analysis of suitable protein–DNA cocrystals is given below.

1. A rotating anode X-ray generator (X-ray source).
2. Goniometer(s) (centering and positioning the crystal in the X-ray beam).
3. A precession camera (photographic interpretation of crystal symmetry [space group], lattice parameters, and crystal quality).
4. Data collection and measurement, e.g., area detector (*see* Chapter 4), 4-circle diffractometer (accurate measurement of diffracted intensities).
5. Computer and molecular graphics system (data analysis and electron density/model display) (*see* Chapter 9).

3. Methods

3.1. Protein Purification and Protein Purity

There are many review articles and books dedicated to protein purification and protein-purification techniques (e.g., *83,84*), and therefore, none of these areas will be covered in this chapter. However, a number of methods have been developed specifically for purifying DNA binding proteins (for a detailed description of methods, *see* ref. *85*). These methods rely on the high-affinity binding of the protein to a sequence-specific DNA site. The biotin–streptavidin affinity technique, for example, utilizes the strong noncovalent interaction between biotin and streptavidin and the ability to end-label DNA oligomers with biotin *(85)*. Another commonly used method is DNA affinity chromatography, whereby catenated DNA binding sites are coupled to cyanogen bromide activated

Sepharose, allowing specific protein–DNA interactions to occur on a column matrix *(86)*. Both of the above methods offer the advantages of sequence-specific protein–DNA interaction. However, proteins that bind DNA nonspecifically may have to be purified using standard purification techniques. A technique applicable to the purification of any DNA binding protein is polyethyleneimine precipitation. This procedure can be useful in removing nonspecific nucleic acids and proteins and has been successfully applied to the purification of the Klenow fragment of DNA polymerase I *(87)*.

It is generally thought that protein purity is a critical factor in crystallization experiments, and it has been suggested that proteins used for crystallization should be as pure as possible and completely homogeneous *(81)*. Furthermore, McPherson has listed a number of factors that contribute to heterogeneity, including denaturation, posttranslational modification, and variation in aggregation state *(81,82)*. All of these factors should be monitored during the purification procedure, and the final purified protein should be fully characterized before attempting crystallization studies. The final homogeneous protein should be judged visually on silver-stained polyacrylamide gels *(88)*, and should be as pure as possible and free from any contaminating material.

3.2. Oligonucleotide Synthesis and Purification

The preferred method for oligonucleotide synthesis is the solid-phase phosphoramidite method, which gives high coupling yields and easily purified products *(89)*. Large-scale synthesis (1/10-µmol scale) can be routinely carried out, and the method offers the advantages of full automation (for a detailed description of methods, *see* ref. *90*).

The phosphoramidite method also employs a capping scheme where noncoupled molecules are acetylated at the 5' terminus (removed with ammonia), and full-length molecules are protected by 5' dimethoxytrityl (DMT) groups (resistant to ammonia treatment). The difference in protection group, among species in the crude reaction mixture, can be exploited by reverse-phase HPLC chromatography *(91)*. The initial purification of the DMT oligomers can be carried out on C4/C8 columns using acetonitrile gradients in 0.1*M* triethylammonium acetate, pH 7.0. The DMT groups can then be removed by treatment with acetic acid, and the oligomers further purified on the same column with different acetonitrile gradients. In order to check the purity and authenticity of a synthesized oligomer, 5'-end labeling, polyacrylamide gel electrophoresis in 7*M* urea, and oligonucleotide sequencing should be carried out (for a detailed description of methods, *see* ref. *92*). It is important that the DNA oligomers for cocrystallization are purified to homogeneity, as impurities can have deleterious effects on cocrystallization *(2)*. The extinction coefficients of the purified oligomers can be calculated directly from the individual coefficients of the constituent nucleotides *(93)*. However, the calculated value should be corrected for the hyperchromic effect, which is caused by base stacking in duplex DNA and results in an apparent decrease in UV absorbance *(94)*.

3.3. Cocrystallization Strategy and Techniques

A survey of all the successful cocrystallization studies (Table 1 *[95,96]*) reveals that there is no general applicable rationale in determining the best conditions for cocrystallizing protein–DNA complexes. However, a number of general observations have been made and are summarized below.

3.3.1. Oligonucleotide Composition and Length

In the design of a cocrystallization experiment, the choice of synthetic oligonucleotide is critical, not only in terms of ease of synthesis and purification, but also in terms of length and biological specificity. Therefore, a full characterization of the DNA-binding properties of the protein of interest should be carried out before attempting cocrystallization. For example, the general region of DNA that the protein binds to should be determined by "footprinting" analysis, which involves binding the protein to DNA and cleaving the DNA with either DNase or chemical reagents, leaving the protected sequence intact *(85)*. To determine specific sequences that are in close contact with the protein, methylation interference assays can be carried out, whereby specific guanines and adenines are methylated and monitored for interference in DNA binding *(85)*. Specific DNA binding can be assayed using low-ionic-strength polyacrylamide gel electrophoresis, commonly termed the "band shift" assay *(85)*. This technique can allow quantitative measurements of the affinity, association/dissociation rate constants, and specificity of binding.

Once the basic biochemical characteristics of the protein–DNA interaction have been ascertained, then oligonucleotides of optimal length and specificity can be synthesized. These oligomers, although representative of the in vivo DNA binding site, do not have to be the exact site, as

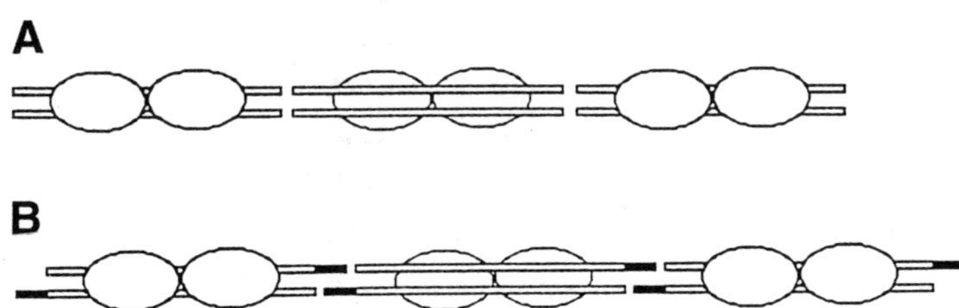

Fig. 1. DNA stacking in protein–DNA Crystals. Protein dimers (ovals) are shown complexed with duplex DNA (▭). The 5'- and complimentary 3'-base overhangs are shown in black (■). **(A)** Blunt-end stacking as observed in cocrystals of 434 and 434-Cro repressor-operator complexes *(12,14)*. **(B)** Stabilization of DNA stacking by single complementary terminal bases as observed in cocrystals of λ repressor–operator complex *(5)*.

in most studies, the protein of interest has been cocrystallized with consensus and/or modified operator sites (Table 1).

In the *Eco*RI–DNA crystal structure and in all of the repressor–operator crystal structures, it has been found that the DNA fragments stack end-to-end forming long pseudo-continuous helical rods, which run through the crystal lattice (Fig. 1). It was therefore suggested that if the DNA were stacked end-to-end in the cocrystal, the length and terminal nucleotides of the DNA would profoundly affect cocrystallization (Fig. 1A; *2,5*). Anderson et al. *(9)* suggested that the oligonucleotide should contain at least one functional protein binding site and have a minimum length of 7*n* bases, where *n* is >1. However, Brennan et al. *(5)*, who also suggested that end-to-end stacking was important for crystallization, concluded that the 7*n* rule was too restrictive and that a number of DNA lengths should be tried. Independently, Jordan et al. *(2)* tested a series of synthetic oligonucieotides of different lengths and end composition, in crystallizing the λ repressor–operator complex, and found that the best crystals were achieved with a 20-bp fragment containing an overhanging 5' dT and corresponding 5' dA on the opposite strand. They concluded that for a cocrystallization experiment of any particular protein, a series of several different DNA fragments should be tried and that complementary overhanging bases could stabilize stacking interactions (*2*; Fig. 1B). However, the best Trp repressor–operator crystals were grown using a self-complementary oligonucleotide that included an unpaired 5' dT overhang *(14)*. An unpaired 5' dT also produced suitable cocrystals between

*Eco*RI and a duplex dodecanucleotide, the unpaired bases stacking end-to-end, similar to other protein–DNA cocrystals *(48,49)*. Overhanging bases have also been found to form Hoogstein base pairs with symmetry-related strands in cases such as MetJ, glucocorticoid receptor, and tramtrack, which help stabilize the packing arrangement *(27,29,31)*. Steitz and colleagues have carried out an extensive survey of crystallization conditions in cocrystallizing catabolite gene activator protein (CAP) with its operator sequence *(19)*. They used 26 different DNA fragments varying in nucleotide sequence, length, and extension of 3' and 5' termini, and found that the best CAP-operator cocrystals were grown using a 31-bp fragment containing a 5' dG at each end *(19)*. By synthesizing a series of half-sites, they were able to generate a wide variety of synthetic DNA fragments, and suggested that their approach overcame the problems of synthesizing and purifying long oligonucleotides *(19)*.

In contrast, the cocrystal structures of Klenow and DNase I show the DNA to be covered mostly by protein, where protein–protein interactions form the majority of crystal lattice contacts. In these specific cases, the choice of oligomers was based on the previously determined three-dimensional structures of the proteins, although a number of different oligomers were tried *(45,51)*.

In summary, therefore, it appears that a number of different duplex oligonucleotides varying in both length and terminal base composition should be tested in a single series of cocrystallization experiments. DNA stacking in some cases appears to be important, and, therefore, the terminal base sequences should be varied. However, depending on the architecture of the protein, the DNA can be completely enveloped, and would therefore have little effect on cocrystallization.

The sequence of the oligonucleotide, however, should be primarily chosen on the basis of both biological specificity and duplex stability. The use of perfect palindromic sequences may give rise to the formation of hairpin secondary structures. These structures may not only interfere in the synthesis and purification of the oligomers *(90)*, but could also affect protein–DNA interaction. The formation of hairpin structures can be prevented by annealing the DNA strands at very high concentrations, where duplex formation is favored over hairpin formation *(14)*. Similarly, sequences that are not exactly twofold symmetric will also prevent hairpin formation *(5)*. In order to promote duplex stability, G-C-rich sequences can be used, although the specificity of the protein for a particular sequence should always be taken into account.

3.3.2. Annealing the DNA Strands

For self-complementary sequences, the DNA strands are dissolved in neutral buffer, heated to 90°C for 10 min, and allowed to cool slowly to 5°C. For palindromic sequences, the strands should be annealed at high concentrations. For duplexes with two different strands, each strand should be dissolved in neutral buffer and then mixed together (1:1 stoichiometry), heated to 90°C for 10 min, and allowed to cool slowly to 5°C.

3.3.3. Cocrystallization

The preferred method for screening crystallization conditions is the vapor diffusion "hanging drop" method *(81,82)*, which is described in detail in Chapter 2 of this volume. Essentially, different concentrations of protein–DNA mixtures (5–20 μL) are placed over reservoirs (1 mL) containing different precipitants and allowed to equilibrate over a period of time. A large number of variables can be tested by this method, including temperature, pH, and precipitation concentration and a number of crystallization protocols have been developed *(97)*.

Table 1 lists some of the crystallization conditions that have produced protein–DNA cocrystals suitable for X-ray structure analysis. Surprisingly, a number of protein–DNA complexes have been crystallized from high salt, although a variety of low-salt precipitants have also been successful. A number of different additives, including spermine, cobalt hexamine, and divalent cations, have also been tried. These additives have been used successfully in crystallizing DNA alone, and have led to the suggestion that neutralization of the polyanionic DNA can allow self-association of DNA molecules, which may lead either to precipitation or crystallization *(98)*. The addition of divalent cations in protein–DNA cocrystallization has also been shown to have dramatic effects on crystallization *(2,14,19)*, although in other cases, no effect is seen *(5)*. It should be noted that in one case, the amount of diffuse scatter and mosaicity of diffraction was reduced by the addition of spermine and *n*-octyl-β-D-glucopyranoside *(19)*. Another important variable in cocrystallization is the protein-to-DNA stoichiometry. In the majority of successful cocrystallizations, changes in the molar ratios of protein:DNA have greatly affected crystal quality. Often a slight molar excess of DNA yields more suitable crystals, such as in the case of the estrogen receptor *(30)*. The integrity of the duplex DNA when complexed to protein should also be considered. For example, in cocrystallization mixtures of Klenow

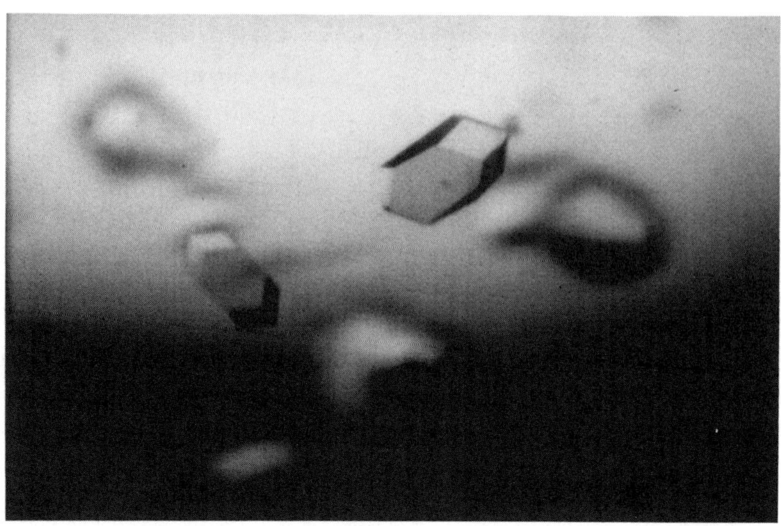

Fig. 2. Cocrystals of the Klenow fragment complexed to duplex DNA. The cocrystals were photographed in plane-polarized light. The largest crystal has dimensions of 0.5 × 0.5 × 0.25 mm. These crystals diffracted to 2.8 Å and were suitable for further crystallographic analysis *(51)*.

and *Eco*RI-DNA, 1 m*M* EDTA was added to inhibit the divalent cation-dependent nuclease activities of both enzymes.

In summary, a wide range of different crystallization conditions should be tried and a number of variables should be tested including the addition of divalent cations and polycations, temperature, the protein-to-DNA stoichiometry, and the use of high- or low-salt precipitants.

3.4. Analysis of Cocrystals

The following two sections will describe the techniques for analyzing potential cocrystals for the presence of protein complexed to DNA, and also for diffraction quality and suitability for further crystallographic analysis.

3.4.1. Biochemical Analysis

From a visual inspection of crystals using a microscope (Fig. 2), it is almost impossible to determine whether the crystals are of protein, protein–DNA complex, or salt. It is therefore necessary to analyze potential cocrystals by the following methods. The crystals should be transferred from "hanging drops" and stabilized in mother liquor solutions, which should contain precipitant at the final concentration that gave crystals, buffer at the crystallization pH, and any extra additives used to promote crystallization. Crystals should then be transferred into several fresh solutions of mother liquor in order to remove any noncrystallized protein or DNA. At this stage, the crystals can be transferred into a microfuge tube, boiled for 10 min to denature the protein–DNA complex, dissolved in a suitable buffer, centrifuged, and the supernatant analyzed by a variety of techniques.

The protein content can be visualized on SDS polyacrylamide gels, stained with Coomassie blue *(99)* or silver stain *(88)*, and the protein concentration determined by a dye binding assay *(100)* or amino acid analysis. The concentration of protein [P] and DNA [D] can also be calculated spectrophotometrically using the following formulae, by measuring the UV absorbance at λ 260 and 280 nm.

$$A_{260} = E_{P260}[P] + E_{D260}[D] \quad (1)$$

$$A_{280} = E_{P280}[P] + E_{D280}[D] \quad (2)$$

and combining (1) and (2):

$$[P] = (A_{280}E_{D260} - A_{260}E_{D280})/(E_{D260}E_{P280} - E_{D280}E_{P260}) \quad (3)$$

$$[D] = (A_{280}E_{P260} - A_{260}E_{P280})/(E_{D280}E_{P260} - E_{D260}E_{P280}) \quad (4)$$

where E_P and E_D are the molar extinction coefficients for the protein and DNA, respectively.

The DNA can also be visualized either complexed to protein or separately on nondenaturing or denaturing polyacrylamide gels stained with methylene blue or ethidium bromide, respectively (*88*; *see* Fig. 2 in ref. *2*). Alternatively, the DNA from dissolved cocrystals can be analyzed by reverse-phase HPLC (Fig. 3). The cocrystals should be carefully washed, especially if excess DNA is used in the cocrystallization mixture, dissolved in buffer and boiled for 10 min to dissociate the protein–DNA complex. Figure 3A shows the HPLC elution profile for the supernatant of Klenow–DNA cocrystals harvested from hanging drops and centrifuged intact. The Klenow:DNA stoichiometry was 1:2 in the cocrystallization mixture, and, therefore, the supernatant from the drop contained at least half the total DNA. Figure 3B and C shows elution profiles for subsequent washes of the cocrystals, and only background amounts of each oligomer can be observed. Figure 3D, however, shows the elution profile of dissolved cocrystals after washing and boiling. The DNA is easily visualized and the ratio of 11 to 8 mer is approx 1:1, indicating

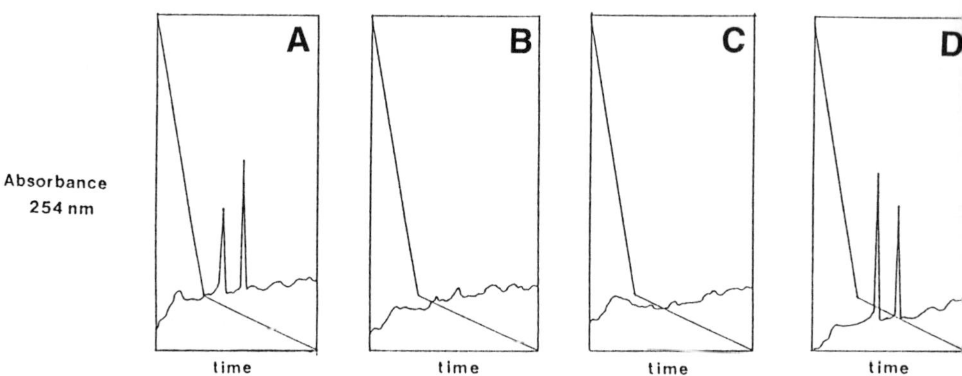

Fig. 3. HPLC Analysis of Klenow–DNA Cocrystals. The cocrystals were harvested from hanging drops, washed thoroughly, dissolved in low ionic strength buffer and boiled for 10 min at 100°C. Fractions from each of these stages were loaded onto a C4 column equilibrated in 0.1*M* triethylammonium acetate, pH 7.0, at a flowrate of 1.0 mL/min, and eluted from the column by the application of acetonitrile gradients. The absorbance was monitored at $\lambda = 254$ nm. Standard solutions of each oligomer (11 and 8 mer) were loaded separately onto the column and used to identify and quantitate peaks. **(A)** Supernatant from freshly harvested cocrystals that were centrifuged intact. **(B)** First wash of intact cocrystals. **(C)** Second wash of intact cocrystals. **(D)** Crystals that were dissolved in low-ionic-strength buffer and boiled for 10 min at 100°C. Elution profiles courtesy of Friedman *(51)*.

possible duplex formation. From measuring the total protein concentration in the dissolved cocrystals and by calibrating the observed oligomer peak heights with standard amounts of oligomers, an approximate protein-to-DNA stoichiometry can be calculated, which was 1:1 for the Klenow–DNA cocrystals. It should be noted that when the dissolved cocrystals were not boiled, the DNA could not be detected by this method.

3.4.2. X-Ray Diffraction Analysis

In order to determine whether protein–DNA cocrystals are suitable for X-ray structure analysis, a series of preliminary diffraction experiments are carried out using a rotating anode generator and precession camera. A cocrystal is mounted into a thin glass capillary (0.8–1.0 mm diameter) and a small volume of mother liquor is placed at one end. The capillary is then attached to a goniometer, which is positioned onto an X-ray generator such that the crystal is aligned in the X-ray beam. The crystal is then exposed to X-rays, and the resultant diffraction pattern is recorded on film. From these preliminary photographs, an estimate of how far the diffraction pattern extends toward the edge of the film can be determined. This is a measure of the resolution limit of the crystals that determines the detail of the electron density map and, thus, the accuracy of the final protein–DNA model. The crystals should diffract to as high a resolution as possible (2–2.5 Å) although, resolution limits of 2.5–3.3 Å have allowed interpretations of a number of protein–DNA crystal structures (Table 1).

Further crystallographic characterization of crystals is carried out by recording diffraction patterns from crystals that are made to precess around the X-ray beam. This allows undistorted images of given reciprocal lattice planes to be recorded on film. From a series of precession photographs of crystals in different orientations, the crystal symmetry (space group) and lattice dimensions can be determined (*see* Chapter 3). At this stage, a judgment of the suitability of crystals for further structural analysis can be made. For example, crystals with long unit cell dimensions will give rise to diffracted intensities that are spatially very close. This may prove problematic in terms of data collection, where measurements of individual intensities could be affected by neighboring reflections. Also, crystals that display anisotropic diffraction, that is, diffraction extending to different resolution limits in different directions, are not ideal. The anisotropy is usually caused by rotational/translational disorder of molecules within the crystal lattice and can result in electron density maps that are not completely interpretable. However, many cocrystal structures have been solved using anisotropic diffracting crystals, although the positions of several amino acid side chains could not be determined *(12)*. Disorder within the crystals can also lead to diffraction patterns that exhibit high levels of mosaicity (broadening of reflections) and diffuse scattering. Another important factor that can be checked at this stage is the stability and lifetime of the crystals in the X-rays beam. Exposure of protein/protein–DNA crystals to X-rays results in damage to the crystal structure, which is observed as a loss of diffracted intensity. The exact mechanism of destruction is unknown, but the rate of diffraction decay can be monitored and corrected for. However, crystals that have limited lifetimes (< 1–2 h) in an X-ray beam may prove difficult in terms of data collection. An approach that has been employed in many recent structural studies is that of "flash freezing" of the crystals. This method involves transferring the crystals into a stabilizing buffer that contains a cryoprotectant, such as glycerol. The crystal is then picked up in a nylon loop (premounted on a goniometer pip) of dimensions comparable with the crystal and is then rapidly transferred into a stream of dry nitrogen, delivered by an appropriate cooling system, at a temperature of around –160°C. This methodology greatly reduces radiation damage to the point where crystal lifetime is almost indefinite, enabling complete data sets to be collected on a single crystal. One drawback of this method is that it usually results in highly mosaic crystals, although current data-processing techniques can tolerate crystals with large mosaicity. The minimum lifetime, which is dependent on the X-ray source, should be at least several hours and ideally 1–2 d.

Finally, a judgment of the homogeneity of the crystals should be made, as crystals that are twinned can be unsuitable for further analysis. Twinning results from the growth of two or more crystals forming a composite crystal and can result in exact superimpositions of different reflections (merohedral twinning). However, methods have been developed to overcome these difficulties, and several structures of proteins have been solved from twinned crystals *(101)*.

From the unit cell dimensions and some knowledge of the molecular weight of the protein–DNA complex, a method for determining the contents of the unit cell has been devised *(102)*. This method involves measuring the density of the cocrystals *(103)*, from which the percentage solvent content (usually in the range 30–80%) and the molecular weight per asymmetric unit (defined as the smallest part of the crystal lattice that contains the complete structure) can be calculated. From these values, the crystal volume per unit of molecular weight (V_m) can be derived and usually lies within the range 1.6–3.6 Å^3/Dalton for globular proteins *(104)*, although anomalously high values have been obtained in protein–DNA cocrystals *(5,9)*. It is important to determine the contents of the asymmetric unit, as this information can be used in the subsequent stages of structural analysis.

In those protein–DNA cocrystals where the DNA forms long pseudo-continuous helices by end-to-end stacking, certain characteristics in the diffraction pattern can provide information concerning the conformation and packing of the DNA within the crystal. For example, strong reflections at 3.4-Å resolution are indicative of the DNA adopting a right-handed B conformation, as the helical rise per base pair for B-type DNA is 3.35 Å. Also, a number of reflections (meridonal layer-line reflections) can be observed in some cases, and relate to the number of helical repeating units per turn of pseudo-superhelix, formed from end-to-end DNA stacking *(2,5,9,12)*.

By combining the space group information, unit cell dimensions, number of molecules per asymmetric unit, and the observed diffraction characteristics, it is possible to determine a protein–DNA packing arrangement within the crystal lattice (Fig. 3 in ref. *5*), which is helpful in the subsequent three-dimensional structure analysis.

3.4.3. Structure Solution

The general crystallographic techniques for solving protein and protein–DNA crystal structures have been reviewed extensively elsewhere *(105)* and in this book. However, one of the major problems in the structure solution of any protein or protein–DNA crystal is in the determination of the phase angles for each of the diffracted intensities. The phase angles in combination with the square root of the intensities (structure factors) produce the electron density image, which in turn give rise to the observed diffraction pattern. A number of methods have been developed for phase determination, the most common being the multiple isomorphous replacement method (reviewed in *105* and Chapter 6). This method involves soaking crystals in solutions of heavy atoms (e.g., mercury, platinum, and silver compounds), measuring the intensity differences between native and soaked crystals, and determining the position of the heavy atoms in the crystal lattice, from which rough estimates of each phase angle can be calculated. However, the method relies on the isomorphous binding of heavy atoms to the protein, producing significant intensity differences without large changes in the cell dimensions, which cannot always be achieved.

In protein–DNA cocrystals, the problem can be partially overcome by directly introducing a heavy atom into the DNA. This can be achieved by replacing deoxythymidine with 5-bromo- or 5-iodo deoxyuracil without affecting the stability or duplex formation of the DNA *(98)*. By cocrystallizing the protein with brominated DNA, the method provides a relatively easy way of obtaining single isomorphous heavy atom derivatives. This method, however, in conjunction with a number of other methods, has been successfully applied to several cocrystal structure solutions

(6,10,12). The use of 5-iodo uridine as a substitute for thymine at multiple sites in the target oligonucleotide can provide enough phasing information for a multiple isomorphous replacement solution of the complex, as has recently been employed for the solution of TBP-TATA box complex *(34)*. It should be noted that heavy-atom soak parameters used in a structure determination of the protein itself may not be applicable in the case of the protein–DNA complex, since specific heavy-atom sites could be involved in protein–DNA interactions.

However, if the structure of the protein is known and the protein–DNA complex crystallizes isomorphously to the protein, then the phases from the protein structure can be used to calculate difference Fourier electron density maps. In these maps, the contribution of the protein has been subtracted, and the remaining electron density should only correspond to the DNA (e.g., *51*). Another technique, which also requires previous knowledge of the protein structure, but can be applied to nonisomorphous crystal forms of the protein–DNA complex, is molecular replacement (reviewed in *106*). In simple terms, the position of the protein within the protein–DNA crystal that best accounts for the observed diffraction pattern can be determined by searching the unit cell with a model of the protein. From the position and orientation of the protein, phases can be calculated and an approximate electron density map produced for the protein–DNA complex. This method, however, relies on the structure of the search protein being very similar to that of the protein complexed to DNA, but has been used successfully in a number of cocrystal structure solutions *(1,6,12,15)*.

The correctness of the molecular replacement solution can be tested by crossdifference Fourier techniques using brominated and nonbrominated protein–DNA complexes. If the molecular replacement phases are correct, then the bromine sites should appear as high peaks in the difference Fourier map, the positions of which should relate to the orientation of the DNA within the lattice. This method has been used in the molecular replacement solutions of 434 repressor–operator and other complexes *(6,30)*. In cocrystals where the asymmetric unit is made up of more than one protein–DNA complex, then it is possible to average the electron density corresponding to each complex within the asymmetric unit. This technique, known as noncrystallographic symmetry averaging, can provide dramatic improvements in the quality of the electron density map and has been used in several cocrystal structure solutions *(12,15)*. The quality of the map can also be improved by applying the method of solvent modification *(107)*. This method relies on differences in the electron density levels between protein/protein–DNA and background. The background or noise can be attributed to solvent and systematic errors in the phases, and is generally lower than the correct protein/protein–DNA density. By determining a boundary around the protein/protein–DNA density and removing weak or negative densities, new phases can be calculated and the process reiterated . This method was used exclusively in the structure solution of a *Eco*RI–DNA complex *(49)*.

All of the above methods have been used at different stages in protein–DNA structure determinations and have led to electron density maps of sufficient quality to determine the positions of the protein–DNA atoms. Because of the well-defined structure of DNA, it is sometimes possible to identify electron density, corresponding to the DNA, in preliminary density maps (Fig. 1 in ref. *12*). From a series of maps, a model of the protein and DNA can be constructed, the atomic positions of which can then be refined by a variety of methods, the most common being rigid body refinement *(108)* and restrained least-squares refinement *(109)*.

4. Notes

An X-ray crystallographic investigation of protein–DNA interactions is a multistep process, the major steps of which are summarized below.

1. Large quantities of purified protein should be available.
2. The final purified protein should be homogeneous and the DNA binding characteristics fully determined.
3. Suitable cocrystals, containing the protein of interest complexed to specific DNA sequences, are required for the crystallographic analysis.
4. The synthetic oligonucleotides for cocrystallization should be chosen on the basis of biological specificity, duplex stability, ease of purification, optimal length, and base sequence.
5. A number of different oligonucleotides, varying in both length and base sequence, terminal base sequence, and extension of 3' and 5' termini should be tested.
6. A wide variety of different cocrystallization conditions and precipitants should be tried, and a number of different additives tested including divalent cations, polycations, and *n*-octyl-β-D-glycopyranoside.
7. A biochemical analysis of the cocrystals should be undertaken and the protein:DNA stoichiometry of the crystals determined.
8. The suitability of cocrystals for full structural analysis should be judged by preliminary diffraction analysis, and be based on the resolution limit of diffraction,

degree of anisotropy, and mosaicity, crystal stability in the X-ray beam, crystal homogeneity, reproducibility of crystals, and size of the asymmetric unit.

References

1. Jordan, S. R. and Pabo, C. O. (1988) Structure of the lambda complex at 2.5 Å resolution: Details of the repressor-operator interactions. *Science* **242,** 893–899.
2. Jordan, S. R., Whitcombe, T. V., Berg, J. M., Pabo, C. O. (1985) Systematic variation in DNA lengths yields highly ordered repressor-operator cocrystals. *Science* **230,** 1383–1385.
3. Beamer, L. J. and Pabo, C. O. (1992) Refined 1.8 Å structure of the λ repressor-operator complex. *J. Mol. Biol.* **227,** 177–196.
4. Brennan, R. G., Roderick, S. L., Takeda, Y., and Matthews, B. W. (1990) Protein–DNA conformational changes in the crystal structure of a λ cro-operator complex. *Proc. Natl. Acad. Sci. USA* **87,** 8165–8169.
5. Brennan, R. G., Takeda, Y., Kim, J., Anderson, W. F., and Matthews, B. W. (1986) Crystallisation of a complex of cro repressor with a 17 base-pair operator. *J. Mol. Biol.* **188,** 115–118.
6. Aggarwal, A. K., Rodgers, D. W., Drottar, M., Ptashne, M., and Harrison, S. C. (1988) Recognition of a DNA operator by the cro repressor of phage 434: a view at high resolution. *Science* **242,** 899–907.
7. Shimon, L. J. W. and Harrison, S. C. (1993) The phage 434 OR2/R1-69 complex at 2.5 Å resolution. *J. Mol. Biol.* **232,** 826–838.
8. Rodgers, D. W. and Harrison, S. C. (1993) The complex between phage 434 repressor DNA-binding domain and operator site O(R)3— structural differences between consensus and non-consensus half-sites. *Structure* **1,** 227–240.
9. Anderson, J. E., Ptashne, M., and Harrison, S. C. (1984) Co-crystals of the DNA-binding domain of phage 434 repressor and a synthetic 434 operator. *Proc. Natl. Acad. Sci. USA* **81,** 1307,1308.
10. Anderson, J. E., Ptashne, M., and Harrison, S. C. (1987) Structure of the repressor–operator complex of bacteriophage 434. *Nature* **326,** 846–852.
11. Wolberger, C. and Harrison, S. C. (1987) Crystallisation and X-ray diffraction studies of a 434 cro–DNA complex. *J. Mol. Biol.* **196,** 951–954.
12. Wolberger, C., Dong, Y.-C., Ptashne, M., and Harrison, S. C. (1988) Structure of a phage 434 cro/DNA complex. *Nature* **335,** 789–795.
13. Mondragón, A. and Harrison, S. C. (1991) The phage 434 cro/OR1 complex at 2.5 Å resolution. *J. Mol. Biol.* **219,** 321–334.
14. Joachimiak, A., Marmorstein, R. Q., Schevitz, R. W., Mandecki, W., Fox, J. L., and Sigler, P. B. (1987) Crystals of the trp repressor–operator complex suitable for X-ray diffraction analysis. *J. Biol. Chem.* **262,** 4917–4921.
15. Otwinowski, Z., Schevitz, R. W., Zhang, R. G., Lawson, C. L. Joachimiak, A., Marmorstein, R. Q., Luisi, B. F., and Sigler, P. B. (1988) Crystal structure of trp repressor/operator complex at atomic resolution. *Nature* **335,** 321–329.
16. Carey, J., Combatti, N., Lewis, D. A. E., and Lawson, C. L. (1993) Co-crystals of *Escherichia coli* trp repressor bound to an alternative operator DNA sequence. *J. Mol. Biol.* **234,** 496–498.

17. Pace, H. C., Ponzy, L., and Lewis, M. (1990) lac repressor: Crystallisation of intact tetramer and its complexes with inducer and operator DNA. *Proc. Natl. Acad. Sci. USA* **87,** 1870–1873.
18. Clark, K. L., Halay, E. D., Lai, E., and Burley, S. K. (1993) Cocrystal structure of the HNF3/forkhead DNA-recognition motif resembles histone H5. *Nature* **364,** 412–420.
19. Schultz, S. C., Shields, G. C., and Steitz, T. A. (1990) Crystallisation of *Escherichia coli* catabolite gene activator protein with its DNA binding site. The use of modular DNA. *J. Mol. Biol.* **213,** 159–166.
20. Schultz, S. C., Shields, G. C., and Steitz, T. A. (1991) Crystal structure of a CAP–DNA complex: The DNA is bent by 90°. *Science* **253,** 1001–1007.
21. Wolberger, C., Pabo, C. O., Vershon, A. K., and Johnson, A. D. (1991) Crystallisation and preliminary X-ray diffraction studies of a MATα2 DNA complex. *J. Mol. Biol.* **217,** 11–13.
22. Wolberger, C., Vershon, A. K., Liu, B., Johnson, A. D., and Pabo, C. O. (1991) Crystal structure of a MATα2 homeodomain-operator complex suggests a general model for homeodomain DNA interactions. *Cell* **67,** 517–528.
23. Liu, B., Kissinger, C. R., Pabo, C. O., Martin-Blanco, E., and Korberg, T. B. (1990) Crystallisation and preliminary diffraction studies of the engraved homeodomain–DNA complex. *Biochim. Biophys. Res. Commun.* **171,** 257–259.
24. Kissinger, C. R., Liu, B., Martin-Blanco, E., Korberg, T. B., and Pabo, C. O. (1990) Crystal structure of engrailed homeodomain–DNA complex at 2.8 Å resolution: a framework for understanding homeodomain–DNA interactions. *Cell* **63,** 579–590.
25. Pavletich, N. P. and Pabo, C. O. (1991) Zinc-finger–DNA recognition: crystal structure of a Zif268-DNA complex at 2.1 Å *Science* **252,** 809–817.
26. Marmorstein, R., Carey, M., Ptashne, M. Harrison, S. C. (1992) DNA recognition by GAL4: Structure of a protein–DNA complex. *Nature* **356,** 408–414.
27. Fairall, L,, Schwabe, J. W. R., Chapman, L., Finch, J. T., and Rhodes, D. (1993) The crystal structure of a two Zinc-finger peptide reveals an extension to the rules for zinc-finger/DNA recognition. *Nature* **366,** 483–487.
28. Pavletich, N. P. and Pabo, C. O. (1993) Crystal structure of a five-finger GLI–DNA complex: New perspectives on Zinc fingers. *Science* **261,** 1701–1707.
29. Luisi, B. F., Xu, W. X., Otwinowski, Z., Freedman, L. P., Yamamoto, K. R., and Sigler, P. B. (1991) Crystallographic analysis of the interaction of the glucocorticoid receptor with DNA. *Nature* **352,** 497–505.
30. Schwabe, J. W. R., Chapman, L., Finch, T., and Rhodes, D. (1993) The crystal structure of the estrogen receptor DNA-binding domain bound to DNA: How receptors discriminate between their response elements. *Cell* **75,** 567–578.
31. Somers, W. S. and Phillips, S. E. V. (1992) Crystal structure of the met repressor operator complex at 2.8 Å resolution reveals DNA recognition by B-strands. *Nature* **359,** 387–393.
32. Raumann, B. E., Rould, M. A., Pabo, C. O., and Sauer, R. T. (1994) DNA recognition by beta-sheets in the Arc repressor-operator crystal-structure. *Nature* **367,** 754–757.
33. Kim, Y., Geiger, J. H., Hahn, S., and Sigler, P. B. (1993) Crystal structure of a yeast TBP/ TATA-box complex. *Nature* **365,** 512–520.

34. Kim, J., Nikolov, D. B., and Burley, S. K. (1993) Co-crystal structure of TBP recognising the minor groove of a TATA element. *Nature* **365,** 520–527.
35. Ferré-D'Amaré, A. R., Prendegast, G. C., Ziff, E. B., and Burley, S. K. (1993) Recognition by Max of its cognate DNA through a dimeric b/HLH/Z domain. *Nature* **363,** 38–45.
36. Ellenberger, T. E., Brandl, C. J., Struhl, K., and Harrison, S. C. (1992) GCN4 Basic region leucine zipper binds DNA as a dimer of uninterrupted a-helices: Crystal structure of the protein–DNA complex. *Cell* **71,** 1223–1237.
37. Ma, P. C. M., Rould, M. A., Weintraub, H., and Pabo, C. O. (1994) Crystal-structure of MyoD bHLH domain-DNA complex—Perspectives on DNA recognition and implications for transcriptional activation. *Cell* **77,** 451–459.
38. Hegde, R. S., Grossman, S. R., Laimins, L. A., and Sigler, P. B. (1992) Structure of the bovine papillomavirus E2 DNA-binding domain bound to its target DNA. *Nature* **359,** 505–512.
39. Klemm, J. D., Rould, M. A., Aurora, R., Herr, W., and Pabo, C. O. (1994) Crystal-structure of the Oct-1 Pou domain bound to an octamer site-DNA recognition with tethered DNA-binding modules. *Cell* **77,** 21–32.
40. Jacob-Molina, A., Clark, A. D., Jr., Williams, R. L., Nanni, R. G., Clark, P., Ferris, A. L., Hughes, S. H., and Arnold, E. (1991) Crystals of a ternary complex of human immunodefficiency virus type 1 reverse transcriptase with a monoclonal antibody Fab fragment and double-stranded DNA diffract X-rays to 3.5 Å resolution. *Proc. Natl. Acad. Sci. USA* **88,** 10,895–10,899.
41. Arnold, E., Jacob-Molina, A., Nanni, R. G., Williams, R. L., Lu, X., Ding, J., Clark, A. D. Jr., Zhang, A., Ferris, A. L., Clark, P., Hizi, H., and Hughes, S. H. (1992) Structure of HIV-1 reverse transcriptase/DNA complex at 7 Å resolution showing active site locations. *Nature* **357,** 85–89.
42. Klimasauskus, S., Kumar, S., Roberts, R. J., and Cheng, X. (1994) Hhal methytransferase flips its target base out of the DNA helix. *Cell* **76,** 357–369.
43. Feng, J.-A., Simon, M., Mack, D. P., Dervan, P. B., Johnson, R. C., and Dickerson, R. E. (1993) Crystallisation and preliminary X-ray analysis of the DNA biding domain of the Hin recombinase with its DNA binding site. *J. Mol. Biol.* **232,** 982–986.
44. Lahm, A. and Suck, D. (1991) DNase λ-induced DNA conformation. 2.0 Å of a DNase I octamer complex. *J. Mol. Biol.* **221,** 645–667.
45. Suck, D., Lahm, A., and Oefner, C. (1988) Structure refined to 2.0 Å of a nicked DNA octanucleotide complex with DNAse I. *Nature* **332,** 464–468.
46. Weston, S. A., Lahm, A., and Suck, D. (1992) X-ray structure of DNase l-d$(GGTATACC)_2$ complex at 2.3 Å resolution. *J. Mol. Biol.* **226,** 1237–1256.
47. Young, T.-S., Modrich, P., Beth, A., Jay, E., and Kim, S. -H. (1981) Preliminary X-ray diffraction studies of *Eco*RI restriction endonuclease–DNA complex. *J. Mol. Biol.* **145,** 607–610.
48. Grable, J., Frederick, C. A., Samudzi, C., Jen-Jacobsen, L., Lesser, D., Greene, P., Boyer, H. W., Itakura, K., and Rosenberg, J. M. (1984) *E. coli* restriction endonuclease *Eco*RI + DNA. *J. Biomol. Struct. Dynam.* **1,** 1149–1160.

49. McClarin, J. A., Frederick, C. A., Wang, B.-C., Greene, P., Boyer, H. W., Grable, J., and Rosenberg, J. M. (1986) Structure of the DNA *Eco*RI endonuclease recognition complex at 3 Å resolution. *Science* **234,** 1526–1541.
50. Kim, Y., Grable, J. C., Love, R., Greene, P. J., and Rosenberg, J. M. (1990) Refinement of *Eco*RI endonuclease crystal structure: a revised protein chain tracing. *Science* **249,** 1307–1309.
51. Freemont, P. S., Friedman, J. M., Beese, L., Sanderson, M. R., and Steitz, T. A. (1988) Cocrystal structure of an editing complex of Klenow fragment with DNA. *Proc. Natl. Acad. Sci. USA* **85,** 8924–8928.
52. Chandrasegaran, S., Smith, H. O., Amzel, M. L., and Ysern, X. (1986) Preliminary X-Ray diffraction analysis of Hhall endonuclease–DNA cocrystals *Proteins: Struct. Funct. Genetics* **1,** 263–266.
53. Winkler, F. K., D'Arcy, A., Blocker, H., Frank, R., and van Boom, J. H. (1991) Crystallisation of complexes of *Eco*RV endonuclease with cognate and non-cognate DNA fragments. *J. Mol. Biol.* **217,** 235–238.
54. Winkler, F. K., Banner, D. W., Oefner, C., Tsernoglou, D., Brown, R. S., Heathman, S. P., Bryan, R. K., Martin, P. D., Petratos, K., and Wilson, K. S. (1993) The crystal structure of EcoRV endonuclease with cognate and non-cognate DNA fragments. *EMBO J.* **12,** 1781–1795.
55. Ghosh, G., Vanduyne, G., Ghosh, S., and Sigler, P. B. (1995) Structure of NF-kappa-B P50 homodimer bound to a Kappa-B site. *Nature* **373,** 303–310.
56. Muller, C. W., Rey, F. A., Sodeoka, M., Verdine, G. L., and Harrison, S. C. (1995) Structure of the NF-kappa-B P50 homodimer bound to DNA. *Nature* **373,** 311–317.
57. Glover, J. N. M. and Harrison, S. C. (1995) Crystal-structure of the heterodimeric βZIP transcription factor C-Fos-C-Jun bound to DNA. *Nature* **373,** 257–261.
58. Xu, W. G., Rould, M. A., Jun, S., Desplan, C., and Pabo, C. O. (1995) Crystal-structure of a paired domain-DNA complex at 2.5-Angstrom resolution reveals structural basis for Pax developmental mutations. *Cell* **80,** 639–650.
59. Cho, Y. J., Gorina, S., Jeffrey, P. D., and Pavletich, N. P. (1994) Crystal-structure of a P53 tumor-suppressor DNA complex-Understanding tumorigenic mutations. *Science* **265,** 346–355.
60. Yang, W. and Steitz, T. A. (1995) Crystal-structure of the site-specific recombinase gamma-delta resolvase complexed with a 34 bp cleavage site. *Cell* **82,** 193–207.
61. Reinisch, K. M., Chen, L., Verdine, G. L., and Lipscomb, W. N. (1994) Crystallisation and preliminary crystallographic analysis of a DNA (cytosine-5)-methyltransferase from *Haemophilus-aegyptius* bound covalently to DNA. *J. Mol. Biol.* **238,** 626–629.
62. Reinisch, K. M., Chen, L., Verdine, G. L., and Lipscomb, W. N. (1995) The crystal-structure of HaeIII methyltransferase covalently complexed to DNA—an extrahelical cytosine and rearranged base-pairing. *Cell* **82,** 143–153.
63. Ellenberger, T., Fass, D., Arnaud, M., and Harrison, S. C. (1994) Crystal-structure of transcription factor E47-E-box recognition by a basic region helix-loop-helix dimer. *Genes & Development* **8,** 970–980.
64. Ferré-D'Amar, A. R., Pognonec, P., Roeder, R. G., and Burley, S. K. (1994) Structure and function of the B/HLH/Z domain of USF. *EMBO J.* **13,** 180–189.

65. Marmorstein, R. and Harrison, S. C. (1994) Crystal-structure of a PPR1-DNA complex: DNA recognition by proteins containing a Zn(2)Cys(6) binuclear cluster. *Genes & Development* **8,** 2504–2512.
66. Strzelecka, T., Newman, M., Dorner, L. F., and Knott, R., Schildkraut, I., and Aggarwal, A. K. (1994) Crystallisation and preliminary X-ray analysis of restriction endonuclease BamHI-DNA complex. *J. Mol. Biol.* **239,** 430–432.
67. Newman, M., Strzelecka, T., Dorner, L. F., Schildkraut, I., and Aggarwal, A. K. (1995) Structure of BamHI endonuclease bound to DNA—Partial folding and unfolding on DNA-binding. *Science* **269,** 656–663.
68. Balendiran, K., Bonventre, J., Knott, R., Jack, W., Benner, J., Schildkraut, I., and Anderson, J. E. (1994) Expression, purification, and crystallisation of restriction endonuclease PvuII with DNA containing its recognition site. *Proteins-Structure Function and Genetics* **19,** 77–79.
69. Cheng, X. D., Balendiran, K., Schildkraut, I., and Anderson, J. E. (1995) Crystal-structure of the PvuII restriction-endonuclease. *Gene* **157,** 139–140.
70. Li, T., Stark, M., Johnson, A. D., and Wolberger, C. (1995) Crystallisation and preliminary X-ray diffraction studies of an A1/Alpha-2/DNA ternary complex. *Proteins-Structure Function and Genetics* **21,** 161–164.
71. Hirsch, J. A. and Aggarwal, A. K. (1995) Purification, crystallisation, and preliminary X-ray diffraction analysis of even-skipped homeodomain complexed to DNA. *Proteins-Structure Function and Genetics* **21,** 268–271.
72. Vassylyev, D. G., Kashiwagi, T., Mikami, Y., Ariyoshi, M., Iwai, S., Ohtsuka, E., and Morikawa, K. (1995) Crystal-structure of T4 endonuclease-V in complex with a DNA substrate. *Protein Engineering* **8,** 64.
73. Nikolov, D. B., Chen, H., Halay, E. D., Usheva, A. A., Hisatake, K., Lee, D. K., Roeder, R. G., and Burley, S. K. (1995) Crystal structure of a TFIIB-TBP-TATA-element ternary complex. *Nature* **377,** 119–128.
74. Brennan, R. G. and Matthews, B. W. (1989) Structural basis of DNA-protein recognition. *TIBS* **14,** 286–290.
75. Freemont, P. S., Lane, A. L., and Sanderson, M. R. (1991) Structural aspects of protein–DNA recognition. *Biochem. J.* **278,** 1–23.
76. Wolberger, C. (1993) Transcription factor structure and DNA binding. *Curr. Opinion in Struct. Biol.* **3,** 3–10.
77. Matthews, B. W. (1988) No code for recognition. *Nature* **335,** 294,295.
78. Richmond, T. J., Finch, J. T., Rushton, B., Rhodes, D., and Klug, A. (1984) Structure of the nucleosome core particle at 7 Å resolution. *Nature* **311,** 532–537.
79. Burlingame, R. W., Love, W. E., Wang, B. C., Hamlin, R., Xuong, N. H., and Moudrianakis, E. N. (1985) Crystallographic structure of the octameric histone core of the nucleosome at a resolution of 3.3 Å. *Science* **228,** 546–553.
80. Richmond, T. J., Rechsteiner, T., and Luger, K. (1993) Studies of nucleosome structure. *Cold Spring Harbor Symposia On Quantitative Biology* **58,** 265–272.
81. McPherson, A. (1982) *The Preparation and Analysis of Protein Crystals.* Wiley, New York.
82. McPherson, A. (1985) Crystallisation of macromolecules. *Methods Enzymol.* **114,** 112–120.

83. Walker, J. M. (ed.) (1984) *Methods in Molecular Biology, vol. 1: Proteins*, Humana, Clifton, NJ.
84. Scopes, R. K. (1987) *Protein Purification: Principles and Practice.* Springer-Verlag, New York.
85. Ausubel, F. M., Brent R., Kingston, R. E., Moore, D. D., Seidman, J. G., Smith, J. A., and Struhl, K. (eds.) (1989) *Current Protocols in Molecular Biology, vol. 2: DNA-Protein Interactions*, Chapter 12, Harvard University Press, Cambridge, MA.
86. Kadonaga, J. T. and Tjian, R. (1986) Affinity purification of sequence-specific DNA binding proteins. *Proc. Natl. Acad. Sci. USA* **83,** 5889–5893.
87. Joyce, C. M. and Grindley, N. D. F. (1983) Construction of a plasmid that overproduces the large proteolytic fragment (Klenow fragment) of DNA polymerase I of *E. coli*. *Proc. Natl. Acad. Sci. USA* **80,** 1830–1834.
88. Wray, W., Boulikas, T., Wray, V. P., and Hancock, R. (1981) Silver staining of proteins in polyacrylamide gels. *Anal. Biochem.* **118,** 197–203.
89. Caruthers, M. H. (1985) Gene synthesis: DNA chemistry and its uses. *Science* **230,** 281–285.
90. Applied Biosystems, Model 380/381 User Bulletin (1987) The evaluation and purification of synthetic oligonucleotides.
91. Becker, C. R., Efcavitch, J. W., Heiner, C. R., and Kaiser, N. F. (1985) Use of a reverse phase column for the HPLC purification of synthetic oligonucleotides. *J. Chromatogr.* **326,** 293–299.
92. Ausubel, F. M., Brent, R., Kingston, R. E., Moore, D. D., Seidman, J. G., Smith, J. A., and Struhl, K. (eds.) (1989) *Current Protocols in Molecular Biology*, vol. 1. Harvard University Press, Cambridge, MA.
93. Fasman, G. D. (ed.) (1975) *Handbook of Biochemistry and Molecular Biology, vol. 1: Nucleic Acids*, 589, 3rd ed., CRC, Boca Raton, FL.
94. Felsenfeld, G. and Sandeen, G. (1962) The dispersion of the hyperchromic effect in thermally induced transitions of nucleic acids. *J. Mol. Biol.* **5,** 587–610.
95. Aggarwal, A. K. (1990) Crystallisation of DNA binding proteins with oligodeoxynucleotides. *Methods* (a companion to *Methods in Enzymol.*) **1,** 83–90.
96. Dock-Bregeon, A. C. and Moras, D. (1992) Crystallisation of nucleic acids and cocrystallisation of proteins and nucleic acids, in *Crystallisation of Nucleic Acids and Proteins* (Ducruix, A. and Giegé, R., eds.), IRL Press at Oxford University Press, Oxford, UK, pp. 145–174.
97. Carter, C. W., Baldwin, E. T., and Frick, L. (1988) Statistical design of experiments for protein crystal growth and the use of a pre-crystallisation assay. *J. Crystal Growth* **90,** 60–73.
98. Holbrook, S. R. and Kim, S. H. (1985) Crystallisation and heavy-atom derivatives of polynucleotides. *Methods Enzymol.* **114,** 167–176.
99. Laemmli, U. (1970) Cleavage of structural proteins during the assembly of the head of bacteriophage T4. *Nature* **227,** 680–685.
100. Bradford, M. M. (1976) A rapid and sensitive method for the quantitation of microgram quantities of protein utilising the principle of protein-dye binding. *Anal. Biochem.* **72,** 248–254.

101. Goldman, A., Ollis, D. L., and Steitz, T. A. (1987) Crystal structure of muconate lactonising enzyme at 3 Å resolution. *J. Mol. Biol.* **194,** 143–153.
102. Matthews, B. W. (1985) Determination of protein molecular weights, hydration, and packing from crystal density. *Methods Enzymol.* **114,** 176–187.
103. Westbrook, E. M. (1985) Crystal density measurements using aqueous ficoll solutions. *Methods Enzymol.* **114,** 187–196.
104. Matthews, B. W. (1968) Solvent content of protein crystals. *J. Mol. Biol.* **33,** 491–497.
105. Wyckoff, H. W., Hirs, C. H. W., and Timasheff, S. N. (eds.) (1985) Diffraction methods for biological macromolecules. *Methods Enzymol.* **115**.
106. Lattman, E. D. (1985) Use of rotation and translation functions. *Methods Enzymol.* **115,** 55–77.
107. Wang, B.-C. (1985) Resolution of phase ambiguity in macromolecular crystallography. *Methods Enzymol.* **115,** 90–112.
108. Sussman, J. L. (1985) Constrained-restrained least squares (CORELS) refinement of proteins and nucleic acids. *Methods Enzymol.* **115,** 271–303.
109. Hendrickson, W. A. (1985) Stereochemically restrained refinement of macromolecular structures. *Methods Enzymol.* **115,** 252–270.

CHAPTER 13

Virus Crystallography

Elizabeth Fry,
Derek Logan, and David Stuart

1. Introduction

Crystallography provides a means of visualizing intact virus particles as well as their isolated constituent proteins and enzymes *(1–3)* at near-atomic resolution, and is thus an extraordinarily powerful tool in the pursuit of a fuller understanding of the functioning of these simple biological systems. We have already expanded our knowledge of virus evolution, assembly, antigenic variation, and host-cell interactions; further studies will no doubt reveal much more. Although the rewards are enormous, an intact virus structure determination is not a trivial undertaking and entails a significant scaling up in terms of time and resources through all stages of data collection and processing compared to a traditional protein crystallographic structure determination. It is the methodology required for such studies that will be the focus of this chapter. The computational requirements were satisfied in the late 1970s, and when combined with the introduction of phase improvement techniques utilizing the virus symmetry *(4,5)*, the application of crystallography to these massive macromolecular assemblies became feasible. This led to the determination of the first virus structure (the small RNA plant virus, tomato bushy stunt virus), by Harrison and coworkers in 1978 *(6)*. The structures of two other plant viruses followed rapidly *(7,8)*. In the 1980s, a major focus of attention was a family of animal RNA viruses; the Picornaviridae. The determination of the structures of a human rhinovirus by Rossmann and colleagues and a poliovirus by Hogle's group shortly

From: *Methods in Molecular Biology, Vol. 56: Crystallographic Methods and Protocols*
Edited by: C. Jones, B. Mulloy, and M. Sanderson Humana Press Inc., Totowa, NJ

afterward were landmarks in macromolecular crystallography *(9,10)*. The 1990s have already led to notable progress and surprises, with results on an RNA bacterial virus *(11)*, a DNA bacterial virus *(12)*, and two DNA animal viruses *(13,14)*.

These structural studies have shown that the arrangement of subunits within a spherical capsid does not adhere strictly to the theory of virus architecture known as quasi-equivalence, put forward by Caspar and Klug *(15)*. This theory allows $60T$ subunits to be accommodated in a spherical shell, where T subunits occupy quasi-equivalent environments and T is subject to some exclusion rules. Thus, we do not have a single mechanism controlling the assembly of these viruses. However, a striking similarity has been observed in the topology and tertiary structure of the coat proteins leading to a more unified view of the evolution of apparently rather disparate groups of animal, insect, and plant RNA viruses. Table 1 provides an overview of the viruses examined crystallographically to date; for most of these, T = 1 or 3. The Picornaviridae are particularly well studied with at least one structure known from four different genera *(9,10,20–25)*. In addition, these have formed the basis for further studies involving, for example, drug binding and mutant virus analyses *(28,29)*, which are relatively easily tackled where isomorphous crystals are obtained (permitting the calculation of difference electron density maps). Future studies will attempt to analyze larger viruses, indeed the structure of simian virus 40 has now been obtained *(14)*; and preliminary analyses have been published for rice dwarf virus *(30)*, blue tongue virus cores *(31,32)*, cauliflower mosaic virus *(33)*, and reovirus type 3 Dearing cores *(34)*. Exciting prospects still remain, such as the determination of structures of assembly intermediates and virus–receptor, virus–antibody complexes.

By now, much of the methodology is relatively straightforward, and we here present an introductory survey of the current "state-of-the-methods," although from a rather personal viewpoint. Someone embarking on the crystallographic analysis of a virus would be well advised to read carefully some of the more detailed methods papers *(35–41)*. Rossmann *(42)* provides a good overview of the molecular replacement phasing method.

An innate feature of many of the simpler viruses that makes them particularly attractive to crystallographers is the symmetry of the protein coat (capsid), which surrounds and protects the nucleic acid *(43)*. This arises from the greater genetic "efficiency" made possible by the use of

Table 1
Icosahedral Virus Capsid Structures Known at Atomic Resolution

	Genome	Symmetry[a]	Approx radius Å	Ref.
Plant viruses				
Tomato bushy stunt virus	(+)ssRNA	T = 3	150	*(6)*
Southern bean mosaic virus	(+)ssRNA	T = 3	150	*(7)*
Satellite tobacco necrosis virus	(+)ssRNA	T = 1	90	*(8)*
Satellite tobacco mosaic virus	(+)ssRNA	T = 1	80	*(16)*
Turnip crinkle virus	(+)ssRNA	T = 3	150	*(17)*
Cowpea mosaic virus	bipartite(+)ssRNA	T = 3	150	*(18)*
Beanpod mottle virus	bipartite(+)ssRNA	T = 3	150	*(19)*
Animal viruses				
Human rhinovirus 14	(+)ssRNA	T = 1(P = 3)	150	*(9)*
Human rhinovirus 1A	(+)ssRNA	T = 1(P = 3)	150	*(20)*
Poliovirus type 1 Mahoney	(+)ssRNA	T = 1(P = 3)	150	*(10)*
Poliovirus type 3 Sabin	(+)ssRNA	T = 1(P = 3)	150	*(21)*
Mengovirus (murine)	(+)ssRNA	T = 1(P = 3)	150	*(22)*
Foot and mouth disease virus	(+)ssRNA	T = 1(P = 3)	150	*(23)*
Theiler virus	(+)ssRNA	T = 1(P = 3)	150	*(24,25)*
Canine parvovirus	ssDNA	T = 1	130	*(13)*
Simian virus 40	dsDNA	T = 7	250	*(14)*
Insect viruses				
Black beetle virus	Two (+)ssRNA	T = 3	150	*(26)*
Flock house virus	Two (+)ssRNA	T = 3	160	*(27)*
Bacterial viruses				
MS2	(+)ssRNA	T = 3	135	*(11)*
φX174	ssDNA	T = 1	165	*(12)*

[a]T-number: the number of triangles into which each triangular face of the icosahedron is divided to form the quasi-equivalent units of the capsid. The number of quasiequivalent subunits is usually T × 60. If the subunits are not chemically identical, then the symmetry is termed pseudo-equivalent, indicated by P = *n*. (+):positive sense nucleic acid.

many copies of just a few proteins. Since the viral genome cannot exhibit the symmetry of the capsid, it is often almost completely invisible in a crystallographic analysis unless, as in the case of bean pod mottle virus *(19)* and canine parvovirus *(13)*, the symmetry of the coat proteins is reflected, in part, in the structure of the nucleic acid of the genome. However, viruses vary considerably in their complexity; the capsid may be spherical, bullet-shaped or rod-shaped, and the genome may consist of one or more segments of single- or double-stranded DNA or RNA. Some viruses also have an outer host derived lipid membrane surrounding the protein coat. Such viruses (the enveloped viruses) have rarely yielded crystals. However, solubilized protein components of these membranes (e.g., the hemagglutinin *(44)*, and neuraminidase *(45,46)* of influenza virus) and individual core proteins *(1)* have been studied with great success. The disk of the rod-shaped tobacco mosaic virus was the first capsid component successfully analyzed crystallographically *(47)*, and remarkable progress has been made in the analysis of these viruses using fibre diffraction by Stubbs and coworkers *(48,49)*. However, since only the isometric viruses have so far yielded to single crystal analysis, we will concentrate on them.

The presence of icosahedral (532) symmetry inevitably leads to a minimum of fivefold noncrystallographic redundancy (for those which crystallize with the virus lying on a point of 23 crystallographic symmetry). This redundancy is of enormous importance in a crystallographic analysis; it provides extremely powerful constraints that facilitate many aspects of the analysis: the location of heavy-atom positions, phase refinement and extension by averaging, model refinement, and the calculation of difference maps from incomplete data sets. The fact that the overall architecture of spherical viruses tends to be closely conserved among related viruses has enabled the successful use of molecular replacement *(17,20–25)* aided by the power of averaging and solvent flattening to extend phases (which can be determined initially often at only low resolution) to near atomic resolution *(50)*. It has even proven possible to derive starting phases from a basic sphere *(51)*, and we believe it is only a matter of time until the first virus structure is determined by entirely *ab initio* phasing *(52,53)*. Although these factors make some aspects of the crystallography more straightforward, there is, inevitably, a price to be paid; to achieve the redundancy, we must do more work, and the size and complexity of the problem grows alarmingly, as we shall see.

Before embarking on a structural study, the logistics of the experiment should be carefully considered. A particular problem can be disease security regulations: Containment facilities may be required for both crystallization and data collection. At present, only one synchrotron station is designed to handle dangerous human pathogens routinely; this is station F1 at CHESS, the Cornell High Energy Synchrotron Source at

Ithaca, NY, although we hope that facilities will shortly become available at the European Synchrotron Radiation Facility (ESRF) at Grenoble, France.

2. Crystallization

As with any crystallographic study, the first hurdle to be overcome is crystallization. With viruses, however, it does seem that highly ordered crystals are often obtained once the optimum crystallization conditions are found. This may be attributable to their isometric nature and is a dramatic reflection of the perfection with which these systems assemble. There is a higher than normal tendency to crystallize in cubic and orthorhombic space groups. Owing to the essentially perfect symmetry of the isometric particles, it is often observed that some of the capsid symmetry axes coincide with those of the crystal *(54)*, e.g., bean pod mottle virus *(19)* and cow pea mosaic virus (hexagonal) *(18)* occupy twofold symmetry axes, whereas FMDV (cubic) *(23)* occupies a site of 23 crystallographic symmetry. In spite of the isometric morphology of the particles, crystals in noncubic space groups may exhibit birefringence.

Crystallization conditions are similar to those used for crystallizing protein molecules (and just as variable). Previously published virus crystallization conditions are listed in Table 2. The use of cloned viral material ensures a genetically homogeneous population. Such material is usually purified by ultracentrifugation using CsCl or sucrose gradients. We find that many viruses may be readily concentrated using 100-kDa microconcentrators. Standard spectrophotometric concentration measurements should be corrected for the nucleic acid content; an approximate formula is:

$$\text{Virus concentration} = (1.55\ A_{280} - 0.76\ A_{260}) \times \text{dilution} \quad (1)$$

Just as with proteins, it is usually best to attempt initial crystallization at virus concentrations in excess of 10 mg/mL. Virus yields vary, and if grown in a monolayer cell culture, a vast number of roller bottles may be required! A concentration of 100 mg/mL has been used to obtain crystals of SBMV *(84)*, but plant viruses can be grown readily in large amounts, and such a method would probably not be the first choice for a virus that is hard to grow, such as hepatitis A. In general, the usual micro methods of crystallization (hanging drop, sitting drop, and microdialysis) work well *(126,127)*. Crystals have been obtained with a wide range of pre-

(text continued on p. 331)

Table 2
Crystallization Conditions for Intact Viruses, Subassemblies, Structural Proteins, and Their Complexes

Virus	Virus, mg/mL	Meth.	Buffer	T, °C	pH	Precipitant	Additives	Habit	Size, mm	Time	Space group	Diff., Å	Unit cell	Ref.
Picornaviruses														
BEV	5	µD	0.1*M* NaP	20	7.6	≈30% NaCl	Trace NaN_3	Hex. plates	0.7	10–14 d	$P2_1$	2.5	338 390 360; 113°	*(55)*
Coxsackie A9	5	µD	0.1*M* NaP	RT	7.6	20–22% AmS	Trace NaN_3	Rhomb. dodec.	0.3	10–14 d	$P4_n22$	3.3	≈500 ≈700	*(56)*
Coxsackie A10	—	B	None	4	—	1% NaCl	None	Dodecahedra	0.1	1–2 wk	—	—	—	*(57)*
Coxsackie A10	—	B	ppt.	4	—	5% AmAc	None	Rect./sq. plates	0.1	—	—	—	—	*(57)*
Coxsackie B1	10	HD	10 m*M* NaC	≈20	5.0	0.1*M* AmH_2PO_4	None	Prisms	1 × 0.2 × 0.05	1 mo	$P2_{(1)}2_{(1)}2_{(1)}$	2.9	323 450 522	*(58)*
Coxsackie B1	10	HD	10 m*M* NaC	≈20	4.0	17–19% AmS	None	Thick plates	0.4 × 0.2 × 0.1	1 mo	—	2.9	all > 300	*(58)*
FMDV $A10_{61}$	5–10	µD	10 m*M* NaP	RT	7.6	20–26% AmS	None	—	0.5	2–4 wk	—	—	—	*(59)*
FMDV $A10_{61}$	5–20	VD/MS	0.1*M* NaP	—	7.6	2.5–3% PEG 20K	2*M* AmCl 3 m*M* NaN_3	Parallelepipeds —	0.4 × 0.4 × 0.2	—	R3	3.0	296; 62.3°	*(59)*
FMDV $A10_{61}$	5–20	VD	0.1*M* NaP	—	7.6	3% PEG 4K	2*M* AmCl 1.5*M* NaCl 3 m*M* NaN_3	Rhomb. dodec.	0.15 × 0.15 × 0.08	—	I23	3.4	347	*(60)*
FMDV $A10_{61}$ (empty)	5–20	µD	0.1*M* NaP	—	7.6	2.5–4*M* AmAc	3 m*M* NaN_3	—	0.25 × 0.25 × 0.13	—	$C222_{(1)}$	3.0	590 560 490	*(60)*
FMDV A22 Iraq 24/64	5–20	VD/MS	0.1*M* NaP	—	7.6	3–5% PEG 4K	4M AmCl 3 m*M* NaN_3	—	0.3 × 0.3 × 0.15	—	I222	3.0	328 342 364	*(60)*
FMDV A22 (empty)	5–20	VD	0.1*M* NaP	—	7.6	2–3% PEG 20K	4*M* AmCl 3 m*M* NaN_3	—	0.3 × 0.3 × 0.15	—	I222	3.0	328 342 364	*(60)*
FMDV A24 Crusiero	5–20	VD/SS	0.1*M* NaP	—	7.6	4–5% PEG 4K	4*M* AmCl 3 m*M* NaN_3	Rhomb. dodec.	0.3 × 0.3 × 0.15	—	$C222_{(1)}$	3.5	950 700 500	*(60)*
FMDV C3	5–10	µD	10 m*M* NaP	RT	7.6	6–15% AmS	None	—	0.3	2–4 wk	—	—	—	*(59)*
FMDV C-S8c1	5–20	VD	0.1*M* NaP	—	7.6	9–12% AmS	3 m*M* NaN_3 10 m*M* DTT	Rhomb. dodec.	0.4 × 0.4 × 0.2	< 10 d	I23	3.5	348	*(60)*
FMDV C-S8c1 (R100)	5–20	VD	0.1*M* Tris or HEPES	—	8.0/ 7.6	13% AmS	3 m*M* NaN_3 10 m*M* DTT	Rhomb. dodec.	0.35 × 0.35 × 0.15	—	—	3.5	—	*(60)*
FMDV O_1BFS	5–10	µD	10 m*M* NaP	RT	7.6	17–19% AmS	Trace NaN_3	Rhomb. dodec.	0.3	2–4 wk	I23	2.3	345	*(59)*
Do.,trypsin treated	5–10	µD	10 m*M* NaP	RT	7.6	15–20% AmS	None	—	0.3	2–4 wk	—	—	—	*(59)*
Do.	5–20	VD	10 m*M* NaP	—	7.6	2.25–2.75% PEG 4K	2*M* AmCl	Rhomb. dodec.	0.15 × 0.15	—	Trigonal/	3.0	635 320	*(60)*

FMDV O_1K (G67)	5–20	µD	0.1*M* NaP	—	7.6	21–23% AmS	3 m*M* NaN_3	Rhomb. dodec.	0.4 × 0.4 × 0.2	—	"Pseudo I432"	2.9	345	*(60)*
FMDV SAT2	5–10	µD	10 m*M* NaP	RT	7.6	27–35% AmS	None	—	0.4	2–4 wk	—	—	—	*(59)*
HRV14	*y*%[a]	VD	0.1*M* NaP	RT	7.2	2.5% AmS	1 m*M* NaN_3	Prisms	0.6	1 wk	$P2_12_12$	3.5	323 358 380	*(61)*
HRV14	5	HD	10 m*M* Tris/HCl	RT	7.2	0.25, 0.5% PEG 8K	20 m*M* $CaCl_2$	Octahedra	0.9	7–10 d	$P2_13$	2.8	445.1	*(62)*
HRV1A	—	B	ppt.	6	—	AmF	None	Hex. prisms	0.3	3–4 d	—	—	—	*(63)*
HRV1A	3–5	µD	10 m*M* Tris	6	7.35	0.15*M* AmF	None	Long hex.	0.6 × 0.6 × 0.2	2 wk	$P6_322$	2.8	341.3 465.9	*(20)*
Mengo	13	SD	10 m*M* NaP[b]	RT	7.8	27.3% AmS[b]	0.1*M* AmF[b]	Oct. tablets	0.6–0.9	Few days	P23	7.0	422	*(64)*
Mengo	5	HD	0.1*M* NaP[b]	RT	7.4	2.8% PEG 8K	None	Orthorhombic	0.8	1–2 d	$P2_12_12_1$	2.5	441.4 427.3 421.9	*(22,65)*
Polio 1	4.17	B	None	4	7.3	None	None	Octahedral	0.4	>1 yr	$P2_12_12$	2.0	353 378 320	*(66,67 68)*
Polio 1	—	D	NaP	—	7.0	NaCl	None	Octahedral	Several tenths	2–3 wk	$P2_12_12$	2.5	359 381 324	*(69)*
Polio 3	≈10	µD[b]	10 m*M* PIPES	4	7.0	≈0.5*M* NaCl	5 m*M* $MgCl_2$ 1 m*M* $CaCl_2$	—	—	—	I222	2.4	320 358 381	*(21)*
Polio 2/1 chimera	10	µD[b]	10 m*M* PIPES	4	7.0	≈0.1*M* NaCl	5 m*M* $MgCl_2$ 1 m*M* $CaCl_2$	Octahedral	—	—	$P2_12_12$	2.6	323.3 358.5 380.5	*(70)*
TMEV (GDVII)	20	HD	40 m*M* NaP	22	9.0	4.5–7% PEG 3350	0.2% (v/v) βME	—	0.6	2 wk	C2	3.0	575.2 324.0 558.4; 108.2°	*(71)*
Plant viruses														
ArMV	—	D	20 m*M* NaP	—	6.5	None	None	Trunc. tet.	0.4	—	$P2_13$	10	387	*(72)*
AMV T=1	12	µD[c]	ppt.	—	7.0→4.6	0.2*M* NaC	None	Hex. plate	0.4 × 1.5	1 wk	$P6_3$	3.4	200 314	*(73)*
AMV T=1	12	µD[c]	ppt.	—	4.6	0.2*M* NaC	None	Hex. pyramid	0.7 × 0.5	—	$P3_{1/2}$	5.0	201 485	*(73)*
BBMV	10	B	—	20	6.5	50% NaC	0.1*M* NaCl	Parallelopipeds, Triang. prisms	0.5	Several days	$P2_1$	—	530 514 275; 114°	*(74)*
BDMV bottom component	12	SD	50 m*M* NaP	—	4.5	10% PEG 6K	None	—	< 3	3–4 wk	R3	3.0	296 729	*(75)*
BDMV top component	25	SD	0.1*M* NaP	—	4.2	10% PEG 6K	None	—	> 3	2–7 d	R3	3.0	296 729	*(75)*
BDMV	30–40	SD	0.1*M* NaC	—	5.6	2.5% PEG 6K	2 m*M* DTT	—	1–1.5	3–7 d	R3	3.5	295.4; 59.9°	*(76)*
BPMV	15	SD	0.1*M* KP	20	7.0	2% PEG 8K	None	Elongated tabular	—	7–10 d	$P22_12_1$	—	311 284 350	*(77)*
CarMV	—	—	0.1*M* Tris-maleate/NaOH	—	6.0	AmS	None	—	—	—	F432	—	482.6	*(78)*

(continued)

Table 2 *(continued)*

Virus	Virus, mg/mL	Meth.	Buffer	T, °C	pH	Precipitant	Additives	Habit	Size, mm	Time	Space group	Diff., Å	Unit cell	Ref.
CCMV	5–21	B	ppt.	20	6–6.5	1.5*M* NaP	2 m*M* $MgCl_2$	cubic, various habits	0.25	—	F432	7.5	510	*(79)*
CCMV	5–21	B	ppt.	20	6–6.5	0.85*M* NaC	2 m*M* $MgCl_2$	Do.	0.25	—	F432	7.5	560	*(79)*
CCMV	5–21	B	ppt.	4	5.5	35.7% AmS	2 m*M* $MgCl_2$	Tet. bipyram.	"Small"	—	—	—	—	*(79)*
CCMV	5–21	B	NaC or None	RT	<6	0.85*M* NaC or 35.7% AmS	2 m*M* $MgCl_2$	Needles	—	—	—	—	—	*(79)*
CCMV	20–50	SD[d]	0.3 *M* succinate	RT	3.3	3.7–4% PEG 8K	1 m*M* NaN_3 1 m*M* EDTA	—	1.5 × 1.2 × 0.7	3–5 d	$P2_12_12_1$	3.1	381.26 381.26;408.59	*(80)*
CMV	15	SD	0.1*M* Tris	RT	6.9	4% PEG 3350	1 m*M* $CaCl_2$ 1 m*M* leupeptin	Hexagonal	0.3–0.5	2–3 mo	—	—	—	*(80)*
CMV	15	SD	0.1*M* Tris	RT	6.9	4% PEG 3350	1 m*M* $CaCl_2$ 0.1 m*M* leupeptin 0.1% βOG	Octahedra	0.3–0.5	3–4 wk	—	27	—	*(80)*
CMV	10	HD	0.1 m*M* Tris	22 + 4	6.8	2% PEG 400	2% AmS (w/v) 1 m*M* $CaCl_2$ 0.1 m*M* leupeptin	Hexagons	0.2	10 d	—	—	—	*(80)*
CpMV	20	HD/SD	ppt.	—	4.9	0.6*M* NaC	None	Hex. bipyram.	1 × 0.8	—	$P6_{1/5}22$	3.0	451 1038	*(81)*
CpMV	20	HD/SD	None	—	4.9	AmS	None	Rhomb. dodec.	1 × 1	—	I23	—	308	*(81)*
CpMV	35	SD	50 m*M* KP	20	7.0	2% PEG 8K	9.9% AmS	Rhomb. dodec.	1 × 1	1–2 wk	I23	—	317	*(77)*
ELV	10–40	VD	None	—	7.0	8% PEG 6K	None	Trunc. tetr.	>1	—	$P2_13$	8.0	414	*(82)*
ELV	10–40	VD	None	—	7.0	8% PEG 6K	None	Rhomb. plates	>1	—	B2	3.7	442 422 387; $\gamma = 95°$	*(82)*
SBMV	"low"	—	—	—	—	"moderate" AmS	None	Rhombs	—	—	R32	—	923 299	*(83,84)*
SBMV	100–110	µD	None	—	—	0.37% AmS	None	Rhombs	<1	1–3 wk	R32	3.6	337 756	*(84)*
SBMV	100–110	µD	None	—	—	0.37% AmS	None	Diamond plates	—	Few days	$C222_1$	3.5	522 330 530	*(84)*
SBMV exp.	32	SD	20 m*M* NaP	4	7.5	8% PEG 6K	40 m*M* EDTA	Orthorhombic	1 × 1 × 0.2	—	—	—	—	(85)
SBMV exp.	40	SD	20 m*M* NaP	4	6.5	6% PEG 6K	40 m*M* EDTA	Orthorhombic	1 × 1 × 0.02	—	—	—	—	*(85)*
SBMV exp.	55	SD	20 m*M* NaP	4	5.0	3% PEG 6K	40 m*M* EDTA	Orthorhombic	0.8 × 0.8 × 0.3	—	$C222_1$	4.0	552 341 551	*(85)*
SBMV exp.	50	SD	20 m*M* NaC	4	4.0	2.5% PEG 6K	40 m*M* EDTA	Orthorhombic	1 × 1 × 0.6	—	$C222_1$	4.0	552 341 551	*(85)*
STMV	20	SD	40 m*M* cacod., NaP or Tris	—	6, 6.5, 7	15% AmS	NaCl or $NaC_2H_4O_2$	"Polyhedral"	—	24 h	I222	2.5	176 192 205	*(86)*

STNV	1.65%	E	None	RT	—	None	1 m*M* MgS	Rhomb. plates or triclinic prisms	0.5 × 0.5 × 0.2	1–2 wk	C2	2.5	319 304 185.5;94.37°	*(88)*
STNV	1.5%	B[e]	None	—	—	None	None	Rhomb. plate	1.5 × 1.3 × 0.3	—	C2	3.0	do.	*(89)*
STNV exp.	1%	D	TNEM	18	6.5	3.5% PEG 6K	None	Octahedra	0.3–0.7	—	$P2_13$	7.5	257.6	*(90,91)*
STNV exp.	1%	D	TNEM	18	5.0	1.5% PEG 6K	20 m*M* MgS	Rhombohedra	—	—	P1	2.5	174.3 178.8 174.2;75.4° 112.9°68.8°	*(90)*
STNV exp.	1%	D	TNEM	18	6.5	3.5% PEG 8K	None	Rhombohedra	—	—	C2	2.5	317.3 304.0 184.6;94.37°	*(90)*
TBSV	30	P	None	4	—	12.3% AmS	None	Rhomb. dodec.	Months	0.3–0.5	I23	2.9	386	*(92,93)*
TCV small particle	8	B	None	RT	—	33% AmS	0.02 mg/mL TCV RNA	Octahedral	0.1	4 wk	$P2_{(1)}3$	—	292	*(94)*
TNV	—	B	None	4	—	None	None	Tricl. prisms, Hex. plates	2 × 1 × 0.5	>1 yr	P_1	2.8	179 219 243; 87.5° 97.5° 97.5°	*(95)*
TNV	10–15	D	0.4*M* NaP	—	6.0	None	None	Dodecahedra	1.0	Few wk	$P4_232$	2.5	338	*(96)*
TRV bottom component	10–15	SD	0.1*M* KP	RT	6.4	1% PEG 6K	10 m*M* EDTA	Rods	1.0	7–14 d	—	Disordered	—	*(97)*
TRV bottom component	7–10	SD	0.1*M* KP	RT	6.4	1% PEG 6K	10 m*M* EDTA	Polyhedral	1.0	2–7 d	$P2_12_1$	3.3	388 396 405	*(97)*
TRV middle component	6–10	SD	0.1*M* KP	RT	6.4	1.2% PEG 6K	10 m*M* EDTA	Rods	1.0	7–14 d	$P2_12_1$	3.3	388 396 405	*(97)*
TSV (M)	10	HD	0.1*M* NaC	25	4.75	4% PEG 8K	1 m*M* NaN_3 2% MPD	Hex. Tablets	0.2 × 0.2 × 0.1	1 wk	Hexagonal	—	—	*(98)*
TYMV	2–6%	B	None	RT	—	17.2–18.5% AmS	None	Octahedral	0.2–0.5	4–12 wk	$F4_132$	20.0	—	*(99,100)*
Other RNA viruses														
BBV	8	HD	50 m*M* NaP	20	6.9–7.2	13.5% AmS	None	Rhomb. dodec.	—	2–3 wk[g]	$P4_232$	—	362	*(77,101)*
FHV	10–20	HD	25 m*M* NaC	20	6.0	13.5% AmS	0.1% βOG	Rhombohedra	—	—	R32	—	330; 63°	*(77)*
FHV	18	SD	10 m*M* Bis-Tris	RT	6.0	2.8% PEG 8K	20 m*M* $CaCl_2$	Rhombohedra	0.7	3 wk	R3(I)[i]	2.8	325.5; 61.8°	*(102)*
FHV	18	SD	10 m*M* Bis-Tris	RT	6.0	2.8% PEG 8K	20 m*M* $CaCl_2$	Rhombohedra	0.7	3 wk	R3(II)[i]	2.8	325.5; 61.8°	*(102)*
FHV	18	SD	10 m*M* Bis-Tris	RT	6.0	2.8% PEG 8K	20 m*M* $CaCl_2$	Orthorhombic	—	—	R3(III)[i]	—	416.7 332.1 351.2	*(102)*
FHV VLP	12	HD	10 m*M* NaP	RT	6.8	5% PEG 8K	20 m*M* $CaCl_2$	Rhombohedra	0.3	4 wk	$P2_1$	3.3	464.8 333.9 325.2; 91.9°	*(103)*
MS2	1%	C	10 m*M* Tris	4	7.6	4% PEG 6K	0.6*M* NaCl 0.5 m*M* spermidine 0.1 m*M* $MgCl_2$ 0.1 m*M* EDTA	Needles	1–2 × 0.1–0.2	—	—	Composite	—	*(104)*

(continued)

Table 2 *(continued)*

Virus	Virus, mg/mL	Meth.	Buffer	T, °C	pH	Precipitant	Additives	Habit	Size, mm	Time	Space group	Diff., Å	Unit cell	Ref.
MS2	1%	HD	ppt.	37	7.4	0.4*M* NaP	1.5% PEG 6K 0.02% NaN_3	Monoclinic	0.35 × 0.35 × 0.15	4 wk	R32	2.6	274 73.4°	*(105)*
MS2	1%	HD	ppt.	37 + 4[h]	7.4	0.4*M* NaP	1.5% PEG 6K 0.02% NaN_3	Needles	1 × 0.1 × 0.2	—	—	15–20	—	*(105)*
MS2	1%	HD	10 m*M* Tris	19	7.4	8% PEG 6K	0.5*M* NaCl 0.1 m*M* MgS 0.01 m*M* EDTA 0.02% NaN_3	Octahedral	0.7 × 0.7 × 0.4	—	Cubic	15–20	—	*(105)*
NDV	10	SD	50 m*M* NaP	20	7.2	2.5% PEG 8K	1 m*M* NaN_3	Monoclinic	—	4–7 d	—	—	—	*(77)*
NβV	8	SD	0.17*M* Tris	20	7.6	2.25% PEG 8K	1*M* CaN	Pseudo-cubic	—	7–10 d	P1	—	405 404 412; 60° 60° 64°	*(77)*
NωV	8–10	SD	75 m*M* MOPS	—	7.0	2% PEG 8K	0.25*M* $CaCl_2$ 1 mM NaN_3	Tabular	—	2–4 wk	P1	2.7	414.0 410.7 420.1;59.1° 58.9° 64.0°	*(106)*
Reovirus cores	3–40	HD/MS	15 m*M* HEPES	RT	7.8	0.8% PEG 20K (v/v)	0.5*M* NaCl 0.2m*M* $MgCl_2$	Cubes	0.2	4–5 wk	F432	8.0	1270	*(34)*
RDV	12	VD	0.1*M* His	20 + 4	6.0	0.4–0.5% PEG 8K	10 m*M* $MgCl_2$	Rhomb. dodec.	0.5	3 wk	I23	6.5	789	*(30)*
Sindbis	2.6–6	SD	50 m*M* PIPES	20	7.2	4.8% PEG 8K	0.24*M* KCl 7.5% glycerol 1 m*M* EDTA	Hex. plates	—	—	R32	30	640 1520	*(107)*
Sindbis nucleocapsid	5–10	SD	50 m*M* Tris	4	7.5	3% PEG 35K	5% glycerol 0.15*M* NaCl 1 m*M* EDTA	Small hex. plates	—	—	—	—	—	*(107)*
Miscellaneous DNA viruses														
CPV	10	HD	10 m*M* Tris	RT	7.5	0.75% PEG 8K	6 m*M* $CaCl_2$	—	0.5	2 wk	$P2_1$	2.8	264.5 350.3 267.8; 90.86°	*(108)*
Polyoma	—	HD	None	—	—	50% NaS	None	Rhomb. dodec.	0.5	1 wk	I23	6.5	568	*(109,110)*
Polyoma	4	VD	20 m*M* Tris	RT	9.5	0.55*M* NaS	None	Rhomb. dodec.	—	1 mo	I23	8	572	*(111)*
SV40	5–10	HD	Tris or NH_4OH	25	7.0–7.5	≈50% AmS		Rhomb. dodec. or tetrahedra	0.15	—	I23	3.8	558	*(112,114)*
ΦX174	8	HD	90–93 m*M* Bis Tris methane	20, 4	6.8	1.5–2% PEG 8K	None	Tabular	0.7	>4 wk	$P2_1$	2.7	306.0 361.1 [illegible]	*(113)*

Adeno 5 hexon	2.3	D	ppt.	4	4.4	0.8*M* KP	None	Tetrahedral	0.1	13 h	$P2_13$	—	213	*(114–116)*
Adeno 2 hexon	—	D	ppt.	—	5.0	0.25–0.55*M* NaAc	None	Tetrahedral	"V. small"	—	—	—	—	*(117)*
Adeno 2 hexon	—	D	ppt.	—	5.0	1.0*M* NaAc	None	Bipyramidal	"V. small"	—	—	—	—	*(117)*
Adeno 2 hexon	—	D	ppt.	—	3.2	NaC	None	Bipyramidal	0.45–0.6	—	$P2_13$	1.6	—	*(117)*
Adeno 2 hexon	—	D	ppt.	—	3.7–4.1	NaC	None	Tetrahedral	—	—	—	—	149.9	*(117)*
Adeno 2 hexon	—	D	ppt.	4	4.4	0.08*M* KP	1 m*M* EDTA	Tetrahedral	0.5	3–5 d	$P2_13$	—	—	*(118)*
TBEV E protein	6	HD	50 m*M* Tris	20	8.2–8.5	PEG 8K	1.4–1.8*M* KCl 0.5% βOG	Hex. rods	1.5 × 0.1 × 0.1	—	—	3.0	—	*(119)*
HIV p24/ Fab	—	HD	40 m*M* BisTris Cl	≈25	6.2–7.5	12–24% PEG 3350	20 m*M* $CaCl_2$ or $MgSO_4$	Cubic-like	0.15	—	—	3.9	—	*(120)*
HIV p24/ Fab	—	HD (/MS)	40 m*M* BisTris Cl	≈25	6.2–7.5	12–24% PEG 3350	20 m*M* $CaCl_2$ or $MgSO_4$	Parallelopiped	0.3 × 0.2 × 0.2	—	$P2_1$	2.7	92.1 85.4 54.0; 90.4°	*(120)*
PRD1 major coat protein	5	HD	25 m*M* NaP	RT	6.0→4.0	4–6% PEG 4K	None	—	0.6 × 0.6 × 0.4	1–2 wk	$P2_12_12_1$	3.0	121.6 123.2 128.6	*(121)*
Sindbis core	5–10	HD	0.1–0.2*M* Tris	RT	8.3	8% PEG 8K	$(pA)_3$	Rect. prism	0.3 × 0.15 × 0.05	Few weeks	$P22_{(1)}2_1$	3.5	37.0 83.8 293.2	*(122)* *(122)*
Sindbis core	5–10	HD	0.1–0.2*M* Tris	RT	8.3	8% PEG 8K	$(pA)_3$	Bipyramids	0.4 × 0.2 × 0.2	Further time	$P4_32_12$	2.5	57 109	*(122)*
Sindbis core	5–10	HD	Near 0.2*M* Tris	RT	8.3	8% PEG 8K	$(pA)_3$	Rhombic	0.3 × 0.3 × 0.3	Further time	$P2_1$	2.5	39 80 61;102.5°	*(122)*
TMV disk	8	B	I = 0.1 Tris	RT	8.0	7.4% AmS	None	Tet. bipyram.	—	2–3 wk	$P22_12_1$	—	228.2 223.9 174.3	*(123)*
TMV disk	50	B	None	20	—	None	None	—	—	2–3 d	—	—	—	*(124)*
TMV disk	50	HD	I = 0.4 Tris	RT	—	10–15% PEG 8K	None	—	0.51 × 0.33 × 0.30	—	—	3.0	—	*(124)*
TSV CP	10	HD	0.1*M* NaC	20	4.9–5.1	18–20% PEG 4K–8K	1 m*M* NaN_3	Hexagonal	—	3–4 d	Hexagonal	—	≈40	*(77)*
TSV (M) CP (trypsinized)	50	SD	0.1*M* NaC	25	5.0	8% PEG 8K	1 m*M* NaN_3	Hex. pyram.	0.7 × 0.25 × 0.25	2 wk	$P6_{2/4}$	4.5	98.3 108.7	*(98)*
TSV (WC) CP (trypsinized)	30	HD	20 m*M* NaAc	—	6.0	1.8*M* AmS	—	—	0.3 × 0.3 × 0.15	2 wk	P1	2.4	60.1 77.5 73.9; 67.8 130 106°	*(98)*

[a]In x% saturated AmS such that $5 < xy < 10$.
[b]Against decreasing concentration of NaCl.
[c]pH gradient.
[d]In darkness.
[e]Seeded in capillary used for eventual data collection.
[f]Grown under microgravity. Octahedral crystals observed in a cell that was ramped from –4 to 17°C.
[g]Grow in 2–3 d if 1% PEG 8000 included in precipitant.

Table 2 *(continued)*

[h]This type of crystal is exclusively present at 4°.

[i]Type I and II crystals differ by a 1.1° particle rotation about the crystallographic threefold axis. The two types never grew together, yet a small percentage of type III crystals grew with each of them.

[j]pH 6.2 crystals show large mosaicity.

Where information is lacking or ambiguous in the cited reference,—appears in the appropriate column. Temperatures separated by a comma indicate that crystallization was initiated at the higher temperature and completed at the lower. Separation by + indicates that crystals were obtained at both temperatures. The size cited is normally the maximum attained, and the time is that taken for full growth. For concision, cell dimensions are not labeled, but follow the normal conventions. Note that occasionally the virus or protein component of the crystallization drop contains a different buffer from that used in the reservoir. Individual papers should be consulted for full details.

Ammonium sulfate concentrations expressed in the source in *M* have been converted to percentages for the sake of convention, based on a molarity of 4.06 for a saturated solution at 20°C. The following solubilities are listed to assist in making up precipitant solutions, and are in g/100 mL of water at the stated temperature.

	Solubility	*T*, °C
Ammonium formate	102	0
Ammonium sulfate	75.6	20
Calcium nitrate tetrahydrate	266	0
Sodium chloride	35.7	0
Sodium citrate dihydrate	72	25
Sodium sulfate heptahydrate	19.5	0

Abbreviations: Viruses: ArMV, arabis mosaic virus; AMV, alfalfa mosaic virus; BBMV, broad bean mottle virus; BBV, black beetle virus; BDMV, belladonna mottle virus; BEV, bovine enterovirus; BPMV, beanpod mottle virus; CarMV, carnation mottle virus; CCMV, cucumber chlorotic mottle virus; CMV, cauliflower mosaic virus; CpMV, cowpea mosaic virus; CPV, canine parvovirus; ELV, Erysimium latent virus; FHV, Flock house virus (VLP, virus-like particle); FMDV, foot-and-mouth disease virus; HIV, human immunodeficiency virus; HRV, human rhinovirus; NβV, *Nudaurelia capensis* β virus; NωV, *Nudaurelia capensis* ω virus; NDV nodamuravirus; RDV, rice dwarf virus; SBMV, southern bean mosaic virus; STMV, satellite tobacco mosaic virus; STNV, satellite tobacco necrosis virus; TBEV, tick-borne encephalitis virus; TBSV, tomato bushy stunt virus; TCV, turnip crinkle virus; TMV, tobacco mosaic virus; TNV, tobacco necrosis virus; TRV, turnip rosette virus; TSV, tobacco streak virus (CP, capsid protein; M, mild strain; WC, wild clover strain); TYMV, turnip yellow mosaic virus.

Methods: μD, microdialysis; B, batch; D, dialysis; HD, hanging drop; LLD liquid/liquid diffusion; MS, macroseeding; SD, sitting drop; SS, streak seeding; VD, vapor diffusion.

Precipitants, buffers, and additives: Am, ammonium; C, citrate; F, formate; N, nitrate; P phosphate; PEG, poly(ethyleneglycol); ppt., precipitant; S, sulfate; βOG, *n*-octyl-β-glucopyranoside; βME, β-mercaptoethanol.

Miscellaneous: unit cell, unit cell dimensions in Å and °; diff., diffraction limit; I, ionic strength; meth., method of crystallisation; RT, room temperature; bipyram., bipyramidal; dodec., dodecahedral; exp., expanded state; hex., hexagonal; rect. rectangular; ref., reference; rhomb., rhombic; sq., square; tet., tetragonal; triang., triangular; tricl., triclinic.

cipitants. It may be that there is a trend toward the use of higher mol-wt polyethylene glycol (PEG) with larger viruses, but this is likely to have more to do with the recently introduced larger PEGs being currently fashionable. As for additives, some detergents have proven useful, perhaps in relieving the aggregation of particles (the aggregation properties of the material may be investigated using an electron microscope). Divalent cations known to interact with the capsid may also be added to some effect. With a macromolecular assembly, there is a possibility of crystallizing a substructure such as an assembly intermediate. The crystallization of intact virus can be verified using SDS-PAGE to demonstrate the presence of the correct complement of proteins, tests to show recovery of infectivity from dissolved crystals, and perhaps by electron microscopic analysis of the crystals. Reproducibility is essential since a large number of virus crystals (sometimes in excess of a hundred) may be required for a *de novo* analysis, since with photographic film, often only one photograph can be obtained from each crystal (although a more effective detector may reduce the number of crystals required by an order of magnitude). Generally speaking, virus crystals are extremely radiation-sensitive, and long exposure times are necessary to record the diffraction, which is rendered weak by the fact that the unit cell is very large. This also fixes a lower size limit (at present) for useful crystals of approx $0.15 \times 0.12 \times 0.08$ mm^3.

The latter points dictate an efficient crystal-mounting and data-collection technique. We prefer to use quartz capillary tubes, since their increased strength makes them less likely to crack in the hurly-burly of it all and they may be required by disease security protocols.

3. Data Collection

The beam geometry, intensity, and wavelength tunability of synchrotron radiation particularly recommend its use with large unit cells. Most synchrotron stations *(128)* need no modification for virus data collection; however, it is well worthwhile carefully checking the alignment of the various components, and in particular making sure that the beam divergence is not excessive and a satisfactory way of marking the direct beam position is available. We have found that wavelengths <1 Å can provide dramatic improvements in the signal-to-noise ratio and crystal lifetime compared to the older standard of approx 1.5 Å. However, when working at X-ray wavelengths of 1 Å or below, photographic film becomes rather inefficient as a detection mechanism. It is then necessary to introduce attenuation beyond that provided by a cassette of three films in order to achieve a reasonable dynamic range. We prefer to do this rather late in the proceedings, i.e., have an A and B film of rather similar intensity to improve the measurement of the majority of the data while introducing attenuation between the B and C films. Imaging plates do not suffer from this problem and are our preferred method of detection. A relative assessment of a selection of recording devices available is given in Table 3.

Very few gas-detector systems can cope with the count rate produced at a synchrotron beam line, and those that can *(129)* cannot record sufficient diffraction orders simultaneously to be really useful for general virus work. The same is true for television area detectors. For many years, film remained the only detector capable of recording and resolving sufficient diffraction orders to enable efficient data collection from large unit cells, hence its previous popularity as a recording medium for such a project. However, the increased availability of imaging plates with increased sensitivity and lower background allowing a reduction in X-ray exposure time, and thus an effective increase in crystal lifetime, has made them the preferred option in many cases particularly coupled with synchrotron radiation. Although they cannot, at present, match the spatial resolution of film, they can be made much larger than the 12 × 12 cm^2 active area usual with film. Current large imaging plates (30 × 30 cm^2) are capable of collecting reasonably high resolution data from primitive cells in excess of 400 Å at synchrotron stations. With the smaller plates (18 × 18 cm^2), it is only possible to collect high-resolution data from such a cell if it is centered. Providing that the data can be resolved *(130)*, imaging plates are much superior to film; we have noted an order of magnitude increase in the number of images from one crystal of FMDV. The Photon Factory (Japan) will soon make available an 80 × 80 cm^2 system (8000 × 8000 pixels) that will be sufficient for the largest cells currently worked on and will enable a good signal-to-noise ratio despite long crystal-to-plate distances *(131)*.

Furthermore, it should be possible to collect very accurate data rapidly from large unit cells "in house" using imaging plates with well-designed focusing mirrors *(132)*, compressing channel-cut monochromators *(133)* or novel focusing systems.

The Laue method has already been used successfully with virus crystals *(134)*. The use of "white" radiation leads to dramatic reduction in

Table 3
A Qualitative Comparison of the Devices Available for Data Collection

	Pixel Size	Size of active area	No. of pixels across detector	No. orders resolvable	Relative disadvantages/advantages
FAST television area detector	$0.1 \times 0.1\ mm^2$	47 × 63 mm	512 × 512	80 orders in long direction of detector	Large point spread function at the outer edge limits the crystal-to-detector distances feasible High data-rate capacity Better signal-to-noise than film
Xentronics gas-filled multi-wire chamber area detector	0.2 mm	12-cm diam.	512 × 512	≈80 orders across detector	Low data-rate capacity Cannot be used with synchrotron radiation Insensitive at short wavelengths Better signal-to-noise than film
Image plates (MAR)	≈150 μ	18-cm diam.	1156 × 1156	≈200 orders across plate	Dead time incurred in data collection owing to scanning Better signal-to-noise than film
R axis	≈150 μ	30-cm diam.	2000 × 2000	≈350 orders across plate	Better detection efficiency than film, particularly for wavelengths < 1.5 Å
Film	50 μ or 25 μ	12-cm diam.	≈2200 or 4400	≈400 orders across film	Approx 70% quantum efficient for 1.5 Å X-rays Background fog Low dynamic range Digitization (50/25μ raster) tedious and time-consuming

exposure times, and in spite of the enormous intensity of incident radiation, it has been claimed that some virus crystals may yield significantly more information with white radiation than with monochromatic illumination. The Laue method represents an efficient method of data collection with viruses, since if the space group is of high symmetry, a high proportion of the unique data can be recorded on a single photograph, whereas in lower symmetry space groups, averaging can be used to compensate for missing data. However, a phenomenal number of data are packed onto the film, creating the need for overlap deconvolution. Given the technical problems, it is a major achievement that the Laue method has been used successfully for the analysis of structural changes relative to a known starting structure *(134)*, but we still await the first *de novo* virus structure determined from Laue data.

The efficiency of practical data collection is limited by the susceptibility of many virus crystals to radiation damage and, as a result of the large unit cell, the need to avoid overlap of reflections on the film or plate. As a direct result of their sensitivity to radiation, virus crystals are normally only aligned optically before data collection (the American method) *(135)*. Such a method is often supposed to be inefficient; we have analyzed the effectiveness of the method, and present some results in Table 4. It will be seen that, providing a somewhat incomplete data set is acceptable, then the method is in fact extremely effective. Thus, for instance, a data set over 85% complete can be collected from only 16° of data for space group I23.

Although in our experience it is often possible to collect more than one photograph per virus crystal when using image plates, there will still be a large number of reflections incompletely recorded at the beginning and end of the rotation range for each crystal. Many of these reflections can in fact be used by careful "postrefinement" of the data. Moderate cooling has been tried to extend crystal lifetime, with occasional success, such as in the case of poliovirus for which data was collected at –15°C *(10)*. Cooling to near liquid nitrogen temperatures has been used successfully in the case of some proteins *(136)*, where the crystal is almost immortalized; however, it often seems to give rise to an increase in mosaic spread. This is a problem that will have to be overcome if flash-freezing techniques are to be generally useful for high-resolution virus data collection; otherwise, there is likely to be an overlap of data at high resolution. A further problem may arise because the currently favored methods involve the crystal being open to the air during X-ray exposure rather

Table 4
A Test of the American Method[a]

Point group	Number of packs					
	5	10	20	40	60	90
1	2.2	4.4	8.5	16.3	23.4	33.0
2	4.3	8.4	16.2	29.4	40.8	54.4
222	8.2	15.7	29.0	48.9	63.5	77.8
4	8.4	16.1	29.6	50.3	64.9	79.2
422	15.1	28.0	48.4	72.6	85.5	94.2
23	22.7	40.2	64.0	86.2	94.6	98.6
432	37.6	60.9	83.8	96.8	99.2	99.8
3	6.5	12.6	23.5	41.3	54.8	69.5
321	12.3	22.8	40.5	63.5	77.8	89.4
6	12.5	23.1	41.0	64.7	78.9	90.2
622	22.4	39.1	62.6	84.6	93.5	98.1

[a]The figures shown are the percentage completeness to 3 Å resolution. These figures were calculated by predicting reflections for 90 film packs with randomly generated missetting angles with a cell a = b = c = 200 Å, $\alpha = \beta = \gamma = 90°$ (or $\alpha = \beta = 90°, \gamma = 120°$ for the trigonal and hexagonal space groups), a wavelength of 0.9 Å, and an oscillation range of 0.4°. Only reflections more than 50% recorded were accepted. These reflections were then reduced to the unique subset for each point group, merged, duplicates removed, and the percentage completeness calculated in ranges of $\sin^2 \theta/\lambda^2$. To a first approximation, these figures will scale directly according to the oscillation range used. The figures are similar for different resolution ranges.

than sealed in a quartz capillary, which will contravene the disease security requirements for some viruses.

The avoidance of overlapped reflections depends on a highly parallel X-ray beam (characteristic of a well-designed synchrotron station), which is usually limited by a collimator of not more than 0.3-mm diameter. For data collection to a resolution of say 3 Å for small viruses (refer to Fig. 1), the oscillation range is usually between 0.3° and 0.5°, depending on the lattice properties. This must be carefully chosen to avoid reflection overlap at high resolution, but using traditional image processing techniques, it is advantageous to collect as large a swept volume on each film as possible to ensure more fully recorded reflections. Most program systems provide an easy means to investigate this. In this respect, a delightful feature of many virus crystals is that the crystal mosaicity is very small (in our experience usually not more than 0.05°, and sometimes considerably less) and coupled with the small beam divergence, which can be achieved at synchrotrons, particularly where

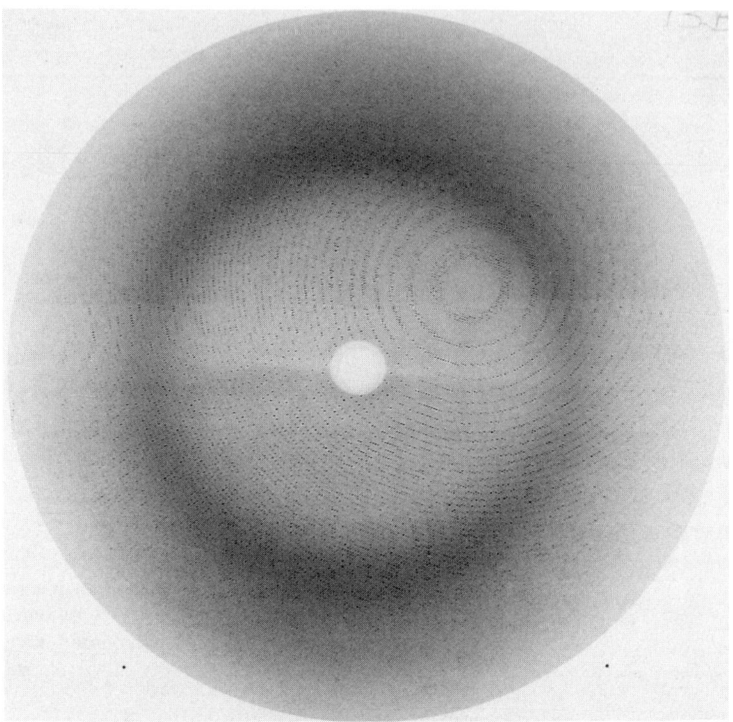

Fig. 1. A 0.5° oscillation photograph from a crystal of the picornavirus Bovine enterovirus, space group $P2_1$, with a = 388 Å, b = 386 Å, c = 358 Å, and β = 113°. This was recorded at the SERC synchrotron radiation facility at Daresbury UK, station 9.6 (wavelength ≈ 0.9 Å) with 0.3-mm diameter collimation and a crystal-to-film distance of 190 mm such that the data at the edge of the film correspond to a resolution of 3.0 Å. Courtesy of Michael Smyth.

undulators are used as insertion devices to reduce the beam divergence in the horizontal direction (e.g., ESRF, Grenoble), it is possible to achieve a good yield of reflections fully recorded on a photograph with an oscillation range as small as 0.3°. Data from viruses of the order of 300-Å diameter are routinely collected and processed to a resolution of 2.5–3.0 Å. At this resolution, a wealth of detail becomes apparent in the electron density maps, as we shall discuss below.

4. Space Group Determination

The classical methods of space group determination (which usually involve a device known as a precession camera) are usually inappropriate for viruses. Since photographs are usually taken in virtually random orientations, the tendency is to extend the American method to its logical conclusion, collect some data first, process initially in a relatively low symmetry point group and determine the space group by analysis of the aggregate of data. The methods for doing this are straightforward. To take a recent example from our laboratory.

Picornavirus C yielded crystals that appeared to be perfect rhombic dodecahedra. The diffraction pattern immediately indicated a primitive unit cell, and measurements of the films and autoindexing calculations suggested that the cell edges were approximately equal with cell angles, all approx 90°. The data were then processed, assuming a primitive orthorhombic cell, and the films merged internally assuming both mmm symmetry and 3m symmetry. The R factors* were indistinguishable showing that the space group was cubic. At this stage, a space group, such as $P2_13$, became the lowest likely symmetry. Correlation coefficient calculations between data from different films with and without index permutation indicated that the reciprocal lattice had a further fourfold symmetry. Given packing considerations (confirmed by the observed systemmatic absences), the space group was identified as $P4_232$. In this case (as in many others), the space group determination immediately provided information on the orientation and position of the virus particle in the unit cell, thus greatly simplifying many of the subsequent stages in the structure determination.

5. Data Processing

Given the sheer volume of the data in virus crystallography and the extremely computationally intensive nature of the phase refinement and extension process, it is essential to have good access to a machine with a fast processor, large memory, and substantial filestore.

In general, the data will be derived from a large number of images from many crystals, with each covering a very narrow angular range. In this situation, it is critical that the observed reflections be accurately assigned as either fully or partially recorded. This requires a precise knowledge of the orientation of the crystal relative to the camera axes together with its unit cell parameters, mosaicity, and the beam characteristics. Once these have been satisfactorily determined, the problem becomes the accurate measurement of the crowded and often weak data.

*$RI = [\Sigma_h \Sigma_i | (I_h - I_{hi}) | / \Sigma_h \Sigma_i I_{hi}] \times 100$

I_h is the weighted mean measured intensity of the observations I_{hi}.

5.1. Determination of Crystal Orientation

Crystal orientations have in the past been determined by hand, but considering the large number of often randomly set crystals, this is a formidable task, since in order to avoid misindexing the closely spaced reflections, the crystal orientation must be known to within a fraction of a degree. Autoindexing procedures *(137,138)*, which initially work with difference vectors derived from the image to render the problem tractable, can cope with data from unit cells on the order of 700 Å and provide very accurate estimates of the missetting angles. In our experience, the errors in the missetting angles are on the order of 0.01° about those axes normal to the X-ray beam. Rossmann *(139)* has introduced a method for refining a rough orientation matrix that involves convoluting the observed and calculated diffraction pattern. This method can correct errors in the missetting angles of 1.0° and, for a large cell, provides results accurate to within 0.01°. All such algorithms and integration of the image require very accurate estimates of the origin of the reciprocal lattice and crystal-to-film distance. With the high spot density of virus films and the mechanical uncertainty often encountered at well-used synchrotrons, the methods used for protein crystallography (such as fiducial marks on photographic film) are rarely satisfactory. It is now usual to work with a semitransparent backstop that provides a weak image of the direct beam on each image. The direct beam position and crystal-to-plate distance may be conveniently calculated from a wax image (showing diffraction rings) recorded prior to data collection when using an image plate (this may need to be repeated after beam refills at synchrotrons).

5.2. Intensity Measurement

The advantages of profile fitting over simple summation of intensities are well documented *(139,140)*, and this is now always used. However, problems remain; given a closely spaced reciprocal lattice, many different lunes may be present on the same image. The consequent variation in spot size, shape, and spacing can make the measurement of the intensity of each spot and the background level in the vicinity of each spot across the whole image problematic if the background level is determined from points in a fixed relationship to the spot. For film data, we overcome this using logical bit-maps representing the scanned image to define (before measurement commences): (1) Pixels within a single spot and (2) within more than one spot. This permits tailoring of the exact area used for individual spots and background areas, and enables overlapped spots to be measured when profile fitting by omitting the regions of overlap. Since we wish to measure both strong low-resolution and very much weaker higher-resolution data from the same film pack, it is important that overloaded reflections can be measured. Again, if we measure these by profile fitting, the intensity may be derived from the nonoverloaded optical densities in the integration area.

5.2.1. Postrefinement

When there are a large number of partially recorded reflections that cannot be combined with their complementary parts on successive images, postrefinement is invaluable, providing improved estimates for the crystal orientation and some beam and crystal parameters. This allows partially recorded reflections to be precisely identified and by using a refined "rocking curve," those that are, say, more than 50% recorded may be scaled up to what they would be if fully recorded. Slight variations on the method originally devised by Schutt and Winkler *(141)* have been used with great success. Orientation, cell, and rocking curve parameters for one or more images are refined by comparing the intensities of partially recorded reflections (whose fraction recordedness may be calculated from the current estimates of the parameters) with a reference intensity (typically the mean of safe fully recorded observations of those reflections). If the procedure reclassifies a large number of spots (e.g., predicted as fully recorded, but found to be partially recorded), then it is really best to remeasure the image. The quality of data is normally judged by the agreement of multiply recorded or symmetry-related data. These R factors (*see* footnote on p. 9) are often worse for virus data than for standard protein molecules (sometimes 50–100% higher than for a good protein crystal). This is largely a consequence of the large number of very weak data usually recorded, the R factors (*see* footnote on p. 9) for the strong data are usually comparable to those seen with smaller proteins. The final merged data set will usually contain from 100,000 to a million or more unique reflections.

Because of the isometric nature of virus particles, it is common for there to be an ambiguity in the relative orientation of the icosahedral axes and the crystallographic axes. For example, in a cubic space group not possessing a fourfold axis, there are two alternative ways of indexing the reciprocal lattice that will not normally be distinguishable until data from different crystals are merged. At that stage, incompatible indexing will result in high merging R factors (*see* footnote on page 9) which can be remedied by the appropriate index conversion. The aforementioned ambiguity may arise within a single crystal with particles in either orientation forming a mosaic of scattering blocks. Data from such crystals have been processed *(142)*.

6. Phase Determination

This divides into two problems: (1) initial phase determination and (2) phase refinement and extension. Because of the inevitable oversampling in crystals of spherical viruses, poor initial phases can, given a sufficient number of accurate native data, be refined to great accuracy against the observed structure factor amplitudes.

6.1. Initial Phase Determination

It is the opinion of some of the authors that in many cases, stage 1 should in principle be achievable by reference only to the observed amplitude and simple physical parameters of the virus. In any event, it will be necessary to determine the precise orientation and position of the virus particle(s) in the unit cell. In some cases, this can be defined trivially because the virus with its very high symmetry lies at a special position in the crystal unit cell. In many cases, information on some of the parameters can be directly derived from packing considerations, and this can greatly simplify the latter stages of the structure determination. If the virus is structurally similar to another of known structure, then the precise particle orientation can be determined using a locked rotation function *(143)*. Alternatively, a self-rotation function can be used. Calculated in polar coordinates, plots of sections at constant κ will show the direction of noncrystallographic symmetry axes. In general, because of the large separation of particles within the cell, few protein interactions will not obey the icosahedral symmetry elements; thus, the peaks tend to be very prominent. The clarity can be further improved by resolution cutoffs applied to the data to reduce virus/RNA and virus/solvent effects. Prior to the combination of calculated phases with the observed amplitudes, a correction for the solvent and nucleic acid scattering may be necessary. Indeed, we find that a simple correction, such as that implemented in XPLOR *(144)* (version 2.1) can produce dramatic improvements at early stages, such as in the calculation of the crossrotation function. Indeed, running on a fast computer, XPLOR provides a simple, reasonably efficient package for rotational and translational searches. We have also used this program to find precise cell dimensions using an iterative rigid-body search procedure with the appropriate probe model and for positional refinement of the correctly oriented model against the observed data to yield improved phases.

With the particles placed in the unit cell, it is time to try to obtain phases for the measured (and often also the unmeasured) structure factor amplitudes. In all cases to date, these have been derived initially either from a model or from heavy-atom derivatives; however, two recent studies came close to *ab initio* determinations, namely MS2 *(11)*, canine parvovirus (CPV) *(38)*, and ΦX174 *(39)* (refer to Table 5). These are fascinating results.

If isomorphous replacement is to be used, the noncrystallographic symmetry can be used in a difference Patterson search for heavy-atom positions and also in the refinement of the heavy-atom parameters *(145)*. Because of the power of these constraints, the heavy-atom data are not required to be complete. Thus, for instance in the recent study of CPV *(38)* only 5% of the unique data to 8 Å were measured.

If one or more structurally similar viruses are available to provide a starting point for phase determination, it is simply a matter of calculating phases for the virus or viruses placed correctly in the cell of the unknown structure. Then by inspecting the correlation coefficient* between the observed and calculated structure factor amplitudes as a function of resolution, it can be seen to what level of detail the homolog will provide useful phase information. It is our experience that summing two or more homologous structures can provide improved phase estimates. Remember that carefully applied phase extension at this stage from 8 Å should be capable, even in the presence of the minimal fivefold noncrystallographic symmetry, of providing ultimately a set of very accurate phases to near-atomic resolution *(37)*.

6.2. Phase Refinement

Once a set of low-resolution phases has been obtained, iterative noncrystallographic symmetry averaging may be used to improve them (refer to Table 5). Arnold and Rossmann *(146)* derived an equation suggesting that the "power," P, of phase determination could be related to the noncrystallographic redundancy, N; the ratio of the volume to be aver-

*$C = \{\Sigma_h (<F_{obs}> - |F_{h,obs}|)(<F_{calc}> - |F_{h,calc}|)/[\Sigma_h (<F_{obs}> - |F_{h,obs}|)^2 \cdot (<F_{calc}> - |F_{h,calc}|)^2]^{1/2}\}$

Table 5
A Summary of the Use of Noncrystallographic Symmetry in Virus Crystallography to Date

	Resolution/Å	Redundancy	Use of noncrystallographic symmetry
TMV	2.8	17	Improvement of phases from a single isomorphous derivative
TBSV	2.9	5	Isomorphous phase improvement,first application of real space averaging to icosahedral viruses—used to improve some 1300 phases
TCV	3.2	15	Phase extension and refinement from 10–3.2 Å using low-resolution TBSV phases
SBMV	2.8	10	30–22.5 Å Phase extension with starting phases from a spherical shell. Phase refinement for the final 'best' multiple isomorphous replacement map
STNV	4	60	10–4 Å Phase extension of structure previously obtained by isomorphous replacement
STMV	3	15	Phase extension and refinement of STNV molecular replacement phases; use of these phases (15–5 Å) to locate eight heavy atom derivatives followed by phase refinement and extension to 3 Å
POLYOMA (murine)	22.5	5	Phase extension with starting phases from a model derived from electron microscopy
HRV 14	2.9	20	Phase refinement and 6–3 Å phase extension
HRV1A	3.2	10	Refinement of phases calculated from HRV14 at 5 Å and extension to 3.2 Å
POLIO (type 1 Mahoney)	2.9	30	Phase refinement and phase extension
POLIO (type 3 Sabin)	2.4	15	Refinement of initial phase estimates from P1/Mahoney
MENGO	2.9	60	Phase refinement and 8–3 Å phase extension with starting phases from a crudely homologous picornavirus structure
BBV	3.0	5	Refinement of 5 Å double-isomorphous replacement phases and extension to 3 Å
FHV	3.0	20	Phase extension and refinement from 4–3 Å with starting phases from BBV and extension to 3 Å

Table 5 *(continued)*

	Resolution/Å	Redundancy	Use of noncrystallographic symmetry
CpMV	2.8	5	Improvement of isomorphous replacement phases
BPMV	2.8	30	Phase extension and refinement with starting phases from CPMV
FMDV	2.9	5	Phase refinement and 8–3 Å phase extension with starting phases based on a combined picornavirus structure
MS2	3.3	10	Phase refinement and phase extension from 13 Å with starting phases from SBMV
CPV	3.3	60	Phase refinement and extension from 8 Å with SIR starting phases
SV40	3.8	5/30	Phase refinement using the fivefold ncs Extension of these and limited high-resolution SIR phases from 6.5 to 5 Å and 5 to 4.5 Å. Phase extension from 4.5 to 3.8 Å averaging the density for both pentamer cores (30-fold averaging for VP1) and fivefold averaging the variable parts
ΦX174	3.4	60	Phase extension, low-resolution structure factors from CpMV used for envelope definition; some heavy atom derivative data

aged U, to the total volume of the unit cell, V; the accuracy of the structure factor amplitudes, R; and the proportion of data measured, f, by the semiquantitative expression:

$$P = (Nf)^{1/2}/[R(U/V)] \tag{2}$$

Thus, to compensate for minimum redundancy, it is advantageous to ensure a high degree of accuracy in the measurement of intensities. The power of this procedure is greatly enhanced in combination with solvent flattening, and we find it beneficial to set the solvent and capsid interior to their respective mean electron densities. The enormous importance of real space averaging can be seen from Table 5, and it is particularly impressive in those solutions employing molecular replacement. In outline, the procedure used is to average the electron density for the noncrystallographically related subunits within the viral envelope, back-transform the average electron density, and then recombine the resultant calculated phases with the original F_{obs}, with suitable weighting, and compute a new electron density map. The region outside the molecular envelope and thus beyond the limits of applicability of the local noncrystallographic symmetry is flattened to represent disordered solvent. In the absence of considerable computer memory, averaging algorithms that are, in principle, simple become complicated by the need to implement sorting procedures before and after the interpolation of noncrystallographic symmetry-related electron densities to avoid the need for random access to a large map. The sorts are time consuming and expensive in terms of file space, making it well worth using as coarse a sampling of the electron density as is possible without introducing serious errors. The original implementation of the software used a linear interpolation for density points not lying exactly on integral grid points. Bricogne *(5)* has shown that with this algorithm, the sampling should be at approx 1/5 of the *d* spacing of the highest resolution data used. However, a sampling twice as coarse as this has proven successful using a quadratic interpolation method (such as that described by Nordman *[147]*), and this is much to be preferred (note, however, that some methods require double interpolation, which will demand finer sampling). If all the noncrystallographic symmetry-related densities can be held simultaneously in memory, the procedure is much less formidable *(148,150)*.

In order to maximize the efficiency and effectiveness of the standard averaging protocol, it has been found important to:

- Fill in the missing F_{obs} by including calculated values (suitably weighted) from the back-transformed electron density map—the phases for the newly phased observed amplitudes should be allowed a few cycles to converge before the unobserved data are added in, to speed convergence.
- Weight reflections according to the likely error in their phases. Sim *(149,150)* and Rayment *(151)* have proposed rather different weighting schemes, and Arnold and Rossmann *(152)* have proposed using a combined weight ($W_{Sim} \times W_{Ray}$). Rayment *(151)* presents evidence for the usefulness of a weighting scheme in facilitating the correction of erroneous phases. If the virus is crystallized in a low-symmetry space group or large unit cell, the increased number of unique data make the computing overheads during averaging very high. If the solution is to be determined by molecular replacement, we have found that it can hasten the process to calculate low-resolution maps, build a crude model, and improve and extend the phases by reciprocal space refinement prior to further phase improvement by averaging.

6.3. Phase Extension

Solvent flattening provides a considerable amount of phasing power for new reflections added at higher resolution. Thus, a protein envelope must allow the maximum solvent volume while not truncating the protein. This is especially important with lower phasing power (as arises for instance when only fivefold noncrystallographic redundancy is available, or if the data are inaccurate or incomplete). Starting envelopes defining the protein shell as the void between an inner and outer sphere have been used with overlap between adjacent viruses avoided by truncating the outer surface by planes normal to the vectors joining the centers of neighboring viruses along the axis of closest approach. An inner boundary can be used, since no more than a small portion of the viral genome normally exhibits icosahedral symmetry. Although such an envelope has in several cases proven to be quite satisfactory for phase extension *(35,36)*, it is our experience that with difficult cases, a more detailed envelope may be crucial to provide extra constraints during phase extension. Thus, a method to tighten the envelope automatically has been developed *(37)* that is similar to the method of Wang *(153)*. In order that a boundary can be clearly defined, the electron density is "smeared" to produce large homogeneous connected volumes of relatively high and relatively low density. Instead of applying the usual conical weighting function in real space *(150)*, a weighting function is applied in reciprocal space allowing the convolution in real space to be replaced by a simple product. Thus, the scattering from an atom is attenuated by a factor corresponding to the Fourier transform of the real-space weighting function. We assume a Gaussian model in real space, which is equivalent to a B factor* in reciprocal space. For example, a B factor of 5000 has been used successfully with data to a maximum resolution of 8 Å. This smearing is performed on structure factors calculated from an electron density map whose low-resolution contrast has been enhanced by setting all pixels with ρ less than zero, to zero. Taking the back-transformed map, a doubly limited envelope is calculated consisting of the formerly defined "spherical" envelope to set outer limits for the protein within which a boundary is defined by reference to the smeared density. To achieve this, the electron density values are accumulated into a histo-

*$B = 8\pi^2 < u^2_{Tot} > /3$

B is the isotropic temperature factor. $< u^2_{Tot} >$ is the total mean square displacement of an atom from its rest position.

gram. The number of "solvent" pixels required to tighten the originally defined envelope to the actual solvent volume is calculated, and from the histogram, the electron density value corresponding to this number of pixels is determined and the envelope determined accordingly. It is useful to recalculate this envelope at intervals during the phase extension, and the number of extra solvent pixels can be adjusted to maximize the phasing power while not truncating ordered protein *(37)*.

Problems have been encountered in which high-resolution phase solutions satisfy the noncrystallographic symmetry constraints, but are anticorrelated with the low-resolution phasing solution *(35)*. These can be avoided if the shell thickness for phase extension is restricted by consideration of the interference function G for the spherical viral particle (of radius R) at a distance of H from the reciprocal space origin *(35)*. H in phase extension is the reciprocal distance between the reflection whose phase is to be determined and its nearest already phased contribution. The argument of G, $H \cdot R$ is given by $(n/a)R$ where n is the limiting number of reciprocal lattice points and a is the cell dimension. The largest values of G will occur within the first positive portion of the function corresponding to $(n/a)R < 0.725$, and thus for a given radius and unit cell, the shell thickness that is likely to contain reliable phase information can be determined (e.g., for R = 150 Å and $a = 400$ Å, $n < 1.9$ reciprocal lattice units). Strategies of applying this vary. For instance, in the HRV14 structure determination *(35)*, the phase extension was done in somewhat thicker shells than this relationship would suggest, but several cycles of refinement were performed between successive extensions. In the determination of FMDV on the other hand, very narrow shells were used, but with continual addition of new data on each cycle *(37)*. Both methods worked satisfactorily.

Poor agreement of the strong low-resolution terms may inhibit phase extension. It may therefore be necessary to impose low-resolution cutoffs on the observed data if, for instance, reflections close to the backstop shadow have not been measured properly.

Series termination errors can be reduced by phase extension for a couple of shells beyond the limiting resolution of the data (i.e., by allowing calculated data to soak up spurious ripples in the map).

Extra phasing power can be obtained by truncating some of the most negative densities within the protein region. However, if the number of points in the shell has been judged accurately, these should represent only a small proportion of the total number of points.

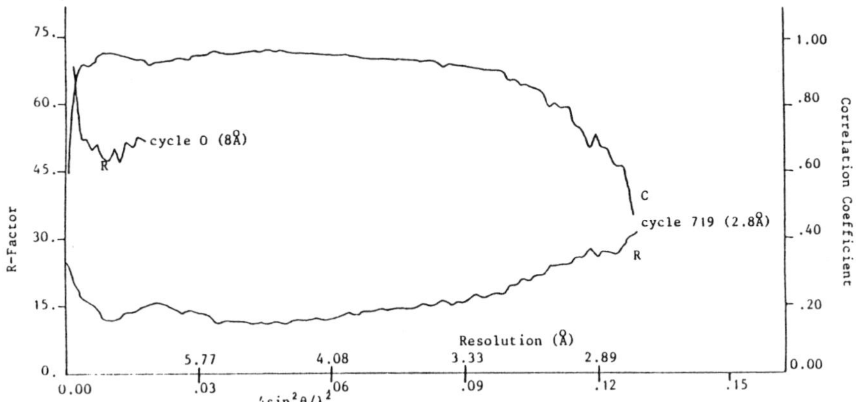

Fig. 2. Correlation coefficients (C) (*see* footnote on p. 13) and R factors (R) (*see* footnote on p. 17) for the phase extension from 8 to 2.8 Å resolution for FMDV. **(A)** Cycle O is starting phases (unaveraged). **(B)** Cycle 719 is the final cycle (the cycle numbering reflects the recalcitrance of the problem; several false starts inflated the value considerably).

The process of phase refinement and extension is normally monitored by reference to averaging R factors* and correlation coefficients (*see* footnote on p. 13), which should be analyzed as a function of resolution (*see* Fig. 2). These have been known to converge to values of 11.79 and 0.961, respectively (these numbers are based on all the data—if one only considers the data above F_{mean}, the R factor drops to 7.89% *[154]*). The correlation coefficient is a more reliable estimate of the success of the phase extension than the R factor. Figure 3 shows the progress of the FMDV phase extension.

7. Map Interpretation—Model Building

A model is normally constructed for a single icosahedral subunit. The quality of the phases produced by the averaging procedure is usually very high, and maps calculated at 3.0-Å resolution clearly show carbonyl bulges and hence the peptide plane. If the virus capsid contains multiple copies of a single protein, then this structure once built for one subunit, can be rotated and adapted to fit the electron density for the symmetrically related subunits. At this stage in the analysis, one reaps the benefits of earlier hard work; enjoy it!

$*R = [\Sigma_h | (| F_{h,obs} | - | F_{h,calc} |) | / \Sigma_h | F_{h,obs} |] \times 100$

F_{obs} is the observed structure amplitude of reflection *h*. F_{calc} is the calculated structure amplitude of reflection *h* as a result of back-transforming the averaged and modified electron density map.

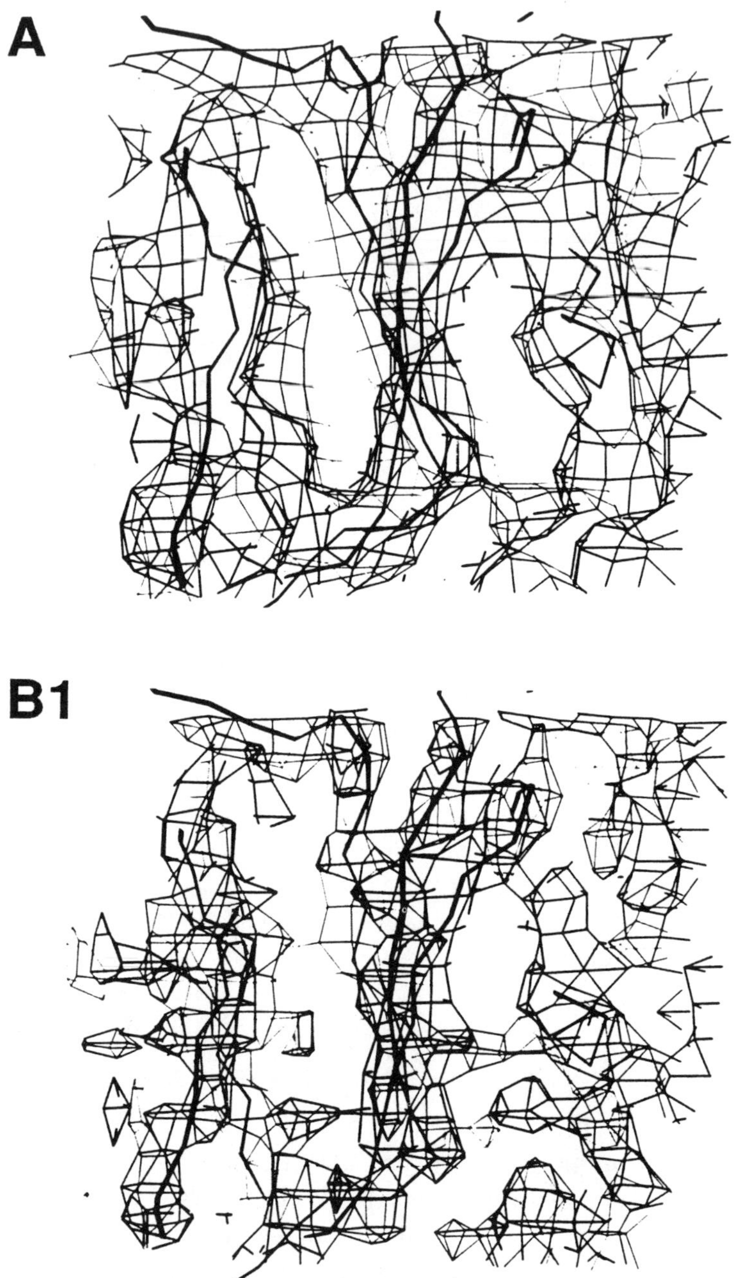

Fig. 3. Identical portions of electron density maps (coefficients $2F_{obs}$–$F_{averaged}$ $\alpha_{averaged}$, contoured at approximately one-tenth of the maximum) calculated at intervals in the phase extension for FMDV. **(A)** 6 Å **(B1)** 4.5 Å. The C_α backbone or all-atom trace is superimposed.

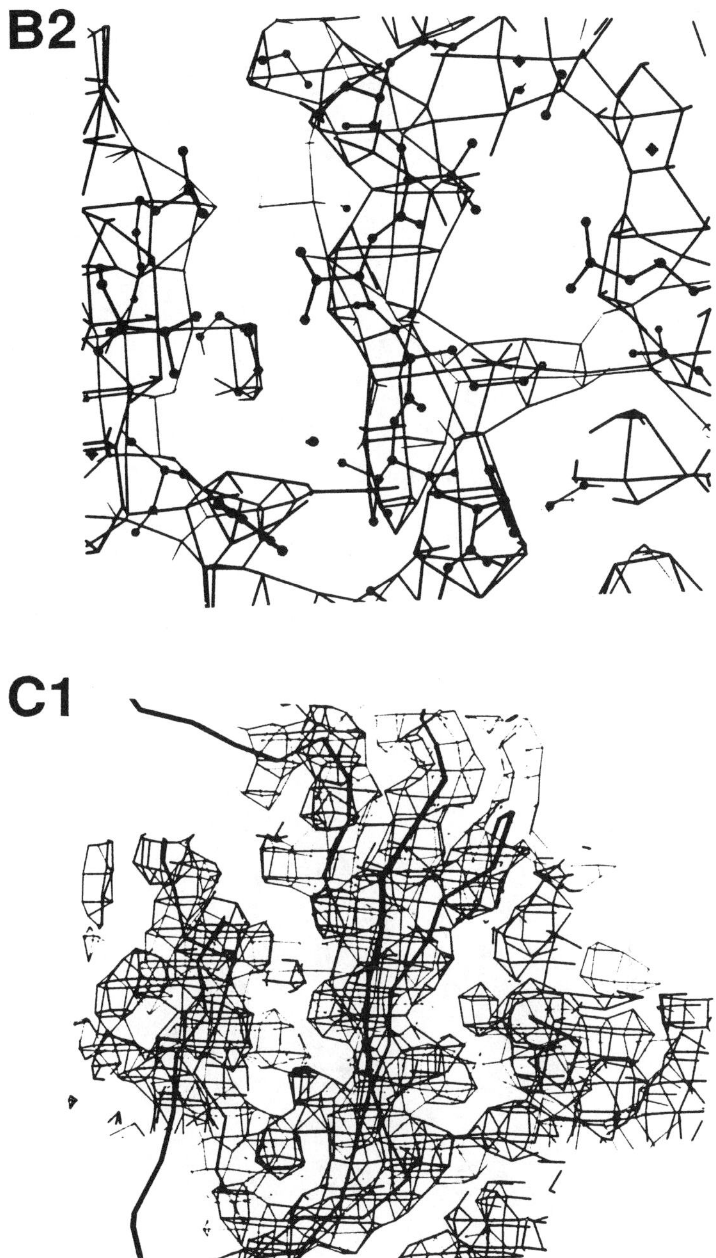

Fig. 3. *(continued on following page)* Identical portions of electron density maps (coefficients $2F_{obs}{-}F_{averaged}\ \alpha_{averaged}$, contoured at approximately one-tenth of the maximum) calculated at intervals in the phase extension for FMDV. **(B2)** 4.5 Å, **(C1)** 3.5 Å. The C_{α} backbone or all-atom trace is superimposed.

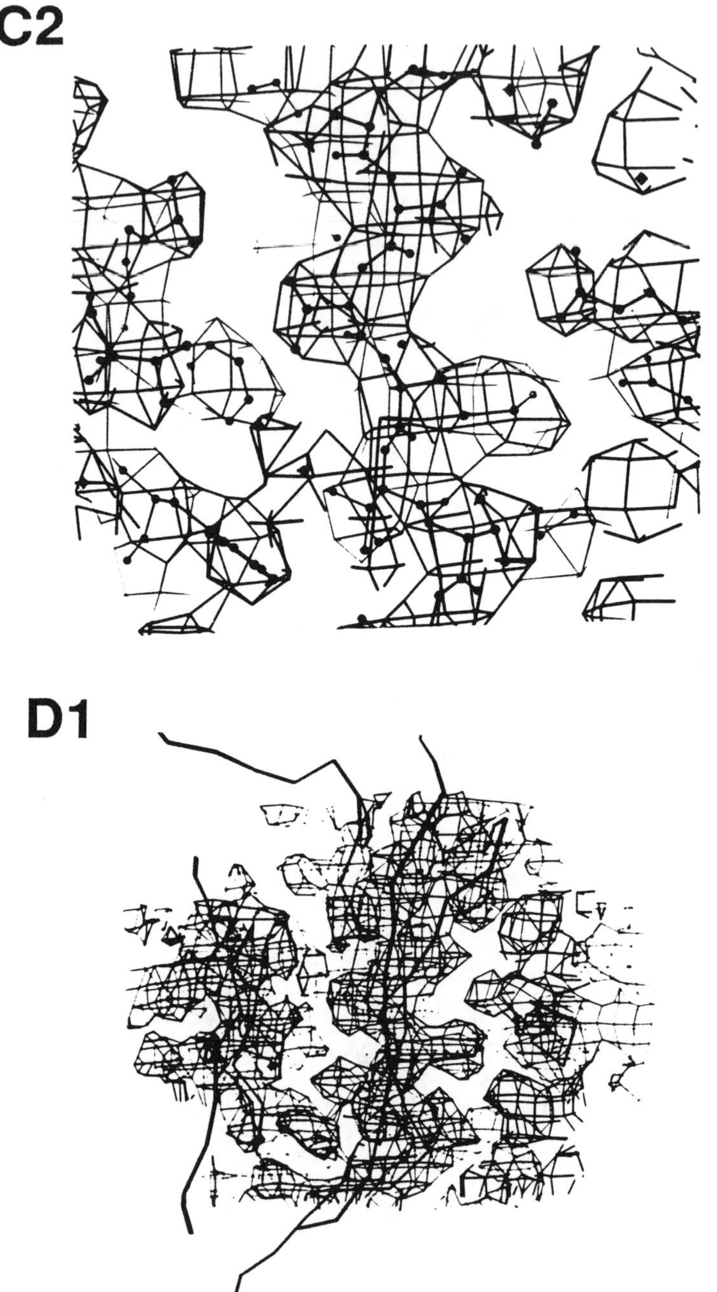

Fig. 3. *(continued from previous page)* Identical portions of electron density maps (coefficients $2F_{obs}–F_{averaged}$ $\alpha_{averaged}$, contoured at approximately one-tenth of the maximum) calculated at intervals in the phase extension for FMDV. **(C2)** 3.5 Å, **(D1)** 2.9 Å. The C_{α} backbone or all-atom trace is superimposed.

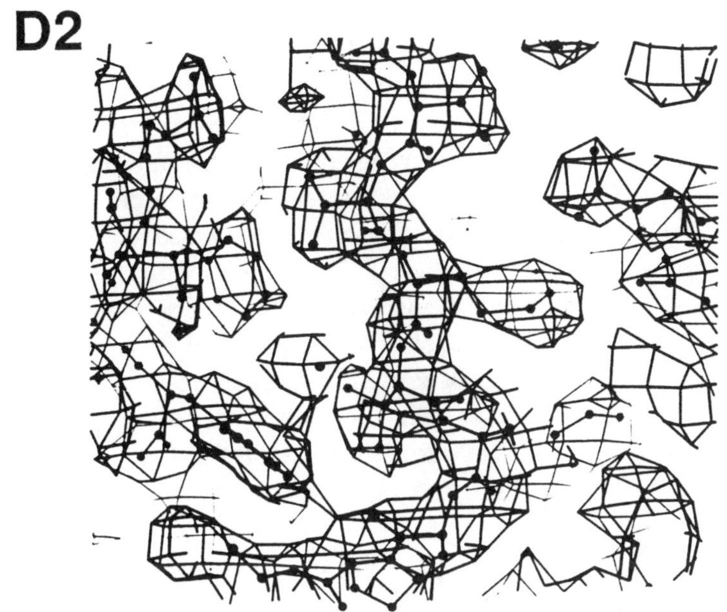

Fig. 3. Identical portions of electron density maps (coefficients $2F_{obs}–F_{averaged}$ $\alpha_{averaged}$, contoured at approximately one-tenth of the maximum) calculated at intervals in the phase extension for FMDV. **(D2)** 2.9 Å. The C_{α} backbone or all-atom trace is superimposed.

8. Refinement

By applying noncrystallographic symmetry constraints, the refinement of the structure becomes, relatively, much better determined than a typical protein refinement at the same resolution. However, it may also be useful, on occasion, to perform a refinement of the entire contents of the crystallographic asymmetric unit in order to observe the agreement of the noncrystallographically related subunits *(24)*. Note that since the phases derived from the averaging process will normally be very accurate the aim of refinement is normally not to improve the phase estimates, but simply to provide an improved atomic model.

Variations on a real-space refinement procedure have been adopted to refine a number of virus structures. The reciprocal space equivalent of this involves minimizing the discrepancy between the real and imaginary parts of the observed structure factors (with the phases derived from the averaging process treated as observations) and the corresponding terms computed from the structural model with geometric and noncrystallographic constraints *(35,36)*. In order to cope with the enormous number of observations, different random subsets of data were taken prior to each stage in the refinement. For STNV, a purely real-space model-fitting refinement procedure was used *(155)*. Simulated annealing has been used very successfully, but the size of the problem necessitates the use of a computer with considerable physical memory. One such refinement program, XPLOR version 2.1 and subsequent versions *(144)*, incorporate a simple solvent correction procedure that we have found can provide a substantially better model at low resolution. Note that XPLOR may be used as a very convenient and extremely efficient general purpose refinement program allowing noncrystallographic symmetry to be incorporated either as restraints or constraints. Typically, an XPLOR refinement protocol without simulated annealing would incorporate:

1. Definition of topology;
2. Assessment of the weighting factor to be applied to the observations;
3. Definition of a solvent mask;
4. Cycles of positional refinement to remove gross inaccuracies;
5. B-factor refinement; and
6. Cycles of positional refinement followed by B-factor refinement with redefinition of the solvent mask as required, in each case allowing the energy gradient to reach a minimum. At this stage, manual rebuilding using $F_{obs} - F_{calc}$ and $2F_{obs} - F_{calc}$ difference maps would be used prior to further refinement.

The refined coordinates obtained will correspond to either the icosahedral asymmetric unit or the crystallographic asymmetric unit, and hence, symmetry operations must be applied to generate the whole capsid. A sample protocol for generating a capsid from the icosahedral asymmetric unit is given in Tables 6 and 7 (for graphics program FRODO *[156]*).

9. Conclusion

The three-dimensional structures of a number of intact viruses have provided explanations for many observations that have arisen from biochemical and biophysical studies. Very recently, three-dimensional reconstructions from cryoelectron microscopy have been used to study complexes between viruses and Fab fragments *(157,158)*, whole antibodies *(159)*, and receptor molecules *(160)* giving low-resolution information on these complexes. However, by combining X-ray data for the individual components that make up such complexes with the cryo-

Table 6
FMDV Symmetry Definition for FRODO (Version of P. Evans)[a]

SAM	–Z,X,–Y
SYMM	–Y,Z,–X
I	–X,–Y,Z
345 345 345 90 90 90	–Z,–X,Y
X,Y,Z	–Y,–Z,X
Z,X,Y	1 0 0 0 1 0 0 0 1 0 0 0
Y,Z,X	.5 –.809 .309 .809 .309 –.5 .309 .5 .809 0 0 0
X,–Y,–Z	.309 –.5 .809 –.5 –.809 –.309 .809 –.309 –.5 0 0 0
Z,–X,–Y	–.5 –.809 .309 –.809 .309 –.5 .309 –.5 –.809 0 0 0
Y,–Z,–X	0 0 1 1 0 0 0 1 0 0 0 0
–X,Y,–Z	

[a]The icosahedron orientation with respect to orthogonal axes is as used in the HRV14 and Mengo virus coordinates *(9,22)*.

Table 7
Symmetry Application[a]

MOL	SYMM S1 SU1
CA	SYMM S1 SU2
NOCO	SYMM S1 SU3
NAM SU1A	SYMM S1 SU4
COL BLUE	SYMM S1 SU5
ZONE 11 2101	NAM SU7
NAM SU1	SYMM N2 SU6
SYMM S1 SU1A	NAM SU8
SYMM S1 SU1A	SYMM N3 SU6
COL GREEN	SYMM N3 SU7
ZONE 52 2182	SYMM S1 SU6
COL RED	SYMM S1 SU7
ZONE 13 2203	NAM SU9
COL YELLOW	SYMM N4 SU8
ZONE 154 394	NAM SU10
ZONE 654 854	SYMM N4 SU9
NAM SU2	NAM SU11
SYMM N1 SU1	SYMM S1 SU8
NAM SU3	SYMM S1 SU9
SYMM N1 SU2	SYMM S1 SU10
NAM SU4	NAM SU12
SYMM N1 SU3	SYMM L1 SU11
NAM SU5	NAM SU 13
SYMM N1 SU4	CELL
NAM SU6	

[a]To display a C_{α} trace of a picornavirus capsid (here using FMDV coordinates) with the individual proteins color-coded according to the standard convention: VP1-Blue, VP2-Green, VP3-Red, VP4-Yellow. Icosahedron orientation as in Table 6.

electron microscopy data, it is possible to obtain rather precise information about the interactions between the components. Although these results are extremely valuable, we also hope that such complexes will be tackled by crystallography in the near future yielding true high-resolution structures.

References

1. Choi, H-K., Tong, L., Minor, W., Dumas, P., Boege, U., Rossmann, M. G., and Wengler, G. (1991) Structure of sindbis virus core protein reveals a chymotrypsinlike serine proteinase and the organisation of thc virion. *Nature* **354,** 37–43.
2. Kohlstaedt, L. A., Wang, J., Friedman, J. M., Rice, P. A., and Steitz, T. A. (1992) Crystal structure at 3.5 Å resolution of HIV-1 reverse transcriptase complexed with an inhibitor. *Science* **256,** 1783–1790.
3. Navia, M. A., Fitzgerald, P. M. D., McKeever, B. M., Leu, C.-T., Heimbach, J. C., Herber, W. K., Sigal, I. S., Darke, P. L., and Springer, J. P. (1989) Three dimensional structure of the aspartyl protease from the human immunodeficiency virus HIV-1. *Nature* **337,** 615–620.
4. Rossmann, M. G. and Blow, D. M. (1962) The detection of subunits within the crystallographic asymmetric unit. *Acta Cryst.* **15,** 24–31.
5. Bricogne, G. (1976) Methods and programs for direct-space exploitation of geometric redundancies. *Acta Cryst.* **A32,** 832–847.
6. Harrison, S. C., Olson, A. J., Schutt, C. E., Winkler, F. K., and Bricogne, G. (1978) Tomato bushy stunt virus at 2.9 Å resolution. *Nature* **276,** 368–373.
7. Abad-Zapatero, C., Abdel-Meguid, S., Johnson, J. E., Leslie, A., G. W., Rayment, I., Rossmann, M. G., Suck, D., and Tsukihara, T. (1980) Structure of southern bean mosaic virus at 2.8 Å resolution. *Nature* **286,** 33–39.
8. Liljas, L., Unge, T., Jones, T. A., Fridborg, K., Lövgren, S., Skoglund, U., and Strandberg, B. (1982) Structure of satellite tobacco necrosis virus at 3.0 Å resolution. *J. Mol. Biol.* **159,** 93–108.
9. Rossmann, M. G., Arnold, E., Erickson, J. W., Frankenberger, E. A., Griffith, J. P., Johnson, J. E., Kamer, G., Luo, M., Mosser, A. C., Rueckert, R. R., Sherry, B., and Vriend, G. (1985) Structure of a human common cold virus and functional relationship to other picornaviruses. *Nature* **317,** 145–153.
10. Hogle, J. M., Chow, M., and Filman, D. J. (1985) Three-dimensional structure of poliovirus at 2.9 Å resolution. *Science* **229,** 1358–1365.
11. Valegård, K., Liljas, L., Fridborg, K., and Unge, T. (1990) The three-dimensional structure of the bacterial virus MS2. *Nature* **344,** 36–41.
12. McKenna, R., Xia, D., Willingham, P., Ilag, L. L., Krishnaswamy, S., Rossmann, M. G., Olson, N. H., Baker, T. S., and Incardona, N. L. (1992) Structure of single-stranded DNA bacteriophage ΦX174 and its functional implications. *Nature* **355,** 137–143.
13. Tsao, J., Chapman, M. S., Agbandje, M., Keller, W., Smith, K., Wu, H., Luo, M., Smith, T. J., Rossmann, M. G., Compans, R. W., and Parrish, C. R. (1991) The three-dimensional structure of canine parvovirus and its functional implications. *Science* **251,** 1456–1464.

14. Liddington, R. C., Yan, Y., Moulai, J., Sahli, R., Benjamin, T. L., and Harrison, S. C. (1992) Structure of simian virus 40 at 3.8 Å resolution. *Nature* **354,** 278–284.
15. Caspar, D. L. D. and Klug, A. (1962) Physical principles in the construction of regular viruses. *Cold Spring Harbor Symp. Quant. Biol.* **27,** 1–24.
16. Larson, S. B., Koszelak, S., Day, J., Greenwood, A., Dodds, J. A., and McPherson, A. (1993) Double-helical RNA in satellite tobacco mosaic virus. *Nature* **361,** 179–182.
17. Hogle, J. M., Maeda A., and Harrison S. C. (1986) Structure and assembly of turnip crinkle virus I. X-ray crystallographic structure analysis at 3.2 Å resolution. *J. Mol. Biol.* **191,** 625–638.
18. Stauffacher, C., Usha, R., Harrington, M., Schidt, T., Hosur, M., and Johnson, J. E. (1987) The structure of cowpea mosaic virus at 3.5 Å resolution, in *Crystallography in Molecular Biology* (Moras, D., Drenth, J., Strandberg, B., Suck, D., and Wilson, K., eds.), Plenum, New York, pp. 293–308.
19. Chen, Z., Stauffacher, C., Li, Y., Schmidt, T., Bomu, W., Kamer, G., Shanks, M., Lomonossoff, G., and Johnson, J. E. (1989) Protein–RNA interactions in an icosahedral virus at 3.0 Å resolution. *Science* **245,** 154–159.
20. Kim, S., Smith, T. J., Chapman, M. S., Rossmann, M. G., Pevear, D. C., Dutko, F. J., Felock, P. J., Diana, G. D., and McKinlay, M. A. (1989) Crystal structure of human rhinovirus type 1A (HRV1A). *J. Mol. Biol.* **210,** 91–111.
21. Filman, D. J., Syed, R., Chow, M., MacAdam, A. J., Minor, P. D., and Hogle, J. M. (1989) Structural factors that control conformational transitions and serotype specificity in type 3 poliovirus. *EMBO J.* **8,** 1567–1579.
22. Luo, M., Vriend, G., Kamer, G., Minor, I., Arnold, E., Rossmann, M. G., Boege, U., Scraba, D. G., Duke, G. M., and Palmenberg, A. C. (1987) The atomic structure of Mengovirus at 3.0 Å resolution. *Science* **235,** 182–191.
23. Acharya, R., Fry, E., Stuart, D. I., Fox, G., Rowlands, D., and Brown, F. (1989) The three-dimensional structure of foot-and-mouth disease virus at 2.9 Å. *Nature* **337,** 709–716.
24. Grant, R. A., Filman, D. J., Fujinami, R. S., Icenogle, J. P., and Hogle, J. M. (1992) Three-dimensional structure of Theiler virus. *Proc. Natl. Acad. Sci. USA* **89,** 2061–2065.
25. Luo, M., He, C., Toth, K. S., Zhang, C. X., and Lipton, H. L. (1992) Three-dimensional structure of Theiler murine encephalomyelitis virus (BeAn strain). *Proc. Natl. Acad. Sci. USA* **89,** 2409–2415.
26. Hosur, M. V., Schmidt, T., Tucker, R. C., Johnson, J. E., Gallagher, T. M., Selling, B. H., and Rueckert, R. R. (1987) Structure of an insect virus at 3.0 Å resolution. *Proteins* **2,** 167–176.
27. Fisher, A. J. and Johnson, J. E. (1993) Ordered duplex RNA controls the capsid architecture of an icosahedral animal virus. *Nature* **361,** 176–179.
28. Badger, J., Minor, I., Kremer, M. J., Oliviera, M. A., Smith, T. J., Griffith, J. P., Guerin, D. M. A., Krishnaswamy, S., Luo, M., Rossmann, M. G., McKinlay, M. A., Diana, G. D., Dutko, F. J., Fancher, M., Rueckert, R. R., and Heinz, B. A. (1988) Structural analysis of a series of antiviral agents complexed with human rhinovirus 14. *Proc. Natl. Acad. Sci. USA* **85,** 3304–3308.

29. Parry, N., Rowlands, D. J., Fox, G., Brown, F., Fry, E., Acharya, K. R., Logan, D. T., and Stuart, D. I. (1990) Structural and serological evidence for a novel mechanism of immune evasion in foot-and-mouth disease virus. *Nature* **347,** 569–572.
30. Mizuno, H., Kano, H., Omura, T., Koisumi, M., Kondoh, M., and Tsukihara, T. (1991) Crystallisation and preliminary X-ray study of a double-shelled spherical virus, rice dwarf virus. *J. Mol. Biol.* **219,** 665–669.
31. Basak, A. K., Grimes, J., Burroughs, J. N., Mertons, P. P. C., Roy, P., and Stuart, D. (1992) Crystallographic studies on the structure of the bluetongue virus inner capsid, in *Bluetongue, African Horse Sickness and Related Orbiviruses* (Walton, T. E. and Osburn, B. I., eds.), CRC, Boca Raton, FL, pp. 483–490.
32. Basak, A. K., Stuart, D. I., and Roy, P. (1992) Preliminary crystallographic study of bluetongue virus capsid protein VP7. *J. Mol. Biol.* **228,** 687–689.
33. Gong, Z. X., Wu, H., Cheng, R. H., Hull, R., and Rossmann, M. G. (1990) Crystallisation of Cauliflower Mosaic virus. *Virology* **179,** 941–945.
34. Coombs, K. M., Fields, B. N., and Harrison, S. C. (1990) Crystallisation of the reovirus type 3 Dearing core. Crystal packing is determined by the λ2 protein. *J. Mol. Biol.* **215,** 1–5.
35. Arnold, E., Vriend, G., Luo, M., Griffith, J. P., Kamer, G., Erickson, J. W., Johnson, J. E., and Rossmann, M. G. (1987) The structure determination of a common cold virus, human rhinovirus 14. *Acta Cryst.* **A43,** 346–361.
36. Luo, M., Vriend, G., Kamer, G., and Rossmann, M. G. (1989) Structure determination of Mengo virus. *Acta Cryst.* **B45,** 85–92.
37. Fry, E., Acharya, R., and Stuart, D. (1993) Methods used in the structure determination of foot-and-mouth disease virus. *Acta Cryst.* **A49,** 45–55.
38. Tsao, J., Chapman, M. S., Wu, H., Agbandje, M., Keller, W., and Rossmann, M. G. (1992) Structure determination of monoclinic canine parvovirus. *Acta Cryst.* **B48,** 75–88.
39. McKenna, R., Xia, Di, Willingmann, P., Ilag, L. L., and Rossmann, M. G. (1992) Structure determination of the bacteriophage ϕX174. *Acta Cryst.* **B48,** 499–511.
40. Zlotnick, A., McKinney, B. R., Munshi, S., Bibler, J., Rossmann, M. G., and Johnson, J. E. (1993) A monoclinic crystal with R32 pseudo-symmetry: a preliminary report of nodamura virus structure determination. *Acta Cryst.* **D49,** 580–587.
41. Wu, H., Keller, W., and Rossmann, M. G. (1993) Determination and refinement of the canine parvovirus empty-capsid structure. *Acta Cryst.* **D49,** 572–579.
42. Rossmann, M. G. (1989) The molecular replacement method. *Acta Cryst.* **A46,** 73–82.
43. Crick, F. H. C. and Watson, J. D. (1956) Structure of small viruses. *Nature* **177,** 473–475.
44. Wilson, I. A., Skehel, J. J., and Wiley, D. C. (1981) Structure of the haemagglutinin membrane glycoprotein of influenza virus. *Nature* **289,** 366–373.
45. Varghese, J. N., Laver, W. G., and Colman, P. M. (1983) Structure of the influenza glycoprotein antigen neuraminidase at 2.9 Å resolution. *Nature* **303,** 35–40.
46. Burmeister, W. P., Ruigrok, R. W. H., and Cusack, S. (1992) The 2.2 Å resolution crystal structure of influenza B neuraminidase and its complex with sialic acid. *EMBO J.* **11,** 49–56.

47. Bloomer, A., Champness, J. N., Bricogne, G., Staden, R., and Klug, A. (1978) Protein disk of tobacco mosaic virus at 2.8 Å resolution showing the interactions within and between subunits. *Nature* **276,** 362–368.
48. Namba, K., Pattanyek, R., and Stubbs, G. (1989) Visualization of protein-nucleic acid interactions in a virus. Refined structure of tobacco mosaic virus at 2.9 Å resolution by X-ray fiber diffraction. *J. Mol. Biol.* **208,** 307–325.
49. Lobert, S. and Stubbs, G. (1990) Fiber diffraction analysis of cucumber green mottle virus using limited numbers of heavy-atom derivatives. *Acta Cryst.* **A46,** 993–997.
50. Rossmann, M. G., Mc Kenna, R., Tong, L., Xia, D., Dai, J.-B., Wu, H., and Choi, H.-K. (1992) Molecular replacement real-space averaging. *J. Appl. Cryst.* **25,** 166–180.
51. Johnson, J. E., Akimoto, T., Suck, D., Rayment, I., and Rossmann, M. G. (1976) Structure of southern bean mosaic virus at 22.5 Å resolution. *Virology* **75,** 394–400.
52. Tsao, J., Chapman, M. S., and Rossmann, M. G. (1992) *Ab initio* phase determination for viruses with high symmetry: a feasibility study. *Acta Cryst.* **A48,** 293–301.
53. Chapman, M. S., Tsao, J., and Rossmann, M. G. (1992) *Ab initio* phase determination for spherical viruses: parameter determination for spherical-shell models. *Acta Cryst.* **A48,** 301–312.
54. Wang, X. and Janin, J. (1993) Orientation of non-crystallographic symmetry axes in protein crystals. *Acta Cryst.* **D49,** 505–512.
55. Smyth, M., Fry, E., Stuart, D., Lyons, C., Hoey, E., and Martin, S. J. (1993) Preliminary crystallographic analysis of bovine enterovirus. *J. Mol. Biol.* **231,** 930–932.
56. Smyth, M., Hall, J., Fry, E., Stuart, D., Stanway, G., and Hyyppiä, T. (1993) Preliminary crystallographic analysis of coxsackievirus A9. *J. Mol. Biol.* **230,** 667–669.
57. Mattern, C. F. T. and duBuy, H. G. (1956) Purification and crystallisation of Coxsackie virus. *Science* **123,** 1037–1038.
58. Li, T., Zhang, A., Iuzuka, N., Nomoto, A., and Arnold, E. (1992) Crystallisation and preliminary X-ray diffraction studies of coxsackievirus B1. *J. Mol. Biol.* **223,** 1171–1175.
59. Fox, G., Stuart, D., Acharya, K. R., Fry, E., Rowlands, D., and Brown, F. (1987) Crystallisation and preliminary X-ray diffraction analysis of foot-and-mouth disease virus. *J. Mol. Biol.* **196,** 591–597.
60. Curry, S., Abu-Ghazaleh, R., Blakemore, W., Fry, E., Jackson, T., King, A., Lea, S., Logan, D., Newman, J., and Stuart, D. (1992) Crystallisation and preliminary X-ray analysis of three serotypes of foot-and-mouth disease virus. *J. Mol. Biol.* **228,** 1263–1268.
61. Erickson, J. W., Frankenberger, E. A., Rossmann, M. G., Shay Fout, G., Medappa, K. C., and Rueckert, R. R. (1983) Crystallisation of a common cold virus, human rhinovirus 14: "Isomorphism" with poliovirus crystals. *Proc. Natl. Acad. Sci. USA* **80,** 931–934.
62. Arnold, E., Erickson, J. W., Fout, G. S., Frankenberger, E. A., Hecht, H. J., Luo, M., Rossmann, M. G., and Rueckert, R. R. (1984) Virion orientation in cubic crystals of the human common cold virus HRV14. *J. Mol. Biol.* **177,** 417–430.

63. Korant, B. D. and Stasny, J. T. (1973) Crystallisation of human rhinovirus 1A. *Virology* **55,** 410–417.
64. Boege, U., Scraba, D. G., Hakayawa, K., James, M. N. J., and Erickson, J. W. (1984) Structure of the Mengo virion. VII. Crystallisation and preliminary X-ray diffraction analysis. *Virology* **138,** 162–167.
65. Kim, S., Boege, U., Krishnaswamy, S., Minor, I., Smith, T. J., Luo, M., Scraba, D. G., and Rossmann, M. G. (1990) Conformational variability of a picornavirus capsid: pH-dependent structural changes of Mengo virus related to its host receptor attachment and disassembly. *Virology* **175,** 176–190.
66. Schaffer, F. L. and Schwerdt, C. E. (1955) Crystallisation of purified MEF-1 poliomyelitis virus particles. *Proc. Natl. Acad. Sci.USA* **41,** 1020–1023.
67. Steere, R. L. and Schaffer, F. L. (1958) The structure of crystals of purified Mahoney poliovirus. *Biochim. Biophys. Acta* **28,** 241–246.
68. Finch, J. T. and Klug, A. (1959) Structure of poliomyelitis virus. *Nature* **183,** 1708–1814.
69. Hogle, J. M. (1982) Preliminary studies of crystals of poliovirus type, I. *J. Mol. Biol.* **160,** 663–668.
70. Yeates, T. O., Jacobsob, D. H., Martin, A., Wychowski, O., Girard, M., Filman, D. J., and Hogle, J. M. (1991) Three-dimensional structure of a mouse-adapted type 2/type 1 poliovirus chimera. *EMBO J.* **10,** 2331–2341.
71. Toth, K. S., Zhang, C. X., Lipton, H. L., and Luo, M. (1993) Crystallisation and preliminary diffraction studies of Theiler's virus (GDVII strain). *J. Mol. Biol.* **231,** 1126–1129.
72. Takemoto, Y., Nagahara, N., Fukuyama, K., Tsukihara, T., and Iwaki, M. (1985) Crystallisation and preliminary characterization of arabis mosaic virus. *Virology* **145,** 191–194.
73. Fukuyama, K., Abdel-Meguid, S. S., and Rossmann, M. G. (1981) Crystallisation of an alfalfa mosaic virus coat protein as a T = 1 aggregate. *J. Mol. Biol.* **150,** 33–41.
74. Finch, J. T., Leberman, R., and Berger, E. (1967) Structure of broad bean mottle virus II. X-ray diffraction studies. *J. Mol. Biol.* **27,** 17–24.
75. Heuss, K. L., Rao, J. K. M., and Argos, P. (1981) Crystallisation of Belladonna mottle virus. *J. Mol. Biol.* **140,** 629–633.
76. Munshi, S. J., Hiremath, C. N., and Murthy, M. R. N. (1987) Symmetry of belladonna mottle virus: rotation function studies. *Acta Cryst.* **B43,** 376–382.
77. Sehnke, P. C., Harrington, M., Hosur, M. V., Li, Y., Usha, R., Tucker, R. C., Bomu, W., Stauffacher, C. V., and Johnson, J. E. (1988) Crystallisation of viruses and virus proteins. *J. Cryst. Growth* **90,** 222–230.
78. Morgunova, E. Y., Urzhumtsev, A. G., Mikchailov, A. M., and Vainshtein, B. K. (1990) Poster abstract at IUCr meeting, Bordeaux.
79. Rossmann, M. G., Smiley, I. E., and Wagner, M. A. (1973) Crystalline cowpea chlorotic mottle virus. *J. Mol. Biol.* **74,** 255,256.
80. Speir, J. A., Munshi, S., Baker, T. S., and Johnson, J. E. (1993) Preliminary X-ray data analysis of crystalline cowpea chlorotic mottle virus. *Virology* **193,** 234–241.
81. White, J. M. and Johnson, J. E. (1980) Crystalline cowpea mosaic virus. *Virology* **101,** 319–324.

82. Colman, P. M., Tulloch, P. A., Shukla, D. D., and Gough, K. H. (1980) Particle and crystal symmetry of Erysimium latent virus. *J. Mol. Biol.* **142,** 263–268.
83. Magdoff, B. S. (1960) Subunits in Southern Bean Mosaic virus. *Nature* **185,** 673.
84. Akimoto, T., Wagner, M. A., Johnson, J. E., and Rossmann, M. G. (1975) The packing of southern bean mosaic virus in various crystal cells. *J. Ultrastruct. Res.* **53,** 306–318.
85. Rayment, I., Johnson, J. E., and Rossmann, M. G. (1979) Metal-free southern bean mosaic virus crystals. *J. Biol. Chem.* **254,** 5243–5245.
86. Koszelak, S., Dodds, J. A., and McPherson, A. (1989) Preliminary analysis of crystals of satellite tobacco mosaic virus. *J. Mol. Biol.* **209,** 323–325.
87. Day, J. and McPherson, A. (1992) Macromolecular crystal growth experiments on International Microgravity Laboratory—1. *Protein Sci.* **1,** 1254–1268.
88. Fridborg, K., Hjertén, S., Höglund, S., Liljas, A., Lundberg, B. K. S., Oxelfelt, P., Philipson, L., and Strandberg, B. (1965) Purification, electron microscopy, and X-ray diffraction studies of the satellite tobacco necrosis virus. *Proc. Nat. Acad. Sci. USA* **54,** 513–521.
89. Åkervall, K. and Strandberg, B. (1971) X-ray diffraction studies of the satellite tobacco necrosis virus. III. A new crystal mounting method allowing photographic recording of 3 Å diffraction data. *J. Mol. Biol.* **62,** 625–627.
90. Unge, T., Montelius, I., Liljas, L., and Öfverstedt, L.-G. (1986) The EDTA-treated expanded satellite tobacco necrosis virus: biochemical properties and crystallisation. *Virology* **152,** 207–218.
91. Montelius, I., Liljas, L., and Unge, T. (1988) Structure of EDTA-treated satellite tobacco necrosis virus at pH 6.5., *J. Mol. Biol.* **201,** 353–363.
92. Harrison, S. C. (1969) Structure of tomato bushy stunt virus I. The spherically averaged electron density. *J. Mol. Biol.* **42,** 457–483.
93. Harrison, S. C. and Jack, A. (1975) structure of tomato bushy stunt virus III. Three-dimensional X-ray diffraction analysis at 16 Å resolution. *J. Mol. Biol.* **97,** 173–191.
94. Leberman, R. and Longley, W. (1970) The structures of turnip crinkle and tomato bushy stunt viruses. I. A small protein particle derived from turnip crinkle virus. Appendix: X-ray diffraction studies on crystals of the small particles. *J. Mol. Biol.* **50,** 213.
95. Crowfoot, D. and Schmidt, G. M. J. (1945) X-ray crystallographic measurements on a single crystal of a tobacco necrosis virus derivative. *Nature* **155,** 504,505.
96. Fukuyama, K., Hirota, S., and Tsukihara, T. (1987) Crystallisation and preliminary X-ray diffraction studies of tobacco necrosis virus. *J. Mol. Biol.* **196,** 961,962.
97. Heuss, K. L., Murthy, M. R. N., and Argos, P. (1981) Crystallisation of tobacco ringspot virus. *J. Mol. Biol.* **153,** 1163–1168.
98. Sehnke, P. C. and Johnson, J. E. (1993) Crystallisation and preliminary X-ray characterization of tobacco streak virus and a proteolytically modified form of the capsid protein. *Virology* **196,** 328–331.
99. Bernal, J. D. and Carlisle, C. H. (1948) Unit cell measurements of wet and dry crystalline turnip yellow mosaic virus. *Nature* **162,** 139.
100. Klug, A., Longley, W., and Leberman, R. (1966) Arrangement of protein subunits and the distribution of nucleic acid in turnip yellow mosaic virus. I. X-ray diffraction strudies. *J. Gen. Virol.* **1,** 403,404.

101. Hosur, M. V., Schmidt, T., Tucker, R. C., Johnson, J. E., Selling, B. E., and Rueckert, R. R. (1984) Black beetle virus—Crystallisation and particle symmetry. *Virology* **113,** 119–127.
102. Fisher, A. J., McKinney, B. R., Wery, J. P., and Johnson, J. E. (1992) Crystallisation and preliminary data analysis of Flock House virus. *Acta Cryst.* **48,** 515–520.
103. Fisher, A. J., McKinney, B. J., Schneemann, A., Rueckert, R., and Johnson, J. E. (1993) Crystallisation of viruslike particles assembled from Flock House virus coat protein expressed in a baculovirus system. *J. Virol.* **67,** 2950–2953.
104. Min Jou, W., Raeymaekers, A., and Fiers, W. (1979) Crystallisation of bacteriophage MS2. *Eur. J. Biochem.* **102,** 589–594.
105. Valegård, K., Unge, T., Montelius, I., Strandberg, B., and Fiers, W. (1986) Purification, crystallisation and preliminary X-ray data of the bacteriophage MS2. *J. Mol. Biol.* **190,** 587–591.
106. Cavarelli, J., Bomu, W., Liljas, L., Kim, S., Minor, W., Munshi, S., Muchmore, S., Schmidt, T., and Johnson, J. E. (1991) Crystallisation and preliminary structure analysis of an insect virus with T = 4 quasi-symmetry: *Nudaurelia capensis* ω virus. *Acta Cryst.* **B47,** 23–29.
107. Harrison, S. C., Strong, R. K., Schlesinger, S., and Schlesinger, M. J. (1992) Crystallisation of Sindbis virus and its nucleocapsid. *J. Mol. Biol.* **226,** 277–280.
108. Luo, M., Tsao, J., Rossmann, M. G., Basak, S., and Compans, R. W. (1988) Preliminary X-ray crystallographic analysis of canine parvovirus crystals. *J. Mol. Biol.* **200,** 209–211.
109. Murakami, W. T. (1963) Crystallization of SE polyoma virus. *Science* **142,** 56,57.
110. Rayment I., Baker, T. S., Caspar, D. L. D., and Murakami, W. T. (1982) Polyoma virus capsid structure at 22.5 Å resolution. *Nature* **295,** 110–115.
111. Adolph, K. W., Caspar, D. L. D., Hollingshead, C. J., Lattmann, E. E., Phillips, W. C., and Murakami, W. T. (1979) Polyoma virion and capsid crystal structures. *Science* **203,** 1117–1120.
112. Lattmann, E. E. (1980) Simian virus 40 crystals. *Science* **208,** 1048–1050.
113. Willingmann, P., Krishnaswamy, S., McKenna, R., Smith, T. J., Olson, N. H., Rossmann, M. G., Stow, P. L., and Incardona, N. L. (1990) Preliminary investigation of the phage ϕX174 crystal structure. *J. Mol. Biol.* **212,** 345–350.
114. Pereira, H. G., Valentine, R. C., and Russell, W. C. (1968) Crystallisation of an adenovirus protein (the hexon). *Nature* **219,** 946,947.
115. MacIntyre, W. M., Pereira, H. G., and Russell, W. C. (1969) Crystallographic data for the hexon of adenovirus type 5. *Nature* **222,** 1165,1166.
116. Franklin, R. M., Harrison, S. C., Pettersson, V., Philipson, L., Brändén, C. I., and Werner, P.-E. (1972) Structural studies on the adenovirus hexon. *Cold Spring Harb. Symp. Quant. Biol.* **36,** 503–510.
117. Franklin, R. M., Pettersson, U., Åkervall, K., Strandberg, B., and Philipson, L. (1971) Structural proteins of adenovirus. V. size and structure of the adenovirus type 2 hexon. *J. Mol. Biol.* **57,** 383–395.
118. Cornick, G., Sigler, P. B., and Ginsberg, H. S. (1971) Characterisation of crystals of type 2 adenovirus hexon. *J. Mol. Biol.* **57,** 397–401.

119. Heinz, F. X., Mandl, C. W., Holzmann, H., Kunz, K., Harris, B. A., Rey, F., and Harrison, S. C. (1991) The flavivirus envelope protein E: isolation of a soluble form from tick-borne encephalitis virus and its crystallisation. *J. Virol.* **65,** 5579–5583.
120. Prongay, A. J., Smith, T. J., Rossmann, M. G., Ehrlich, L. S., Carter, C. A., and McClure, J. (1990) Preparation and crystallisation of a human immunodeficiency virus p24-Fab complex. *Proc. Natl. Acad. Sci. USA* **87,** 9980–9984.
121. Stewart, P. L., Ghosh S., Bamford, D. H., and Burnett, R. M. (1993) Crystallisation of the major coat protein of PRD1, a bacteriophage with an internal membrane. *J. Mol. Biol.* **230,** 349–352.
122. Boege, U., Cygler, M., Wengler, G., Dumas, P., Tsao, T., Luo, M., Smith, T. J., and Rossmann, M. G. (1989) Sindbis virus core protein crystals. *J. Mol. Biol.* **208,** 79–82.
123. Leberman, R., Finch, J. T., Gilbert, P. F. C., Witz, J., and Klug, A. (1974) X-ray analysis of the disk of tobacco mosaic virus protein. I. Crystallisation of the protein and of a heavy-atom derivative. *J. Mol. Biol.* **86,** 179–182.
124. Raghavendra, K., Kelly, J. A., Khairallah, L., and Schuster, T. M. (1988) Structure and function of disk aggregates of the coat protein of tobacco mosaic virus. *Biochemistry* **27,** 7583–7588.
125. Warburg, O. and Christian, W. (1941) Isolierung und kristallisation des gärungsferments enolase. *Biochem. Z.* **310,** 384–421.
126. Harlos, K. (1992) Micro-bridges for sitting drop crystallizations. *J. Appl. Cryst.* **25,** 536–538.
127. McPherson, A. (1985) The crystallization of macromolecules: General principles, in *Methods in Enzymology: Diffraction Methods.* (Hirs, M., Timasheff, S. N., and Wyckoff, H., eds.), Academic, New York, vol. 114, pp. 112.
128. Helliwell, J. R. (1992) Global instrumentation survey: macromolecular crystallography. *Synchrotron Radiation News* **5,** 22–27.
129. Kahn, R., Fourme, R., Bosshard, R., and Saintage, V. (1986) An area-detector diffractometer for the collection of high-resolution and multiwavelength anomalous diffraction data in macromolecular crystallography. *Nucl. Instr. and Methods* **A246,** 596–603.
130. Morgunova, E. Y., Mikchailov, A. M., Dauter, Z., and Wilson, K., unpublished results.
131. Sakabe, N. personal communication.
132. Arndt, U. (1990) Focusing optics for laboratory sources in X-ray crystallography. *J. Appl. Cryst.* **22,** 53–60.
133. Wilkins S., personal communication.
134. Hajdu, J. and Johnson, L. N. (1990) Progress with laue diffraction studies on protein and virus crystals. *Biochemistry* **29,** 1669–1678.
135. Rossmann, M. G. and Erickson, J. W. (1983) Oscillation photography of radiation-sensitive crystals using a synchrotron source. *J. Appl. Cryst.* **16,** 629–636.
136. Hope, H., Frolow, F., von Boehlen, K., Makowski, I., Kratky, C., Halfon, Y., Danz, H., Webster, P., Batels, K. S., Wittmann H. G., and Yonath, A. (1989) Cryocrystallography of ribosomal particles. *Acta Cryst.* **B45,** 190–199.

137. Kabsch, W. (1988) Automatic indexing of rotation diffraction patterns. *J. Appl. Cryst.* **21,** 67–71.
138. Kim, S. (1989) Auto-indexing oscillation photographs. *J. Appl. Cryst.* **22,** 53–60.
139. Rossmann, M. G. (1979) Processing oscillation diffraction data for very large unit cells with an automatic convolution technique and profile fitting. *J. Appl. Cryst.* **12,** 225–238.
140. Greenhough, T. J. and Suddath, F. L. (1986) Oscillation camera data processing. 4. Results and recommendations for the processing of synchrotron radiation data in macromolecular crystallography. *J. Appl. Cryst.* **19,** 400–409.
141. Schutt, C. E. and Winkler, F. K. (1977) The oscillation method for very large unit cells, in *The Rotation Method in Crystallography* (Arndt, U. V. and Wonacott, A. J., eds.), Elsevier, North-Holland, Amsterdam, pp. 173–186.
142. Lea, S. and Stuart, D. (1994) Deconvolution of fully overlapped reflections from crystals of foot and mouth disease virus O_1 G67. *Acta Cryst.* **D50,** 160–167.
143. Tong, L. and Rossmann, M. G. (1990) The locked rotation function. *Acta Cryst.* **A46,**783–792.
144. Brünger, A. T., Kuriyan, J., and Karplus, M. (1987) Crystallographic R factor refinement by molecular dynamics. *Science* **235,** 458–460.
145. Tong, L. and Rossmann, M. G. (1993) Patterson-map interpretation with non-crystallographic symmetry. *J. Appl. Cryst.* **26,** 15–21.
146. Arnold, E. and Rossmann, M. G. (1986) Effect of errors, redundancy and solvent content in the molecular replacement procedure for the structure determination of biological macromolecules. *Proc. Natl. Acad. Sci. USA* **83,** 5489–5493.
147. Nordman, C. E. (1980) Procedures for detection and idealisation of noncrystallographic symmetry with application to phase refinement of the satellite tobacco necrosis virus structure. *Acta Cryst.* **A36,** 747–754.
148. Grimes, J. and Stuart, D. (1992) General Averaging Program (GAP).
149. Sim, G. (1959) The distribution of phase angles for structures containing heavy atoms II—A modification of the normal heavy-atom method for noncentrosymmetric structures. *Acta Cryst.* **12,** 813–815.
150. Sim, G. (1960) A note on the heavy atom method. *Acta Cryst.* **13,** 511,512.
151. Rayment, I. (1983) Molecular replacement method at low resolution: optimum strategy and intrinsic limitations as determined by calculations on icosahedral virus models. *Acta Cryst.* **A39,** 102–116.
152. Arnold, E. and Rossmann, M. G. (1988) The use of molecular replacement phases for the refinement of the human rhinovirus 14 structure. *Acta Cryst.* **A44,** 270–282.
153. Wang, B. C. (1985) Resolution of phase ambiguity in macromolecular crystallography. *Methods Enzymol.* **115,** 90–111.
154. Morgunova, E. Y., Dauter, Z., Fry, E., Stuart, D. I., Stel'Mashchuk, V. Y., Mikhailov, A. M., Wilson, K. S., Vainshtein, B. K. (1994) The atomic structure of carnation mottle virus capsid protein. *FEBS Lett.* **338,** 267–271.
155. Jones, T. A. and Liljas, L. (1984) Crystallographic refinement of macromolecules having non-crystallographic symmetry. *Acta Cryst.* **A40,** 50–57.
156. Jones, T. A. (1985) Interactive computer graphics: FRODO, in *Methods in Enzymology*, vol. 115 (Wyckoff, H. W., Hirs, C. H. W., and Timasheff, S. N., eds.), Academic, Orlando, FL, pp. 157–171.

157. Wang, G., Porta, C., Chen, Z., Baker, T., and Johnson, J. E. (1992) Identification of a Fab interaction footprint site on an icosahedral virus by cryo electron microscopy and X-ray crystallography. *Nature* **355,** 275–278.
158. Smith, T. J., Olson, N. H., Cheng, R. H., Liu, H., Chase, E. S., Lee, W. M., Lieppe, D. M., Mosser, A. G., Rueckert, R. R., and Baker, T. S. (1993) Structure of human rhinovirus complexed with Fab fragments from a neutralizing antibody. *J. Virology* **67,** 1148–1158.
159. Smith, T. J., Olson, N. H., Cheng, H., Chase, E. S., and Baker, T. S. (1993) Structure of a human rhinovirus-bivalently bound antibody complex: implications for viral neutralization and antibody flexibility. *Proc. Natl. Acad. Sci. USA* **90,** 7015–7018.
160. Olson, N. H., Kolatkar, P. R., Oliveira, M. A., Cheng, R. H., Greve, J. M., McClelland A., Baker T. S., and Rossmann, M. G. (1993) Structure of a human rhinovirus complexed with its receptor molecule. *Proc. Natl. Acad. Sci. USA* **90,** 507–511.

CHAPTER 14

Crystallization and Structure Analysis of Membrane Proteins

Richard Newman

1. Introduction

In recent years, there has been great progress in the determination of high-resolution three-dimensional (3D) structures of membrane proteins. The first major breakthrough came with the crystallization *(1)* and X-ray crystallography *(2,3)* of the bacterial photosynthetic reaction center (*see* refs. *4* and *5* for reviews). The structure of another, entirely different membrane protein, the bacterial outer membrane porin from *Rhodobacter capsulatus*, has now been determined by X-ray crystallography *(6)*. Recent results by electron crystallography of two-dimensional (2D) crystals have been most encouraging. The high-resolution 3D structure of bacteriorhodopsin *(7)* plant light-harvesting complex *(8)* and projection maps of several other membrane proteins at similar resolutions *(9–11)* have been obtained by this technique. Electron crystallography seems particularly appropriate for membrane proteins that are prone to form 2D crystals, and it is hoped that many more structures will be determined in this way.

For the foreseeable future, the structure analysis of membrane proteins at atomic or near-atomic resolution will depend on crystallographic techniques. (Recent progress by crystallographic and noncrystallographic methods has been concisely reviewed by Pattus *[12]*.) NMR spectroscopy, which has been so successful in elucidating the structure of some small soluble proteins (*see* Chapter 2 by Neuhaus and Evans in Vol. 17 of this series), may not be suitable for studying membrane proteins for

From: *Methods in Molecular Biology, Vol. 56: Crystallographic Methods and Protocols*
Edited by: C. Jones, B. Mulloy, and M. Sanderson Humana Press Inc., Totowa, NJ

two reasons: First, the detergent required to keep the protein in solution adds substantially (15–20 kDa) to the molecular mass; second, the majority of biologically active membrane proteins are large assemblies of several subunits. Most detergent-solubilized membrane proteins therefore fall outside the size range of 10–20 kDa that is currently accessible to solution NMR spectroscopy. Solid-state NMR techniques are useful for investigating specific aspects of membrane protein structure *(13)*, but at present, the determination of entire 3D structures by this method seems a remote prospect.

Secondary structure prediction algorithms for membrane proteins (reviewed by Fasman and Gilbert *[14]* and Jähnig *[15]*), based largely on hydrophobicity of amino acid side chains, are popular and widely applied to amino acid sequences of putative membrane proteins. However, since the number of known structures that can serve as guidelines is still very small, results may not be reliable and are, in any case, no substitute for experimentally determined structures.

For the time being, therefore, the crystallization of membrane proteins remains a prerequisite for structure analysis. 3D crystallization has been reviewed recently *(16–17)*, and a brief update is included in this chapter. An introduction to X-ray crystallography can be found in Chapter 1 in this volume. 2D crystallization of membrane proteins and proteins associated with biological membranes is described in Section 3. of this chapter. The requirements and experimental techniques for studying the structure of 2D crystals by electron crystallography are discussed briefly in Section 4.

2. 3D Crystallization of Membrane Proteins: An Update

2.1. 3D Crystals for X-ray Crystallography

A detailed study of the crystallization of OmpF porin from *E. coli* with a variety of detergents and short-chain phospholipids has been carried out by Eisele and Rosenbusch *(18)*. Even though crystals of some *E. coli* porins diffract X-rays rather well, the structure analysis by multiple isomorphous replacement proved difficult. The problem was finally solved by using protein isolated from a different species. The crystallization *(19)* and structure determination *(6)* of porin from the outer membrane of the purple bacterium, *Rhodobacter capsulatus*, has been a major recent achievement.

The quality of two other 3D crystals of membrane proteins has now been improved to the point where data collection by X-ray crystallography is feasible. Crystals of the photosystem I reaction center of cyanobacteria have been grown by several groups *(20,21)*. The best crystals, diffracting to 4 Å *(21)*, were obtained by a method that differed significantly from the one previously reported *(22)*, in particular with respect to protein concentration, which was higher by a factor of 100. Crystals of bacterial light-harvesting complex from *R. acidophila* diffract X-rays to 3.5 Å *(23)*. The improvement of crystal quality was brought about by changing the detergent from LDAO to OG previous to crystallization, and by using benzamidine-HCl instead of heptane triol as an additive. More recently crystal structures have been obtained, at or near atomic resolution, of the photosynthetic reaction center from Rhodobacter sphaeroides at 2.65 Å resolution *(24)*, bacterial porin, refined at 1.8 Å *(25)*, prostaglandin, H2 synthase-1 *(26)*, light-harvesting complex from photosynthetic bacteria *(27)*, and structures of bacterial *(28)* and bovine cytochrome c oxidase *(29)*.

The structure of the detergent in 3D crystals of the bacterial reaction center from *Rhodopseudomonas viridis* has been determined at low resolution by neutron diffraction *(30)*. The detergent surrounding the hydrophobic surface areas forms a convoluted 3D fretwork complementary to the structure of the protein. This suggests that the detergent micelles surrounding the protein come into contact and may even merge during crystallization, highlighting the particular importance of detergent properties for membrane protein crystallization.

2.2. Thin 3D Crystals for Electron Crystallography

Some membrane proteins form microcrystals more readily than large 3D crystals. These can be useful for structure determination by electron microscopy and image analysis (*see* Section 3.), in particular, if they consist of 2D crystals stacked in register. Electron crystallography of frozen-hydrated, thin 3D crystals of Ca^{2+}ATPase *(31)* has yielded a 6-Å projection map of the complex *(32)*. No 3D protein structure has yet been determined from thin 3D crystals of membrane proteins because of technical difficulties in collecting and combining diffraction data.

3. 2D Crystals

Specimens for electron crystallography should be no more than a few hundred Å thick to avoid multiple scattering and loss of image contrast caused by inelastically scattered electrons. 2D crystals of membrane proteins and of proteins forming on lipid monolayers are therefore ideal objects for electron crystallography.

3.1. 2D Crystallization of Membrane Proteins

3.1.1. Naturally Occurring 2D Crystals

A few membrane proteins form 2D crystals in the membranes of living cells. A well-known example is the purple membrane from *Halobacterium halobium*, a natural, almost perfect 2D crystal of bacteriorhodopsin *(33)*. In the membrane, crystalline patches measure around 1 μm across *(34)*. Fusion of native purple membranes produces large 2D crystals up to 20 μm in diameter *(35)*. 3D electron diffraction data have been collected from fused purple membranes *(36)* and the 3D structure of bacteriorhodopsin recently has been determined at high resolution by electron crystallographic techniques *(7)*.

Unfortunately, the high degree of order of purple membrane seems to be unique among naturally occurring 2D crystals. All other known examples, such as photoreceptor units of certain purple bacteria *(37)* and some bacterial porins *(38,39)*, are less well ordered.

3.1.2. 2D Crystallization Within Isolated Membranes

Occasionally, membrane proteins arrange themselves on a 2D lattice during the isolation of membranes, either on extraction of other membrane components or by specific induction. The quality of the resulting arrays depends to a large extent on the size and composition of the isolated membranes. Acetylcholine receptor *(40)* and Ca^{2+}ATPase *(41,42)* form tubular crystals. 2D crystals of cytochrome oxidase *(43–45)*, rhodopsin *(46,47)*, and photosystem II reaction center *(48)* are isolated as collapsed vesicles, whereas gap junction channels *(49)*, mitochondrial outer membrane channels *(50,51)*, and Na^{+}/K^{+}ATPase *(52,53)* are arranged in planar sheets.

3.1.3. 2D Crystallization from Solution

A more systematic approach to 2D crystallization is possible with detergent-solubilized, purified membrane proteins. 2D crystals tend to form as the protein molecules insert into reconstituted lipid bilayers. Protein and lipid may be combined in detergent solution, and the lipid is removed by dialysis or absorption. 2D crystals of cytochrome reductase *(54,55)* mitochondrial complex I *(56)*, various bacterial porins *(57–59)*, and photosystem I reaction center complex *(60)* have been produced in this way. The lipids used for reconstitution are either pure, synthetic compounds, such as dimyristoyl phosphatidyl choline (DMPC) *(57,59,60)*, or lipid extracts of natural membranes *(58)*. 2D crystals of bacterial phosphoporin *(59)* and of outer membrane porin *(57)* prepared in this way were particularly well ordered and have yielded high-resolution projection maps *(10,11)*. A 3D set of electron diffraction data has been collected of 2D crystals of phosphoporin *(61)*. The phosphoporin crystals had been treated with phospholipase A after reconstitution into DMPC bilayers to improve the crystallinity *(59)*.

Some isolated, detergent-solubilized membrane proteins carry enough lipid with them to recombine into membrane-like structures without the addition of extraneous lipid. Bacteriorhodopsin from detergent-solubilized purple membrane recrystallizes into extensive, well-ordered 2D crystals of a different plane group *(62)*. Bacterial reaction center complex also forms 2D crystals without lipid addition *(63)*, and small 2D arrays of photosystem II reaction centers have been obtained by a similar method *(64)*.

LHC-II forms 2D crystals with *(65)* or without *(66)* lipid addition. Crystals produced by the latter method are highly ordered and have yielded a high-resolution projection map of the complex *(9)*. The 2D crystallization of LHC-II has been investigated in some detail *(67)*. It seems that the crystals grow from merging detergent/protein/lipid micelles, enabling hydrophobic interactions between neighboring molecules that presumably are the major forces in the formation of 2D crystals of membrane proteins. Crystallization is made more reproducible by the presence of microcrystalline aggregates acting as seeds. The size of the crystals can be increased by adding a second detergent. The crystals grow in three stages: first, patches of 0.5–1 µm form during incubation at 25°C for 12 h. These merge into larger, mosaic-like arrays after another 36 h at the same temperature. Finally, during incubation at 40°C, the mosaic lattices rearrange to form extensive, well-ordered single sheets measuring up to 10 µm in diameter.

3.2. 2D Crystallization on Lipid Monolayers

Crystallization of protein subunits into a 2D lattice is a cooperative phenomenon. If crystallization is to occur at an interface, then molecules must be present at that interface in sufficient concentrations for spontaneous nucleation. Further, there must be molecular mobility to enable assembly of molecules with identical orientations. However, when molecules are irreversibly adsorbed at solid interfaces, random binding prevents close association of the molecules and long-range order, so crystallization does not occur.

The ordering of proteins below a lipid layer has been used to produce 2D crystals suitable for structure analysis by electron microscopy and image analysis. So far, crystallization by this technique has relied mainly on specific binding of normally water-soluble macromolecules to ligands attached to or incorporated into lipid monolayers. This approach has the advantages of rapidity and requirement for only microgram quantities of protein. Its disadvantage is that it requires the presence of a lipid-bound ligand. More recently, however, the direct interaction of macromolecules with lipid films has been demonstrated.

Lipid monolayers at the air-water interface have been used as a simple model for biological membranes *(68)*. In this technique, the lipid is deposited on the surface of a liquid (the hypophase) contained in a shallow hydrophobic Langmuir trough *(69)*. The lipid spreads and eventually forms a monolayer. The thermodynamic properties of such lipid layers can be determined by film balance measurements, and the monolayer may then be compressed laterally by a moveable bar so that a predetermined surface pressure is obtained. The surface pressure can be determined by the change in surface tension as measured by a Wilhelmy-plate balance. Surface pressure-area isotherms show up to three-phase transitions of the lipid. The transitions are from a "gaseous-expanded" phase, through a "fluid" phase to, in some cases, a "gel-crystalline" phase, and are dependent on the acyl chain length and head group of the phospholipid *(70)*. Information about specificity of binding of protein to lipid has been derived from measurements of surface pressure where an increase in surface pressure at constant surface area has been interpreted as penetration of the monolayer by whole or part of the macromolecule (for a review of protein–lipid interactions in monolayers, *see* Verger and Pattus *[71]*). The electrostatic interactions of proteins with polar head groups can also have an effect on surface pressure, but can be discriminated by varying acyl chain length and ionic conditions.

In order to facilitate crystallization, Uzgiris and Kornberg *(72)* proposed the use of a lipid monolayer derivatized with a ligand as a crystallization interface. They argued that the lateral mobility of the lipid-ligand molecule at the air–water interface would fulfill the requirement of molecular orientation, whereas the relatively small size of the lipid head group would allow for close packing of most bound molecules so that crystallization could occur. The exploitation of this technique and its extension to molecules that interact with lipid directly was a logical one, and the successful crystallization of various macromolecules has been described. Recently, the mechanism of 3D crystal growth of streptavidin from 2D crystals on lipid layers has been described *(73)*.

The number of crystallizations using lipid monolayers and analyzed by electron microscopy is shown in Table 1. There are two main categories, the first being those macromolecules crystallized via a lipid-linked ligand *(72,74)* or receptor *(75–77,81)*. The second category represents a more direct interaction with the film, either in the presence of amine to induce a net positive charge *(79,80,83–85)*, or perhaps mediated by a Ca^{2+} bridge *(82)*.

The problems associated with this technique are:

1. Obtaining the correct crystallization conditions;
2. Harvesting of the crystalline arrays onto electron microscope grids; and
3. Obtaining sufficiently large undistorted crystalline arrays to achieve the resolution limit imposed by negative staining (approx 20 Å).

3.2.1. Protein Concentration

Proteins for crystallization studies have to be as pure as possible. In order to obtain 2D crystals with lipid monolayers, final protein concentrations of between 0.15 and 500 μg/mL have been used at protein:lipid ratios of from 2:1 to 20:1 (Table 1).

3.2.2. Lipids

For the majority of successful crystallizations, lipids have been obtained commercially and used without further purification. In some cases, the electrostatic charge in the head group plane was made positive by the addition of amines, whereas in other cases, derivatized lipid ligands were used or specific receptors added to authentic lipids. In all cases, crystals only formed when lipids were in the fluid state (Table 1).

3.2.3. Crystallization Conditions

With one exception, monolayer crystals have been formed on small, typically 20 μL, drops of protein in buffer, under a lipid film, when incubated for the necessary time. During the incubations, the drops were kept in humid conditions, usually at room temperature (Table 1). In the case of Annexin VI, crystallization also occurred in a Langmuir trough, where

Table 1
Crystallizations Using Monolayers and Analyzed by Electron Microscopy

Protein	Lipid species	Ratio Protein/ lipid, w/w	Crystallization conditions: Protein conc./buffer	Temp.	Time	Resolution Å	Year/ reference
IgG (anti-DNP)	DNP-caproyl-PE	5–10:1	50 μg/mL, 150 m*M* NaCl 50m*M* Tris, pH 8.1	Amb.	1–3 h	60 Å	1983/*(72)*
C1q-complement	DNP-PE DNP-PE	5–10:1	50 μg/mL (as for IgG above)			60 Å	1983/*(72)*
IgG	DNP-PE	5–10:1	50–100 μg/mL, 150 m*M* NaCl, 50 m*M* Tris, pH 7.4	Amb.	1–3 h	20 Å	1985/*(65)*
Cholera toxin B-subunit	PC+ dioleoyl-PE +10% GM1	20:1	250 μg/mL, 150 m*M* NaCl, 50 m*M* Tris, pH 7.3	Amb.	4–48 h	17 Å	1986/*(75)*
Cholera toxin	PC + GM1	Na	100 μg/mL, 200 m*M* NaCl, 3 m*M* NaN_3, 1 m*M* EDTA, 50 m*M* Tris-HCl, pH 7.5	Amb.	0.5–0.75 h	30 Å	1987*(76)* 1988[a]/*(77)*
Ribonucleotide reductase	dATP-amino -caproyl-PE:PC (1:10)	20:1	250 μg/mL 100 m*M* NaCl, 5–15 m*M* $MgCl_2$, 2–10 m*M* spemidine 2 m*M* dTTP,15 m*M* Tris-HCl, pH 7.6	Amb.	24–48 h	18 Å	1987/*(78)*
RNA-polymerase holoenzyme	Octa-decylamine +DMPC	2.5:1	250 μg/mL, 20 m*M* $MgC1_2$, 1 m*M* spermidine, 6 m*M* K_2PO_4, 30 m*M* KCl,0.15 m*M* dithiothreitol, 15*M* EDTA,7.5% w/v glycerol, 20 m*M* Tris, pH 7.2–7.5	Amb.	5–10 h	30 Å	1988/*(79)* 1989[a]*(80)*

Tetanus toxin	PC:ganglioside GM 1	2:1	50–100 μg/mL Citrate-phosphate pH 5.5–6.0	4°C	Weeks	14–20 Å	1988/*(81)*
Annexin VI	DMPE	6:1	0.15 μg/mL, 100 m*M* Na-phosphate, pH 6.8 20 m*M* NaCl, 3 μ*M* Ca^{2+}	37°C	4 h	50 Å	1989/*(73)*
Actin	PC/DSPC+ stearylamine	10:1	100–500 μg/mL phosphate buffer, pH 7.0, 100 m*M* KCl, 2 m*M* $MgCl_2$, 20 m*M* EGTA, 0.01% 1, 2-mercaptoethanol	Amb.	0.5 h	59 Å	1989/*(83)*
RNA-polymerase II	DMPC-octa-decylamine	Na	150 μg/mL, 10 m*M* Tris-HCl, pH 7.5 50 m*M* ammonium sulfate, 10.2 m*M* EDTA, 2 m*M* spermidine		1–20 h	30 Å	1990/*(84)*
DNA gyrase B subunit	Novobiocin-phospholipld		250 μg/mL	Amb.	24 h	27 Å	1990/*(85)*

[a]3D reconstruction from 2D crystals produced by same method.

Abbreviations: PE, phosphatidylethanolamine; DMPE, dimyristoylphosphatidylethanolamine; PC, phosphathylcholine; DMPC, dimyristoylphosphatidylcholine; DSPC, distearoylphosphatidylcholine.

the surface pressure of the lipid film and its state could be determined *(82)*. The correct crystallization conditions need to be obtained on a trial-and-error basis with special attention given to the protein:lipid ratio.

The size of crystalline areas is the major problem to be solved and needs more research. At present, most monolayer crystals do not extend over large enough areas (or contain sufficient unit cells) to give high-quality diffraction patterns. If larger crystalline areas can be produced, it is possible to use techniques described in Sections 4.1.2. and 4.1.3. to extend the resolution considerably beyond that obtainable in negative stain *(86)*.

4. Structure Analysis of 2D Crystals by Electron Microscopic Techniques

The regular arrangement of molecules on lattice positions means that powerful crystallographic techniques can be brought to bear on electron micrographs and electron diffraction patterns of 2D crystals to determine the 3D structure of the protein. For an excellent introduction to crystallographic image processing, *see* the review by Amos et al. *(87)*.

4.1. Specimen Preparation

The preservation of crystalline order and, hence, the success of the structure analysis depend to a large extent on the method of specimen preparation. Several different techniques are in use, depending on requirements.

2D crystals of membrane proteins are applied to a carbon-coated, specimen-support grid in suspension, and then contrasted or otherwise treated as described in more detail below. 2D crystals on lipid monolayers need special preparation techniques, because the monolayers formed at the air–water interface have their hydrocarbon tails in the air, so that picking up such films requires a hydrophobic surface. Generally, the method of choice has been to use a freshly coated carbon/plastic electron microscope grid made from silver or nickel to pick up the lipid monolayer with bound protein. Silver or nickel grids are used to avoid corrosion problems associated with incubations in salt-containing buffers. The grids are prepared by covering one side with a Formvar or nitrocellulose film, and then coating that surface with a thin layer of evaporated carbon. The carbon surface generally remains hydrophobic for only 12 h, so that grids should usually be used almost immediately. In the case of actin *(83)*, perforated plastic films were prepared as described by Fukami and Adachi *(88)*, and then the crystals harvested and carbon coated. Picking up crystals can present problems because of the possible variation in hydrophobicity of the carbon-coated electron microscope grids.

4.1.1. Negative Stain

The quickest and most convenient way to examine 2D crystals is by electron microscopy of negatively stained specimens (*see* ref. *89* for detailed instructions). All negative stains (uranyl acetate and sodium phosphotungstate are commonly used) are heavy metal compounds that form a glassy cast around macromolecular assemblies on drying. This cast is fairly insensitive to radiation damage, which makes direct observation of coarse structural features possible. Frequently, the lattice of negatively stained 2D crystals is visible on the screen of the electron microscope. Negatively stained specimens are therefore most useful for assessing the progress of 2D crystallization experiments.

The structure of biological specimens is conserved to about 20 Å resolution; high-resolution internal detail is lost. Ionic stains may or may not penetrate the hydrophobic lipid bilayer. This can complicate the interpretation of low-resolution 3D maps of membrane proteins. A small amount of detergent, added to the staining solution *(66)*, helps to make stain penetration more reproducible and, thus, to increase the contrast within the bilayer. Specimens that are sensitive to the conditions prevailing in a film of drying, ionic negative stain may be contrasted with aurothioglucose instead. This medium also helps to avoid artifacts resulting from specific staining of charged sites. The disadvantage of aurothioglucose is its sensitivity to electron irradiation.

Nearly all 2D crystals of membrane proteins and of proteins crystallized on lipid monolayers have first been examined in negative stain. Although not ideal, this is a very useful medium for establishing low-resolution detail, such as molecular shape, symmetry, and in some cases, subunit composition of large protein complexes.

Negatively stained specimens of 2D crystals on lipid monolayers are generally prepared as follows: The grid with the adsorbed protein/lipid arrays is removed from the incubation chamber, touched briefly on the side of a filter paper to remove excess liquid (and in some cases, washed with distilled water), and floated on a drop of 1% (w/v) uranyl acetate for 20 s. Excess stain is blotted, and the grid is dried in air prior to examination in a transmission electron microscope.

4.1.2. Frozen-Hydrated Specimens

A more recent, and more sophisticated, method of specimen preparation is rapid freezing in a thin film of aqueous buffer. Very briefly, the specimen is applied to holey carbon film mounted on a support grid. Excess buffer is blotted with filter paper, and the specimen is injected into liquid ethane. At the high cooling rates achieved in this way, water does not crystallize and the specimen is embedded in a glass-like film of vitrified buffer. The grid is then placed into a cryo-transfer holder cooled with liquid nitrogen and inserted into the electron microscope. The temperature should not rise above –140°C during the entire process, because ice crystals form at this point, causing structural damage. The method of preparing frozen-hydrated specimens has been extensively and excellently reviewed by Dubochet et al. *(90)*.

The advantages are obvious: Biological specimens are preserved in their native, aqueous environment. The specimen is not exposed to high concentrations of ions as in negative stain. Artifacts arising from positive staining can be ruled out. However, the inherently low contrast among water, lipid, and protein means that thin, more or less uniform objects, such as 2D crystals, may be difficult to detect. Frozen-hydrated specimens are highly susceptible to radiation damage and, therefore, cannot be inspected visually in the electron microscope, but only on micrographs recorded with minimal electron dose. High-resolution detail may be obscured by electrons scattered by the vitrified buffer, which tends to be several times thicker than a 2D protein crystal. Finally, charging and beam-induced movement may present further obstacles to high-resolution imaging of frozen-hydrated specimens.

Nevertheless, the 3D structures of several membrane proteins, including gap junctions *(91)*, acetylcholine receptor *(92,93)*, and cytochrome oxidase *(94)*, have been determined by image processing of frozen-hydrated crystalline arrays at 15–20 Å resolution and of Ca^{2+} ATPase at 35 Å resolution *(95)*. Projection maps have been obtained from frozen-hydrated *(22)* 2D crystals of LHC-II *(96)*. 2D crystals of LHC-II in vitreous buffer diffract electrons to 3.2 Å resolution *(67)*. At a resolution of 3.4 Å, the projected structure of LHC-II is the same, regardless of whether the crystals are preserved in vitrified buffer, glucose, or tannin *(67)*. However, the latter medium proved to be by far the easiest to work with and is being used for data collection.

4.1.3. Specimen Preparation for High-Resolution Electron Crystallography

An important prerequisite for high-resolution electron crystallography is a preparation technique that maintains the specimen in its native, hydrated state when examined in the high vacuum of the electron microscope. It would seem, therefore, that vitrified water is the best medium for this purpose. In practice, however, few frozen-hydrated 2D crystals have been studied at high resolution because of the difficulties mentioned in the previous section. Other media, such as thin films of glucose *(97)*, trehalose *(10)*, and tannin *(67)*, may be preferable for this purpose. The sugar compounds glucose and trehalose are rich in hydroxyl groups and, thus, able to mimic the native aqueous environment on dehydration. They work best with fairly robust 2D crystals, such as purple membrane and phosphoporin, which are not damaged by air-drying in sugar solutions. Tannin, a gallic acid derivative of glucose, seems to have an additional, mild protective function, making it particularly suitable as a medium for delicate specimens, such as LHC-II, which are disordered by other preparation techniques.

Another important consideration is the influence of the carbon support film on the specimen. In principle, it would be desirable to do without a support film altogether and suspend the specimen instead in a thin film of vitrified buffer. The difficulties of preparing sufficiently thin films reproducibly and the unresolved problem of charging of unsupported specimens in the electron beam make carbon films essential for the time being.

Purple membrane is sensitive to the degree of hydrophilicity of the carbon film *(7,35)*. Freshly prepared hydrophilic films tend to disorder the crystals. Aged hydrophobic carbon films preserve the order, but few membranes adhere to them. The right balance of hydrophilic and hydrophobic properties can be found by aging carbon-coated grids for a few days before use *(35)*. 2D crystals of LHC-II and phosphoporin do not show the same sensitivity to hydrophilic films.

The curvature of the support film needs to be as small as possible to minimize blurring of reflections far from the tilt axis at high-tilt angles *(35)*. The "lens technique" of preparing specimens has been found to yield a reasonable proportion of flat 2D crystals *(67)*. By this technique, the specimen is prepared on the wet surface of a small piece of carbon film floated off on water or glucose solution, and picked up with the grid. The crystal suspension is applied to the small lens of liquid that adheres to it, and the grid is blotted through the grid bars. With 2D crystals of LHC-II, it was found that the surface roughness of the support film determined the quality of diffraction patterns. Isotropically sharp diffraction spots of highly tilted 2D crystals of LHC-II could be recorded only with perfectly smooth, flat carbon films *(98)*.

4.2. High-Resolution Electron Crystallography

One of the attractions of electron crystallography is the direct determination of the phases of structure factors from the Fourier transforms of electron micrographs. The resolution to which phases can be measured depends on the resolution of the image. The problem of phase determination is therefore reduced to the problem of recording high-resolution images. Although not trivial, this problem is not prohibitive, provided that the crystals themselves are sufficiently well ordered and that the instrument is capable of recording high-resolution detail at the required conditions *(see below)*. To date, the highest resolution obtained with a biological specimen (purple membrane) is 2.8 Å *(99)*.

4.2.1. Radiation Damage and Specimen Temperature

Electrons interact strongly with matter and cause severe damage to biological specimens. Most modern electron microscopes are equipped with a minimum dose facility for focusing images (which requires a high dose) on an area adjacent to the area of interest. However, this is not sufficient for high-resolution imaging, because, at room temperature, all fine detail is destroyed by the number of electrons required to record even a very faint image of the specimen. The critical dose (which reduces the intensity of an electron diffraction spot to 1/e of its original value) for 2D crystals of purple membrane at room temperature is 0.5 electrons/Å^2 *(97)*. High-resolution reflections tend to be weak and are therefore the first to be lost as a result of radiation damage. At temperatures between –120 and –180°C, readily achieved by cooling with liquid nitrogen, radiation damage takes about five times longer to manifest itself than at room temperature *(100,101)*, so that images and diffraction patterns of 2D protein crystals may be recorded with a correspondingly higher dose, resulting in a better signal-to-noise ratio.

4.2.2. Electron Diffraction

An excellent method for assessing the suitability of 2D crystals for high-resolution work is electron diffraction. The prerequisite is, of course, a specimen preparation technique that preserves high-resolution detail. The intensity of diffraction spots depends on the number of unit cells in the beam path. As a rule, 20,000–40,000 unit cells are required for recording a diffraction pattern with a reasonable signal-to-noise ratio. 2D crystals for electron diffraction therefore need to be fairly large. Purple membranes of about 1 μm diameter are adequate because the unit cell is fairly small, but the quality of diffraction patterns increases with the size of the crystal *(35)*. For lattices with a larger unit cell, 2D crystals need to be proportionally larger *(59,67)*.

Conditions for recording high-resolution electron diffraction patterns are much less stringent than for imaging. Specimen movements of a few 100 nm can be tolerated because, unlike the phases, the amplitudes of structure factors are not sensitive to position. Electron diffraction amplitudes are more accurate than amplitudes taken from Fourier transforms of images. Electron diffraction is therefore the method of choice for measuring 3D structure factor amplitudes. Programs for processing electron diffraction patterns and for merging electron diffraction data are available *(35)*.

4.2.3. High-Resolution Imaging

Most modern electron microscopes can achieve resolutions of 2 Å or better. However, resolution tests are normally performed with conditions that are quite unsuitable for imaging biological specimens (magnification about 500,000×, high dose, room temperature, specimen insensitive to radiation damage). At the conditions of biological electron crystallography (magnification 40,000×–60,000×, low temperature, low-dose, radiation-sensitive specimen), the instrument performance cannot be taken for granted.

The necessity to cool a part of the instrument (the specimen holder, the specimen stage, or the entire objective lens assembly) to a temperature below –120°C presents special problems. The resulting mechanical movement (thermal drift and vibrations caused by boiling coolant) needs to be kept to the absolute minimum for recording high-resolution images. The magnitude of the problem is best appreciated by considering that, in order to achieve a 3 Å resolution image, the specimen must not move by more than about 1 Å while the image is being recorded (which normally takes several seconds). Currently, there are three electron microscopes with liquid-nitrogen-cooled *(102)* or liquid-helium-cooled *(103,104)* specimens, which are used routinely to record images of the required high quality.

An additional difficulty arises from the fact that electron irradiation causes small, random movements of the specimen that tend to blur high-resolution detail. This beam-induced movement can be minimized by recording images in "spotscan" mode, whereby small areas of the specimen are irradiated in sequence *(105–107)*. The spotscan method also helps to reduce the effect of specimen drift.

4.3. High-Resolution Image Processing

For computer processing, the electron micrographs need to be digitized with a microdensitometer at a suitable step size (typically 7–10 μm for images recorded at magnifications around 50,000×). The signal-to-noise ratio of peak amplitudes increase with the size of the processed area, which should therefore be as large as possible.

Even with all the precautions of minimizing radiation damage and specimen movement, Fourier transforms of most images of 2D protein crystals show, at best, a few reflections beyond about 7 Å. This is because of small deviations of unit cells from a perfect 2D lattice, for example, caused by interaction of the crystal with the support film. Additional electron optical distortions are introduced by imperfections of the magnetic lenses. Lattice distortions tend to blur high-resolution reflections in the Fourier transform and, therefore, reduce them close to the noise level.

Programs to detect and correct lattice distortions of 2D crystals by image processing have been developed by Henderson and colleagues *(108)*. With these programs, it has been possible to determine the structure of bacteriorhodopsin at high resolution in 3D. The same methods are being used in two other laboratories to determine the structures of LHC-II and phosphoporin.

5. Conclusions and Outlook

Several structures of membrane proteins have now been determined at high resolution by X-ray or electron crystallography. Progress with 3D crystallization of membrane proteins has been slower than anticipated a few years ago, but 2D crystallization provides a viable alternative. Many membrane proteins tend to form 2D rather than 3D crystals. With the improvement of 2D crystallization techniques and a better understanding of the processes involved, a larger number of well-ordered 2D crystals of membrane proteins should become available for examination by electron crystallography.

The 2D crystallization of proteins on lipid monolayers seems a promising general technique for the 2D crystallization of soluble proteins that can be derivatized to associate specifically with membrane lipids or that do so under the appropriate charge conditions. Proteins that associate naturally with lipids provide a particularly attractive proposition and, furthermore, would provide structural information under in vivo conditions. The potential for 2D crystallization of membrane proteins by this technique is largely unexplored.

X-ray crystallography of 3D crystals is now almost routine, and a large number of highly developed instruments for data collection are available. By comparison, protein electron crystallography is still in its infancy. Methods and computer programs for structure determination at high resolution have now been worked out, but only a few electron microscopes are currently equipped for this type of work. Large improvements in instrumentation for recording high-resolution images still seem possible *(109)*. Progress in electron crystallography of membrane proteins will therefore continue as more 2D crystals and new instruments become available.

References

1. Michel, H. (1982) Three-dimensional crystals of a membrane protein complex. The photosynthetic reaction center from *Rhodopseudomonas viridis*. *J. Mol. Biol.* **158,** 567–572.
2. Deisenhofer, J., Epp, O., Miki, K., Huber, R., and Michel, H. (1984) X-ray structure analysis of a membrane protein complex. Electron density map at 3Å resolution and a model of the chromophores of the photosynthetic reaction center from *Rhodopseudomonas viridis*. *J. Mol. Biol.* **180,** 385–398.
3. Deisenhofer, J., Epp, O., Miki, K., Huber, R., and Michel, H. (1985) X-ray structure analysis at 3Å resolution of a membrane protein complex: Folding of the protein subunits in the photosynthetic reaction center from *Rhodopseudomonas Viridis*. *Nature* **318,** 618–624.
4. Feher, G., Allen, J. P., Okamura, M. Y., and Rees, D. C. (1989) Structure and function of bacterial photosynthetic reaction centers. *Nature* **339,** 111–116.
5. Deisenhofer, J. and Michel, H. (1989) The photosynthetic reaction center from the purple bacterium *Rhodopseudomonas viridis*. *Science* **245,** 1463–1473.
6. Weiss, M. S., Wacker, T., Weckesser, J., Welte, W., and Schulz, G. E. (1990) The three-dimensional structure of porin from *Rhodobacter capsulatus* at 3 Å resolution. *FEBS Lett.* **267,** 268–272.
7. Henderson, R., Baldwin, J. M., Ceska, T. A., Zemlin, F., Beckmann, E., and Downing, K. H. (1990) Model for the structure of bacteriorhodopsin based on high-resolution electron cryo-microscopy. *J. Mol. Biol.* **213,** 899–929.
8. Kühlbrandt, W., Wang, D. N., and Fujiyoshi, Y. (1994) Atomic model of plant light-harvesting complex by electron crystallography. *Nature* **367,** 614–621.
9. Kühlbrandt, W. and Downing, K. H. (1989) Two-dimensional structure of plant light-harvesting complex at 3.7 Å resolution by electron crystallography. *J. Mol. Biol.* **207,** 823–828.
10. Jap, B. K., Downing, K. H., and Wallian, P. J. (1990) Structure of PhoE porin in projection at 3.5 Å resolution. *J. Struct. Biol.* **103,** 57–63.
11. Sass, H. J., Beckmann, E., Zemlin, F., van Heel, M., Zeitler, E., Rosenbusch, J. P., Dorset, D. L., and Massalski, A. (1989) Densely packed β-structure at the protein-lipid interface of porin is revealed by high-resolution cryo-electron microscopy. *J. Mol. Biol.* **209,** 171–175.

12. Pattus, F. (1990) Membrane protein structure. *Curr. Opinion in Cell Biol.* **2,** 681–685.
13. Smith, R., Thomas, D. E., Separovic, F., Atkins, A. R., and Cornell, B. A. (1989) Determination of the structure of a membrane-incorporated ion channel. Solid-state nuclear magnetic resonance studies of gramicidin A. *Biophys. J.* **56,** 307–314.
14. Fasman, G. D. and Gilbert, W. A. (1990) The prediction of transmembrane protein sequences and their conformation: an evaluation. *TIBS* **15,** 89–92.
15. Jähnig, F. (1990) Structure predictions of membrane proteins are not that bad. *TIBS* **15,** 93–95.
16. Kühlbrandt, W. (1988) Three-dimensional crystallization of membrane proteins. *Q. Rev. Biophys.* **21,** 429–477.
17. Michel, H. (1991) General and practical aspects of membrane protein crystallization, in *Crystallisation of Membrane Proteins*, 9th ed. (Michel, H., ed.) CRC, Boca Raton, FL, pp. 73–88.
18. Eiselé, J. -L. and Rosenbusch, J. P. (1989) Crystallization of porin using short chain phospholipids. *J. Mol. Biol.* **206,** 209–212.
19. Nestel, U., Wacker, T., Woitzik, D., Weckeser, J., Kreutz, W., and Welte, W. (1989) Crystallization and preliminary X-ray analysis of porin from *Rhodobacter capsulatus*. *FEBS Lett.* **242,** 405–408.
20. Ford, R. C., Picot, D., and Garavito, R. M. (1987) Crystallization of the photosystem I reaction center. *EMBO J.* **6,** 1581–1586.
21. Witt, I., Witt, H. T., Di Fiore, D., Rögner, M., Hinrichs, W., Saenger, W., Granzin, J., Betzel, Ch., and Dauter, Z. (1988) X-Ray characterization of single crystals of the reaction center I of water splitting photosynthesis. *Ber. Bunsenges. Phys. Chem.* **92,**1503–1506.
22. Witt, I., Witt, H. T., Gerken, S., Saenger, W., Dekker, J. P., and Rögner, M. (1987) Crystallization of reaction center I of photosynthesis. *FEBS Lett.* **221,** 260–264.
23. Papiz, M. Z., Hawthornthwaite, A. M., Cogdell, R. J., Woolley, K. J., Wightman, P. A., Ferguson, L. A., and Lindsay, J. G. (1989) Crystallization and characterization of two crystal forms of the B800-850 light-harvesting complex from *Rhodopseudomonas acidophila* strain 10050. *J. Mol. Biol.* **209,** 833–835.
24. Ermler, U., Fritzsch, G., Buchanan, S. K., and Michel, H. (1994) Structure of the photosynthetic reaction centre from Rhodobacter sphaeroides at 2.65 Å resolution: cofactors and protein-cofactor interactions. *Structure* **2,** 925–936.
25. Weiss, M. S. and Schulz, G. E. (1992) Structure of porin refined at 1.8 Å resolution. *J. Mol. Biol.* **227,** 493–508.
26. Picot, D., Loll, P. J., and Garavito, R. M. (1994) The X-ray crystal structure of the membrane protein prostaglandin H_2 synthase-1. *Nature* **367,** 243–249.
27. McDermott, G., Prince, S. M., Freer, A. A., Hawthorthwaite-Lawless, A. M., Papiz, M. Z., Cogdell, R. J., and Isaacs, N. W. (1994) Crystal structure of an integral light-harvesting complex from photosynthetic bacteria. *Nature* **374,** 517–521.
28. Iwata, S., Ostermeier, C., Lugwig, B., and Michel, H. (1995) Structure at 2.8 Å resolution of cytochrome c oxidase from Paracoccus denitrificans. *Nature* **376,** 660–669.
29. Tsukihara, T., Aoyama, H., Yamashita, E., Tomizaki, T., Yamaguchi, H., Shinzawa-Itoh, K., Nakashima, R., Yaona, R., and Yoshikawa, S. (1995) Structure of metal sites of oxidised bovine heart cytochrome c oxidase at 2.8 Å. *Science* **268,** 1069–1074.

30. Roth, M., Lewit-Bentley, A., Michel, H., Deisenhofer, J., Huber, R., and Oesterhelt, D. (1989) Detergent structure in crystals of a bacterial photosynthetic reaction center. *Nature* **340,** 659–662.
31. Stokes, D. L. and Green, N. M. (1990) Three-dimensional crystals of CaATPase from sarcoplasmic reticulum. Symmetry and molecular packing. *Biophys. J.* **57,** 1–14.
32. Stokes, D. L. and Green, N. M. (1990) Structure of CaATPase: Electron microscopy of frozen-hydrated crystals at 6 Å resolution in projection. *J. Mol. Biol.* **213,** 529–538.
33. Henderson, R. (1975) The structure of the purple membrane from *Halobacterium halobium*: Analysis of the X-ray diffraction pattern. *J. Mol. Biol.* **93,** 123–138.
34. Henderson, R. and Unwin, P. N. T. (1975) Three-dimensional model of purple membrane obtained by electron microscopy. *Nature* **257,** 28–32.
35. Baldwin, J. and Henderson, R. (1984) Measurement and evaluation of electron diffraction patterns from two-dimensional crystals. *Ultramicroscopy* **14,** 319–336.
36. Ceska, T. A. and Henderson, R. (1990) Analysis of high-resolution electron diffraction patterns from purple membrane labelled with heavy-atoms. *J. Mol. Biol.* **213,** 539–560.
37. Kuhlbrandt, W. (1987) Photosynthetic membranes and membrane proteins, in *Electron Microscopy of Proteins, vol. 6: Membranous structures* (Harris, J. R. and Horne, R. W., eds.), Academic, London, pp. 155–207.
38. Kessel, M., Brennan, M. J., Trus, B. L., Bisher, M. E., and Steven, A. C. (1988) Naturally crystalline porin in the outer membrane of *Bordetella pertussis*. *J. Mol. Biol.* **203,** 275–278.
39. Rachel, R., Engel, A. M., Huber, R., Stetter, K.-O., and Baumeister, W. (1990) A porin-type protein is the main constituent of the cell envelope of the ancestral eubacterium *Thermotoga maritima*. *FEBS Lett.* **262,** 64–68.
40. Brisson, A, and Unwin, P. N. T. (1984) Tubular crystals of acetylcholine receptor. *J. Cell Biol.* 99, 1202–1211.
41. Dux, L., Pikula, S., Mullner, N., and Martonosi, A. (1986) Crystallization of Ca^{2+}-ATPase in detergent-solubilized sarcoplasmic reticulum. *J. Biol. Chem.* **262,** 6439–6442.
42. Taylor, K. A., Dux, L., Varga, S., Ting-Beall, H. P., and Martonosi, A. (1988) Analysis of two-dimensional crystals of Ca^{2+}-ATPase in sarcoplasmic reticulum. *Methods Enzymol.* **157,** 271–289.
43. Vanderkooi, G., Senior, A. E., Capaldi, R. A., and Hayashi, H. (1972) Biological membrane structure. III. The lattice structure of membranous cytochrome oxidase. *Biochim. Biophys.* Acta **274,** 38–48.
44. Frey, T. G., Chan, S. H. P., and Schatz, G. (1978) Structure and orientation of cytochrome c oxidase in crystalline membranes. *J. Biol. Chem.* **253,** 4389–4395.
45. Fuller, D. S., Capaldi, R. A., and Henderson, R. (1979) Structure of cytochrome c oxidase in deoxycholate-derived two-dimensional crystals. *J. Mol. Biol.* **134,** 305–327.
46. Corless, J. M., McCaslin, D. R., and Scott, B. L. (1982) Two-dimensional rhodopsin crystals from disk membranes of frog retinal rod outer segments. *Proc. Natl. Acad. Sci. USA* **79,** 1116–1120.

47. Dratz, E. A., Van Breemen, J. F. L., Kamps, K. M. P., Keegstra, W., and Van Bruggen, E. F. J. (1985) Two-dimensional crystallization of bovine rhodopsin. *Biochim. Biophys. Acta* **832,** 337–342.
48. Bassi, R., Magaldi, A. G., Tognon, G., Giacometti, G. M., and Miller, K. R. (1989) Two-dimensional crystals of the photosystem II reaction center complex from higher plants. *Eur. J. Cell Biol.* **50,** 84–93.
49. Zampighi, G. and Unwin, P. N. T. (1979) Two forms of isolated gap junctions. *J. Mol. Biol.* **135,** 451–464.
50. Mannella, C. A. (1984) Phospholipase-induced crystallization of channels in mitochondrial outer membranes. *Science* **224,** 165,166.
51. Mannella, C. A. (1989) Fusion of the mitochondrial outer membrane: use in forming large, two-dimensional crystals of the voltage-dependent, anion-selective channel protein. *Biochim. Biophys. Acta* **981,** 15–20.
52. Mohraz, M., Yee, M., and Smith, P. R. (1985) Novel crystalline sheets of Na,K-ATPase induced by phospholipase A_2. *J. Ultrastruct. Res.* **93,** 17–26.
53. Skriver, E., Maunsbach, A. B., Hebert, H., and Jørgensen, P. L. (1988) Crystallization of membrane-bound Na^+,K^+-ATPase in two dimensions. *Methods Enzymol.* **156,** 80–87.
54. Wingfield, P., Arad, T., Leonard, K., and Weiss, H. (1979) Membrane crystals of ubiquinone: cyochrome c reductase from *Neurospora* mitochondria. *Nature* **280,** 696,697.
55. Hovmöller, S., Slaughter, M., Berriman, J., Karlsson, B., Weiss, H., and Leonard, K. (1983) Structural studies of cytochrome reductase. Improved membrane crystals of the enzyme complex and crystallization of a subcomplex. *J. Mol. Biol.* **165,** 401–406.
56. Leonard, K. (1987) Three-dimensional structure of NADH: ubiquinone reductase (complex I) from *Neurospora* mitochondria determined by electron microscopy of membrane crystals. *J. Mol. Biol.* **194,** 277–286.
57. Dorset, D. L., Engel, A., Häner, M., Massalski, A., and Rosenbusch, J. P. (1983) Two-dimensional crystal packing of matrix porin. A channel forming protein in *Escherichia coli* outer membranes. *J. Mol. Biol.* **165,** 701–710.
58. Lepault, J., Dargent, B., Tichelaar, W., Rosenbusch, J. P., Leonard, K., and Pattus, F. (1988) Three-dimensional reconstruction of maltoporin from electron microscopy and image processing. *EMBO J.* **7,** 261–268.
59. Jap, B. K. (1988) High-resolution electron diffraction of reconstituted PhoE porin. *J. Mol. Biol.* **199,** 229–231.
60. Ford, R. C., Hefti, A., and Engel, A. (1990) Ordered arrays of the photosystem I reaction centre after reconstitution: projections and surface reliefs of the complex at 2 nm resolution. *EMBO J.* **9,** 3067–3075.
61. Walian, P. J. and Jap, B. K. (1990) Three-dimensional electron diffraction of PhoE porin to 2.8 Å resolution. *J. Mol. Biol.* **215,** 429–438.
62. Michel, H., Oesterhelt, D., and Henderson, R. (1980) Orthorhombic two-dimensional crystal form of purple membrane. *Proc. Natl. Acad. Sci. USA* **77,** 338–342.
63. Miller, K. R. and Jacob, J. S. (1983) Two-dimensional crystals formed from photosynthetic reaction centers. *J. Cell Biol.* **97,** 1266–1270.

64. Dekker, J. P., Betts, S. D., Yocum, C. F., and Boekema, E. J. (1990) Characterization by electron microscopy of isolated particles and two-dimensional crystals of the CP47-D1-D2-cytochrome b-559 complex of photosystem II. *Biochemistry* **29,** 3220–3225.
65. Li, J. (1982) Formation of crystalline arrays of chlorophyll *a/b*-light-harvesting protein by membrane reconstitution. *Biophys. J.* **37,** 363–370.
66. Kühlbrandt, W. (1984) Three-dimensional structure of the light-harvesting chlorophyll *a/b*-protein complex. *Nature* **307,** 478–480.
67. Wang, D. N. and Kühlbrandt, W. (1991) High-resolution electron crystallography of light-harvesting chlorophyll *a/b*-protein complex in three different media. *J. Mol. Biol.* **217,** 691–699.
68. Phillips, M. C. and Chapman, D. (1968) Monolayer characteristics of saturated 1,2-diacylphosphatidylcholine (lecithin) and phosphatidylethanolamines at the air-water interface. *Biochim. Biophys. Acta* **163,** 301–313.
69. Albrecht, O., Gruler, H., and Sackmann, E. (1978) Polymorphism of phospholipid monolayers. *J. de Phys.* **39,** 301–313.
70. Gaines, G. L. (1966) *Insoluble Monolayers at the Liquid–Water Interface.* Wiley Interscience, New York.
71. Verger, R. and Pattus, F. (1982) Protein-lipid interactions in monolayers. *Chem. Phys. Lipids* **30,** 189–227.
72. Uzgiris, E. E. and Kornberg, R. D. (1983) Two-dimensional crystallization technique for imaging macromolecules, with application to antigen-antibody-complement complexes. *Nature* **310,** 134–136.
73. Hemming, S. A., Bochkarev, A., Darst, S., Kornberg, R. D., Ala, P., Yang, D. S. C., and Edwards, A. M. (1995) The mechanism of protein crystal growth from lipid layers. *J. Mol. Biol.* **246,** 308–316.
74. Kornberg, R. D. and Ribi, H. O. (1987) Formation of two-dimensional crystals of proteins on lipid layers, in *Protein Structure, Folding and Design 2* (Oxender, D. L., ed.), Alan R. Liss, New York, pp. 175–186.
75. Ludwig, D. S., Ribi, H. O., Schoolnik, G. K., and Kornberg, R. D. (1986) Two-dimensional crystals of cholera toxin B-subunit-receptor complexes: Projected structure at 17-Å resolution. *Proc. Natl. Acad. Sci. USA* **83,** 8585–8588.
76. Reed, R. A., Mattai, J., and Shipley, G. G. (1987) Interaction of cholera toxin with ganglioside G_{M1} receptors in supported lipid monolayers. *Biochemistry* **26,** 824–832.
77. Ribi, H. O., Ludwig, D. S., Mercer, L. K., Schoolnik, G. K., and Kornberg, R. D. (1988) Three-dimensional structure of cholera toxin penetrating a lipid membrane. *Science* **239,** 1272–1276.
78. Ribi, H. O., Reichard, P., and Kornberg, R. D. (1987) Two-dimensional crystals of enzyme—effector complexes: Ribonucleotide reductase at 18Å resolution. *Biochemistry* **26,** 7974–7979.
79. Darst, S. A., Ribi, H. O., Pierce, D. W., and Kornberg, R. D. (1988) Two-dimensional crystals of *Escherichia coli* RNA-polymerase holoenzyme on positively charged lipid layers. *J. M. Biol.* **203,** 269–273.
80. Darst, S. A., Kubalek, E. W., and Kornberg, R. D. (1989) Three-dimensional structure of *Escherichia coli* RNA polymerase holoenzyme determined by electron crystallography. *Nature* **340,** 730–732.

81. Robinson, J. P., Schmid, M. F., Morgan, D. G., and Chiu, W. (1988) Three-dimensional structural analysis of tetanus toxin by electron crystallography. *J. Mol. Biol.* **200,** 367–375.
82. Newman, R., Tucker, A., Ferguson, C., Tsernoglou, D., Leonard, K., and Crumpton, M. J. (1989) Crystallisation of p68 on lipid monolayers and as three-dimensional single crystals. *J. Mol. Biol.* **206,** 213–219.
83. Ward, J. R., Menetret, J., Pattus, F., and Leonard, K. (1990) Method for forming two-dimensional paracrystals of biological filaments on lipid monolayers. *J. Electr. Microsc. Tech.* **14,** 335–341.
84. Edwards, A. E., Darst, S., Feaver, W. J., Thompson, N. E., Burgess, R. R., and Kornberg, R. D. (1990) Purification and lipid-layer crystallisation of yeast RNA polymerase II. *Proc. Natl. Acad. Sci.* **87,** 2122–2126.
85. Lebeau, L., Regnier, E., Schultz, P., Wang, J. C., Mioskowski, C., and Oudet, P. (1990) Two-dimensional crystallisation of DNA gyrase B subunit on specifically designed lipid monolayers. *FEBS Lett.* **267,** 38–42.
86. Stewart, M., and Vigers, G. (1986) Electron microscopy of frozen-hydrated biological material. *Nature* **319,** 631–636.
87. Amos, L. A., Henderson, R., and Unwin, P. N. T. (1982) Three-dimensional structure determination by electron microscopy of two-dimensional crystals. *Prog. Biophys. Mol. Biol.* **39,** 183–231.
88. Fukami, A. and Adachi, K. (1965) A new method of preparation of a self-perforating micro-plastic grid and its application. *J. Electron Microsc.* **14,** 112–118.
89. Haschemeyer, R. H. and Myers, R. J. (1972) Negative staining, in *Principles and Techniques of Electron Microscopy* (Hayat, M. S., ed.), Van Nostrand Reinhold, New York, pp. 101–150.
90. Dubochet, J., Adrian, M., Chang, J.-J., Homo, J.-C., Lepault, J., McDowall, A. W., and Schultz, P. (1988) Cryo-electron microscopy of vitrified specimens. *Q. Rev. Biophys.* **21,** 129–228.
91. Unwin, P. N. T. and Ennis, P. D. (1984) Two configurations of a channel-forming membrane protein. *Nature* **307,** 609–612.
92. Brisson, A., and Unwin, P. N. T. (1985) Quaternary structure of the acetylchoiine receptor. *Nature* **315,** 474–477.
93. Toyoshima, C., and Unwin, N. (1988) Ion channel of acetylcholine receptor reconstructed from images of postsynaptic membranes. *Nature* **336,** 247–250.
94. Valpuesta, J. M. and Henderson, R. (1990) Electron cryo-microscopic analysis of crystalline cytochrome oxidase. *J. Mol. Biol.* **214,** 237–251.
95. Taylor, K. A., Ho, M.-H., and Martonosi, A. (1986) Image-analysis of the Ca^{2+}ATPase from *Sarcoplasmic reticulum. Ann. NY Acad. Sci.* **483,** 461–489.
96. Lyon, M. K. and Unwin, P. N. T. (1988) Two-dimensional structure of the light-harvesting chlorophyll *a/b*-complex by cryoelectron microscopy. *J. Cell Biol.* **106,** 1515–1523.
97. Unwin, P. N. T. and Henderson, R. (1975) Molecular structure determination by electron microscopy of unstained crystalline specimens. *J. Mol. Biol.* **94,** 425–440.
98. Butt, H.-J., Wang, D. N., Hansma, P. K., and Kühlbrandt, W. (1991) Effect of surface roughness of carbon support films on high-resolution electron diffraction of two-dimensional protein crystals. *Ultramicroscopy*, submitted.

99. Baldwin, J. M., Henderson, R., Beckman, R., and Zemlin, F. (1988) Images of purple membrane at 2.8 Å resolution obtained by cryo-electron microscopy. *J. Mol. Biol.* **202,** 585–591.
100. Hayward, S. B. and Glaeser, R. M. (1979) Radiation damage of purple membrane at low temperature. *Ultramicroscopy* **4,** 201–210.
101. Lamvik, M. K., Kopf, D. A., and Davilla, S. D. (1987) Mass loss rate in collodion is greatly reduced at liquid helium temperature. *J. Microsc.* **148,** 211–217.
102. Hayward, S. B. and Glaeser, R. M. (1980) High resolution cold stage for the JEOL 100B and 100C electron microscopes. *Ultramicroscopy* **5,** 3–8.
103. Lefranc, G., Knapek, E., and Dietrich, I. (1982) Superconducting lens design. *Ultramicroscopy* **10,** 111–124.
104. Fujiyoshi, Y. (1989) High resolution cryo-electron microscopy for biological macromolecules. *J. Electron Microsc.* **38,** 97–101.
105. Downing, K. H. and Glaeser, R. M. (1986) Improvement in high resolution image quality of radiation-sensitive specimens achieved with reduced spot size of the electron beam. *Ultramicroscopy* **20,** 269–278.
106. Bullough, P. and Henderson, R. (1987) Use of spot-scan procedure for recording low-dose micrographs of beam-sensitive specimens. *Ultramicroscopy* **21,** 223–230.
107. Downing, K. (1988) Observations of restricted beam-induced specimen motion with small-spot illumination. *Ultramicroscopy* **24,** 387–398.
108. Henderson, R., Baldwin, J. M., Downing, K. H., Lepault, J., and Zemlin, F. (1986) Structure of purple membrane from *Halobacterium halobium*: Recording, measurement and evaluation of electron micrographs at 3.5 Å resolution. *Ultramicroscopy* **19,** 147–178.
109. Henderson, R. and Glaeser, R. M. (1985) Quantitative analysis of image contrast in electron micrographs of beam-sensitive crystals. *Ultramicroscopy* **16,** 139–150.

Index